T0213079

Lecture Notes in Computer Science   10521

Commenced Publication in 1973
Founding and Former Series Editors:
Gerhard Goos, Juris Hartmanis, and Jan van Leeuwen

More information about this series at http://www.springer.com/series/7407

Michael Floater · Tom Lyche
Marie-Laurence Mazure · Knut Mørken
Larry L. Schumaker (Eds.)

Mathematical Methods for Curves and Surfaces

9th International Conference, MMCS 2016
Tønsberg, Norway, June 23–28, 2016
Revised Selected Papers

 Springer

Editors
Michael Floater
University of Oslo
Oslo
Norway

Tom Lyche
University of Oslo
Oslo
Norway

Marie-Laurence Mazure
Université Joseph Fourier
Grenoble
France

Knut Mørken
University of Oslo
Oslo
Norway

Larry L. Schumaker
Vanderbilt University
Nashville, TN
USA

ISSN 0302-9743　　　　　　　　ISSN 1611-3349　(electronic)
Lecture Notes in Computer Science
ISBN 978-3-319-67884-9　　　　ISBN 978-3-319-67885-6　(eBook)
https://doi.org/10.1007/978-3-319-67885-6

Library of Congress Control Number: 2017956079

LNCS Sublibrary: SL1 – Theoretical Computer Science and General Issues

Printed on acid-free paper

This Springer imprint is published by Springer Nature
The registered company is Springer International Publishing AG
The registered company address is: Gewerbestrasse 11, 6330 Cham, Switzerland

Preface

The 9th International Conference on Mathematical Methods for Curves and Surfaces took place during June 23–28, 2016, in Tønsberg, Norway. The earlier conferences in the series took place in Oslo (1988), Biri (1991), Ulvik (1994), Lillehammer (1997), Oslo (2000), Tromsø (2004), Tønsberg (2008), and Oslo (2012). This conference series is integrated with the French conferences Curves and Surfaces organized by SMAI-SIGMA (Signal-Image-Géométrie-Modélisation-Approximation), and the last conference in France was in Paris in 2014.

The conference gathered 140 participants from 27 countries who presented a total of 115 talks. This includes nine invited talks and seven mini-symposia. This book contains 17 original articles based on talks presented at the conference. The topics range from mathematical theory to industrial applications. The papers have been subject to the usual peer-review process, and we thank both the authors and the reviewers for their hard work and helpful collaboration. We wish to thank all those who supported and helped organize the conference. It is a pleasure to acknowledge the generous financial support from the Research Council of Norway and the Department of Mathematics at the University of Oslo. We also thank Elisabeth Seland, Espen Sande, and Espen Lybekk for their help with administrative and technical matters, and Anne Berg Floater for help with the registration.

June 2017

Michael Floater
Tom Lyche
Marie-Laurence Mazure
Knut Mørken
Larry L. Schumaker

Organization

Organizing Committee

Morten Dæhlen	University of Oslo, Norway
Michael Floater	University of Oslo, Norway
Tom Lyche	University of Oslo, Norway
Marie-Laurence Mazure	UJF Grenoble, France
Knut Mørken	University of Oslo, Norway
Larry L. Schumaker	Vanderbilt University, Nashville, TN, USA

Invited Speakers

Gershon Elber	Technion, Israel
Greg Fasshauer	Colorado School of Mines, USA
Leif Kobbelt	RWTH Aachen University, Germany
Helmut Pottman	Technical University Vienna, Austria
Ulrich Reif	Technical University of Darmstadt, Germany
Giancarlo Sangalli	University of Pavia, Italy
Scott Schaefer	Texas A&M University, USA
Hal Schenk	University of Illinois, USA
Wenping Wang	University of Hong Kong, SAR China

Mini-Symposia Organizers

Heidi Dahl	SINTEF, Norway
Carlotta Giannelli	University of Florence, Italy
Ying He	Nanyang Technological University, Singapore
Gitta Kutyniok	Technical University of Berlin, Germany
Jean-Louis Merrien	University of Rennes 1, France
Tomas Sauer	University of Passau, Germany
Grady Wright	Boise State University, USA

Contents

Computational Assessment of Curvatures and Principal Directions of Implicit Surfaces from 3D Scalar Data

Eric Albin[1]([✉]), Ronnie Knikker[1], Shihe Xin[1], Christian Oliver Paschereit[2], and Yves D'Angelo[3]

[1] Univ. Lyon, CNRS, INSA-Lyon, Université Claude Bernard Lyon 1, CETHIL UMR5008, F-69621 Villeurbanne, France
eric.albin@univ-lyon1.fr
[2] Institute of Fluid Dynamics and Technical Acoustics, Hermann-Föttinger-Institut., Technische Universität Berlin, Müller-Breslau-Str. 8, 10623 Berlin, Germany
[3] Laboratoire de Mathématiques J.A. Dieudonné, CNRS UMR 7351, Université de Nice Sophia-Antipolis, Parc Valrose, 06108 Nice, France

Abstract. An implicit method based on high-order differentiation to determine the mean, Gaussian and principal curvatures of implicit surfaces from a three-dimensional scalar field is presented and assessed. The method also determines normal vectors and principal directions. Compared to explicit methods, the implicit approach shows robustness and improved accuracy to measure curvatures of implicit surfaces. This is evaluated on simple cases where curvature is known in closed-form. The method is applied to compute the curvatures of wrinkled flames on large triangular unstructured meshes (namely a 3D isosurface of temperature).

Keywords: Implicit surface · Curvature · Principal directions · Isosurface · 3D scalar field · Combustion analysis

Introduction

Curvatures play an important role in many areas of physics where interfaces are encountered [13,20]. For instance, local curvatures can modify combustion speed [1,8,9,38], surface tension [4] and evaporation speed [7]. Accurate methods for determining curvatures can then be of paramount importance for flow analysis. In the recent study of Yu and Bai [38], local mean and Gaussian curvatures are computed to analyse deflagration fronts during an auto-ignition. Strong saddle front and small sphere front are shown to play an important role during the auto-ignition process. However, numerical techniques used to estimate curvatures are not detailed and plots with curvatures show a large scattering of data. The present work proposes implicit methods to improve such analyses. These methods measure curvatures and principal directions of isosurfaces extracted from 3D data. Many methods use an explicit representation of surfaces, where

M. Floater et al. (Eds.): MMCS 2016, LNCS 10521, pp. 1–22, 2017.
https://doi.org/10.1007/978-3-319-67885-6_1

the surface is discretized by elementary 2D cells like triangles, but very few make use of values of the scalar function that defines implicitly the surface.

The visualization of implicit surfaces is however commonly encountered in magnetic resonance, tomography [19] or in the post-processing of three dimensional data of numerical simulation [31]. The implicit description of surfaces is also used in level set methods [26] where a scalar function $\phi(x, y, z)$ is introduced for computational purpose. The use of such an implicit representation can handle complex changes of topology that may be encountered e.g. in primary break-up of sprays [17] or in turbulent flame propagation [36]. The dynamics of explicit three-dimensional flame surfaces can be computed, but requires complex treatment of intersections (coalescence, break-up) [9]. Whereas the mean curvature is often deduced from the divergence of $\nabla\phi$ [26], the Gaussian curvature is usually not computed. However, a similar expression for the Gaussian curvature for implicitly defined surfaces, provided in the recent studies of Trott [35] and Goldman [14], is of interest to deduce minimal and maximal curvatures for flow analysis. Note that Goldman also extends curvature formulas to higher dimension than 3D but does not discuss about principal directions. In [18], Lehmann and Reif detail how to build the curvature tensor. No explicit formulas are given for principal curvatures or principal directions but their methodology allows to compute them independently of a given parametrization by searching eigenvalues and eigenvectors. In our study, a scalar formulation of Goldmann's expression and an algorithm derived from [18] are validated through numerical tests with high order diffuse approximation and Lagrange differentiation schemes.

Implicit surfaces of 3D scalar data are commonly extracted by a marching cube method to be visualized or analysed. This method has been introduced by Lorensen and Cline [19] to extract an explicit unstructured mesh of the isosurface using triangles including topological information. The obtained isocontour is a three-dimensional surface. It is for instance used for flow visualization in the open-source software ParaView [31] or in the GTS library [29, 34]. Various algorithms are available to determine curvatures from a triangle mesh [10, 15, 22, 24, 27, 28]. Such algorithms may be used to compute curvatures of the explicit mesh representing the isosurface. We call them by *"explicit methods"* in opposition to *"implicit methods"* that use 3D scalar data. A comparison between many usual methods can be found in Gatze et al. [12]. For instance, local fitting methods of analytical surface with a least square approximation allow to approximate mean and Gaussian curvatures [12, 22]. Meyer et al. [24] define several discrete operators to compute normal vectors and curvatures. If discrete methods [11, 24] are appealing from a computational point of view compared to fitting methods, they suffer from the noise and regularity issues [12]. In this paper, we show that the present 'implicit method' to determine curvatures is more accurate and less sensitive to surface discretization errors than these 'explicit methods' to analyse isosurfaces from 3D scalar data.

In Sect. 1, we gather some formula of the literature allowing to compute the principal curvatures of implicit surfaces and propose an algorithm to determine corresponding principal directions. High order differentiation schemes used in 'implicit methods' are briefly presented in Sect. 2. The accuracy of the proposed implicit methods to determine curvatures, normals and principal directions is assessed in Sect. 3. The method is also tested to measure curvatures of isocontours extracted from 3D scalar fields and compared to 'explicit methods' in terms of accuracy and speed. It is then applied on large data of Direct Numerical Simulations of expanding and imploding flames in Sect. 4. Concluding remarks end the paper in Sect. 5.

1   Curvatures and Principal Directions of Implicit Surfaces

In level set methods, it is common to use a 3D scalar function $\phi(x, y, z)$ to implicitly define an interface. The interface is then a 3D surface defined as the isocontour $\phi(x, y, z) = \phi_0 = \text{cst}$. Some geometrical properties like normals $\boldsymbol{n}$ and mean curvatures κ_H are deduced from ϕ using formulas (1) and (2):

$$\boldsymbol{n} = \frac{\boldsymbol{\nabla}\phi}{\|\boldsymbol{\nabla}\phi\|} = \frac{\boldsymbol{\nabla}\phi}{[(\partial\phi/\partial x)^2 + (\partial\phi/\partial y)^2 + (\partial\phi/\partial z)^2]^{1/2}} \tag{1}$$

$$\kappa_H = \frac{\boldsymbol{\nabla}\cdot\boldsymbol{n}}{2} = \frac{\phi_x^2(\phi_{yy} + \phi_{zz}) + \phi_y^2(\phi_{xx} + \phi_{zz}) + \phi_z^2(\phi_{xx} + \phi_{yy})}{2\cdot(\phi_x^2 + \phi_y^2 + \phi_z^2)^{3/2}}$$

$$- \frac{\phi_x\phi_y\phi_{xy} + \phi_x\phi_z\phi_{xz} + \phi_y\phi_z\phi_{yz}}{(\phi_x^2 + \phi_y^2 + \phi_z^2)^{3/2}} \tag{2a}$$

$$= \frac{\phi_{uu} + \phi_{vv}}{2\cdot|\phi_n|} \tag{2b}$$

The two-dimensional version of Eq. (2a) was first used in the context of the level set method by Chang et al. [6]. The three-dimensional version can be found for example in Osher and Fedkiw [26, p. 12]. In an orthonormal frame $(\boldsymbol{n}, \boldsymbol{u}, \boldsymbol{v})$ where $\boldsymbol{u}$ and $\boldsymbol{v}$ are arbitrary vectors defining the tangent plane to the surface, the formula (2a) reduces to a simpler expression (2b).

A similar expression for the Gaussian curvature κ_K is derived in Goldman (see Eq. (4.1) in [14]) where κ_K is expressed as a function of the gradient $\boldsymbol{\nabla}\phi$, the Hessian matrix $H(\phi)$ and its adjugate matrix. A scalar formulation is also derived using the Mathematica software in Trott [35, pp. 1285–1286]. The formulation (3a) used in this work is a scalar formulation similar to [35] but slightly reformulated to prevent a division by zero when ϕ_z vanishes:

$$\kappa_K = 2\frac{\phi_x\phi_y(\phi_{xz}\phi_{yz} - \phi_{xy}\phi_{zz}) + \phi_x\phi_z(\phi_{xy}\phi_{yz} - \phi_{xz}\phi_{yy}) + \phi_y\phi_z(\phi_{xy}\phi_{xz} - \phi_{yz}\phi_{xx})}{(\phi_x^2 + \phi_y^2 + \phi_z^2)^2}$$

$$+ \frac{\phi_x^2\cdot(\phi_{yy}\phi_{zz} - \phi_{yz}^2) + \phi_y^2\cdot(\phi_{xx}\phi_{zz} - \phi_{xz}^2) + \phi_z^2\cdot(\phi_{xx}\phi_{yy} - \phi_{xy}^2)}{(\phi_x^2 + \phi_y^2 + \phi_z^2)^2} \tag{3a}$$

$$= \frac{\phi_{uu}\phi_{vv} - \phi_{uv}^2}{\phi_n^2} \tag{3b}$$

Expressions (2b) and (3b) are also the eigenvalues of the curvature tensor as explained in [18] and shown in the annex A. The annex A also shows how to derive expressions in the (x, y, z) basis from ϕ_{uu}, ϕ_{uv} and ϕ_{vv} using change of basis (15). Since $\kappa_K = \kappa_{min} \cdot \kappa_{max}$ and $\kappa_H = (\kappa_{min} + \kappa_{max})/2$, the minimal and maximal curvatures are then deduced from the mean and Gaussian curvatures:

$$\kappa_{min} = \kappa_H - \sqrt{|\kappa_H^2 - \kappa_K|} \tag{4a}$$

$$\kappa_{max} = \kappa_H + \sqrt{|\kappa_H^2 - \kappa_K|} \tag{4b}$$

With the chosen convention of orientation, normal vectors point to large values of ϕ and curvatures are negative when normals converge towards high ϕ values.

At umbilical points where $\kappa_{min} = \kappa_{max}$, any direction is a principal direction. However, directions of the minimal and maximal curvatures of implicit surfaces are single solutions at non-umbilical points and may be computed by searching eigenvectors of the curvature tensor, as detailed in Lehmann and Reif [18]. Rather than using an eigenvector solver, we derive in annex A expressions to directly compute principal directions t_1, t_2 corresponding to principal curvatures $\kappa_1 = \kappa_H - \sqrt{|\kappa_H^2 - \kappa_K|} \cdot \zeta$, $\kappa_2 = \kappa_H + \sqrt{|\kappa_H^2 - \kappa_K|} \cdot \zeta$:

$$t_1 = \begin{bmatrix} 0 \\ \phi_{uv} \\ \kappa_1\phi_n - \phi_{uu} \end{bmatrix}_{n,u,v} = \begin{bmatrix} (\kappa_1\phi_n - \phi_{uu}) \cdot v_x + \phi_{uv}u_x \\ (\kappa_1\phi_n - \phi_{uu}) \cdot v_y + \phi_{uv}u_y \\ (\kappa_1\phi_n - \phi_{uu}) \cdot v_z + \phi_{uv}u_z \end{bmatrix}_{x,y,z} \tag{5a}$$

$$t_2 = \begin{bmatrix} 0 \\ \kappa_2\phi_n - \phi_{vv} \\ \phi_{uv} \end{bmatrix}_{n,u,v} = \begin{bmatrix} (\kappa_2\phi_n - \phi_{vv}) \cdot u_x + \phi_{uv}v_x \\ (\kappa_2\phi_n - \phi_{vv}) \cdot u_y + \phi_{uv}v_y \\ (\kappa_2\phi_n - \phi_{vv}) \cdot u_z + \phi_{uv}v_z \end{bmatrix}_{x,y,z} \tag{5b}$$

Principal direction coordinates into brackets are given both in the (n, u, v) and (x, y, z) bases. The subscript notations in (5a) indicate the coordinate system used for the vector decomposition. These expressions are quite similar to those found in the literature dealing with the computation of crest lines [5,25,33], except we introduce a sign function $\zeta = \pm1$ to circumvent degeneracies (see case studies proposed at the end of annex A). Because a wrong choice of u, v may yield to find $t_1 = t_2 = 0$, we propose a criterion on ζ to ensure that the found vectors are non-null for any arbitrary choice of u, v. Since Eq. (2b) implies $|\kappa_{min}\phi_n - \phi_{uu}| = |\kappa_{max}\phi_n - \phi_{vv}|$, we propose the following algorithm in order to compute t_{min} and t_{max} associated to κ_{min} and κ_{max}:

choose arbitrarily u, v from n;
compute ϕ_{uu} and ϕ_{vv} with system (15);
if $|\kappa_{min}\phi_n - \phi_{uu}| \geqslant |\kappa_{min}\phi_n - \phi_{vv}|$ **then**
| choose $\zeta = +1$ to avoid $\kappa_1\phi_n - \phi_{uu} = -(\kappa_2\phi_n - \phi_{vv}) = 0$;
| compute $t_{min} = t_1$ and $t_{max} = -t_2$ with $\kappa_{min} = \kappa_1$ and $\kappa_{max} = \kappa_2$;
else
| choose $\zeta = -1$ since $|\kappa_{max}\phi_n - \phi_{uu}| > |\kappa_{max}\phi_n - \phi_{vv}| \geqslant 0$;
| compute $t_{min} = -t_2$ and $t_{max} = t_1$ with $\kappa_{max} = \kappa_1$ and $\kappa_{min} = \kappa_2$;
end

This algorithm requires to arbitrarily choose (u, v) and compute ϕ_{uu} and ϕ_{vv} to determine κ_H, κ_K, t_1 and t_2 with explicit expressions (2b), (3b), (5a) and (5b). An alternative choice would be to build the curvature tensor in standard coordinates [18] and use a specific solver for searching eigenvalues and eigenvectors. If principal directions are not required, curvatures are computed by the *intrinsic* formulas (2a) and (3a) that do not require any choice of u, v or eigen solver. Some exercises are also proposed at the end of annex A to better understand the following algorithm. This algorithm implies that the basis (t_{min}, t_{max}, n) is direct and:

$$\kappa_{min} = \kappa_1 \frac{1+\zeta}{2} + \kappa_2 \frac{1-\zeta}{2} \quad \kappa_{max} = \kappa_1 \frac{1-\zeta}{2} + \kappa_2 \frac{1+\zeta}{2} \tag{6a}$$

$$t_{min} = t_1 \frac{1+\zeta}{2} + t_2 \frac{1-\zeta}{2} \quad t_{max} = t_1 \frac{1-\zeta}{2} - t_2 \frac{1+\zeta}{2} \tag{6b}$$

2 High Order Differentiation Schemes for Implicit Methods

In order to compute curvatures and principal directions using Eqs. (1) to (5), it is necessary to compute second-order cross derivatives of ϕ at a given point $P(x, y, z)$ from neighbor points P_i where ϕ is discretized. The point P where curvatures have to be computed is not necessarily lying on one of the nodes P_i. Figure 1 shows a schematic discretization of ϕ.

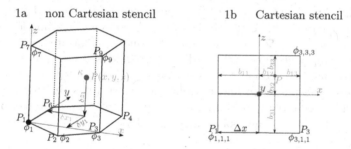

Fig. 1. Schematic representation of a point P where curvatures are computed from a 3D discrete scalar field ϕ_j. Coefficients $(\delta x_j, \delta y_j, \delta z_j)$ or b_{ij} represent the position of the neighbor discrete points P_j to P.

2.1 Diffuse Approximation (DA)

Diffuse approximations (DA) allow to compute all derivatives of ϕ at P from the values ϕ_i known at points P_i of coordinates $(x + \delta x_i, y + \delta y_i, z + \delta z_i)$ with any kind of mesh [30]. Coefficients δx_i, δy_i and δz_i are then small real numbers defining the position of the ith neighbor point P_i relatively to P (see Fig. 1a). The value of the scalar at a point P_i is estimated by a Taylor expansion $\phi_i^* = \sum_{j=1}^{N_u} P_{ij} \alpha_j$. The matrices P and α are given as an example for a second-order

Taylor expansion ($N = 2$, $N_u = 10$) but can be similarly constructed for a fourth order $N = 4$:

$$\boldsymbol{P} = [P_{ij}] = \left[1, \ \delta x_i, \ \delta y_i, \ \delta z_i, \ \delta x_i^2, \ \delta y_i^2, \ \delta z_i^2, \ \delta x_i \delta y_i, \ \delta x_i \delta z_i, \ \delta y_i \delta z_i \right] \quad (7a)$$

$$\boldsymbol{\alpha} = [\alpha_j] = \left[\phi, \ \phi_x, \ \phi_y, \ \phi_z, \ \frac{\phi_{xx}}{2}, \ \frac{\phi_{yy}}{2}, \ \frac{\phi_{zz}}{2}, \ \phi_{xy}, \ \phi_{xz}, \phi_{yz} \right] \quad (7b)$$

If a number N_n of neighbor nodes are used around P, the method consists in minimizing the quadratic error $I(\boldsymbol{\alpha}) = \sum_{i=1}^{N_n} \omega_i (\phi_i - \phi_i^*)^2$ weighted by $\omega_i = e^{-(\delta x_i^2 + \delta y_i^2 + \delta z_i^2)/\Delta x^2}$. This minimization yields a linear system $\boldsymbol{A} \cdot \boldsymbol{\alpha} = \boldsymbol{B}$ to solve with:

$$\boldsymbol{A} = [A_{kj}] = \left[\sum_{i=1}^{N_n} \omega_i \cdot P_{ik} \cdot P_{ij} \right] \qquad \text{with} \quad j \in [1; N_u] \qquad (8a)$$

$$\boldsymbol{B} = [B_k] = \left[\sum_{i=1}^{N_n} \omega_i \cdot P_{ik} \cdot \phi_i \right] \qquad (8b)$$

$\boldsymbol{P}$ is a matrix of size $N_n \times N_u$. The searched vector $\boldsymbol{\alpha}$ of size N_u is then computed after inverting the matrix $\boldsymbol{A}$ of size N_u^2. Minimizing the quadratic error with a Taylor expansion of order $N + 1$ implies a Nth-order accuracy for the first derivatives and an order $N - 1$ for the second derivatives.

Note that Marchandise et al. [21], in the context of two-phase flows, already used a least-squares method (2nd-order DA with no weighting function: $\omega_i = 1$) and equation $\kappa_H = -\boldsymbol{\nabla} \cdot \boldsymbol{n}$ to compute the mean curvature with a slightly different two-step procedure. A 2nd-order DA method is also used in the coupled VOF-level set method proposed by Sussmann et al. [23,32], but only to compute the normal of the piecewise linear surface reconstructions. In this paper, the DA method is used directly to compute principal curvatures and directions with higher order accuracy at any point on the interface.

2.2 Lagrange Differentiation (LD)

High-order differentiation with Lagrange polynomials are also tested to compute principal curvatures and directions if the implicit function ϕ is discretized on a Cartesian grid (Fig. 1b shows a Cartesian stencil as an example with $n = 3$). First, second and cross derivatives like ϕ_x, ϕ_{zz} and ϕ_{xy} are computed by weighting neighbour values of the implicit function over the Cartesian stencil:

$$\left.\frac{\partial \phi}{\partial x_1}\right|_P = \phi_x = \sum_{i=1}^{n} \sum_{j=1}^{n} \sum_{k=1}^{n} \dot{\bar{\beta}}_{1i} \cdot \bar{\beta}_{2j} \cdot \bar{\beta}_{3k} \cdot \phi_{i,j,k} \qquad (9a)$$

$$\left.\frac{\partial \phi}{\partial x_3^2}\right|_P = \phi_{zz} = \sum_{i=1}^{n} \sum_{j=1}^{n} \sum_{k=1}^{n} \bar{\beta}_{1i} \cdot \bar{\beta}_{2j} \cdot \ddot{\bar{\beta}}_{3k} \cdot \phi_{i,j,k} \qquad (9b)$$

$$\left.\frac{\partial^2 \phi}{\partial x_1 \partial x_2}\right|_P = \phi_{xy} = \sum_{i=1}^{n} \sum_{j=1}^{n} \sum_{k=1}^{n} \dot{\bar{\beta}}_{1i} \cdot \dot{\bar{\beta}}_{2j} \cdot \bar{\beta}_{3k} \cdot \phi_{i,j,k} \qquad (9c)$$

Weighting coefficients $\bar{\beta}_{ij}$, $\dot{\beta}_{ij}$ or $\ddot{\beta}_{ij}$ correspond respectively to an interpolation, a first differentiation or a second differentiation in the x_i direction. Their expression are similar to Lagrange basis polynomials [16] and their derivatives, they are obtained from the inversion of the Vandermonde matrix [2]:

$$\bar{\beta}_{ij} = \left(\prod_{\substack{k=1 \\ k \neq j}}^{n} b_{ik} \right) / A_{ij} \quad \text{with} \quad A_{ij} = \prod_{\substack{k=1 \\ k \neq j}}^{n} (b_{ik} - b_{ij}) \tag{10a}$$

$$\dot{\beta}_{ij} = -\left(\sum_{\substack{k=1 \\ k \neq j}}^{n} \prod_{\substack{l=1 \\ l \neq j,k}}^{n} b_{il} \right) / (A_{ij} \cdot \Delta x) \tag{10b}$$

$$\ddot{\beta}_{ij} = \left(\sum_{\substack{k=1 \\ k \neq j}}^{n} \sum_{\substack{l=1 \\ l \neq j,k}}^{n} \prod_{\substack{m=1 \\ m \neq j,k,l}}^{n} b_{im} \right) / (A_{ij} \cdot \Delta x^2) \tag{10c}$$

These coefficients can be directly computed from b_{ij} that represents the position of the point P in the x_i direction within the stencil (see Fig. 1b). As an example, the reader may compute $\bar{\beta}_{ij}$, $\dot{\beta}_{ij}$ or $\ddot{\beta}_{ij}$ for $j \in [1, 2, n = 3]$ with $b_{i1} = -\Delta x$, $b_{i2} = 0$, $b_{i3} = \Delta x$ to recover well-known coefficients of finite differences.

3   Test of Implicit Methods

3.1   Accuracy Assessment of Implicit Methods

A spherical surface with known curvature is used to test the accuracy of the method. The radius of this sphere centered at the point C is denoted R and the parameter δ controls the stiffness of the scalar function using a hyperbolic tangent profile:

$$\phi(x, y, z) = \tanh\left((r - R) / \delta \right) \tag{11}$$
$$r = \sqrt{(x - x_C)^2 + (y - y_C)^2 + (z - z_C)^2}$$
$$x_C = x_O + R \sin\phi \cos\theta$$
$$y_C = y_O + R \sin\phi \sin\theta$$
$$z_C = z_O + R \cos\phi$$

Figure 2 shows a schematic representation of this spherical function around a cubic stencil of center O. In the present analysis, the stencils are defined on Cartesian grids, with $N_n = n^3$ points in order to allow comparisons with LD and DA differentiation schemes. The implicit spherical function is characterized by its radius R, angles (θ, ϕ) and stiffness $1/\delta$ in Eq. (11).

The accuracy of the implicit method for computing curvatures is investigated by varying the surface normal via (θ, ϕ) and the spherical function via R and δ. Figure 3a shows the relative error on the mean-curvature $\Delta\kappa_H / \kappa_H = \kappa_H \cdot R - 1$ determined at point O when varying the normal angles ($\theta \in [0; 2\pi]$, $\phi \in [0; \pi]$,

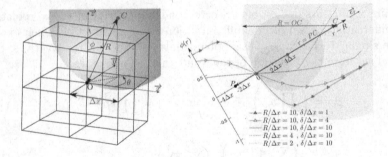

Fig. 2. Representation of a volumetric spherical function with known curvatures.

$R/\Delta x = 10$, $\delta/\Delta x = 2$) with the second-order diffuse approximation method ($N = 2$, $N_n = 5^3$). Because the expressions (2) and (3) and the stencil present a symmetry towards planes $(O, \vec{x})$, $(O, \vec{y})$ and $(O, \vec{z})$, the error is periodical along θ and ϕ directions and the study of errors could be limited to one eighth of the domain. The maximal error is obtained at some positions of the curvature center and can be extracted from these curves; it will be denoted $\max|\Delta \kappa_H / \kappa_H|$. The maximal error on this domain is subsequently plotted for various radius R and stiffness $1/\delta$ in Fig. 3b. The error is higher for stiff or highly-curved functions. The stiffness and the curvature have a similar influence on the error that is then controlled by the most critical parameter between high curvatures and high stiffness.

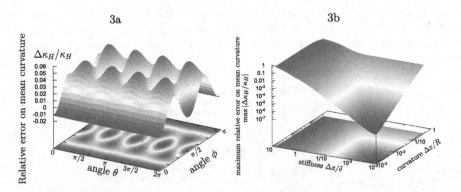

Fig. 3. Measurements of the relative error when computing the mean curvature at the center O of a stencil of size $\pm 2\Delta x$ with a second-order diffuse approximation.

Figure 4 shows the maximum relative errors on the mean curvatures computed at the center of the stencil as a function of the critical parameters (stiffness or curvature) to compare the effect of the numerical schemes on the error. The obtained error on Gaussian curvatures are really similar to mean curvature errors and are not shown for brevity. Errors decrease for smooth function and

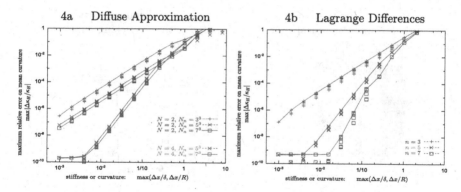

Fig. 4. Relative error on mean curvatures for various stencil sizes ($N_n = n^3$ with $n \in \{3; 5; 7\}$) and numerical schemes (accuracy order: N for DA and n for LD schemes).

weakly curved surfaces. Implicit methods with Lagrange differentiation (LD) and diffuse approximation (DA) present similar accuracies. For a second-order DA ($N = 2$) with $N_n = 3^3$ points, the error is similar to the 2nd-order LD ($n = 3$). However, this error decreases when the stencil is larger with second-order DA ($N_n = 5^3$ or $N_n = 7^3$). Accuracies on curvatures increase for high-order LD or DA numerical schemes. They are similar with 4th order schemes (DA with $N = 4$, LD with $n = 5$). The implicit method with a 6th-order LD scheme for first-derivatives ($n = 7$) reaches lowest errors in this test where no interpolation inside stencils is performed.

3.2 Accuracy Comparison Using a Marching Cube Extraction of Isocontours

The accuracy of proposed implicit methods is now compared to standard explicit methods when computing mean and Gaussian curvatures of a sphere with varying sphere resolution $R/\Delta x$. In this test, the initial data is for all cases an implicit 3D spherical function defined by Eq. (11) and centered at $x_C = y_C = z_C = 0$ in a cubic box of size $[-\frac{N}{2}\Delta x; \frac{N}{2}\Delta x]$ large enough to contain the radius R and stencil points; the stiffness is chosen equal to $\frac{\Delta x}{R}$. As shown in Fig. 5a, an explicit irregular triangulated surface of this sphere of radius R is extracted using a marching cube method [19]. Figure 5b illustrates how triangle meshes are generated from linear interpolation on edges of the Cartesian grid with a marching cube method (cf. edges P_1P_2, P_1P_3, P_5P_6 and P_5P_7).

Mean and Gaussian curvatures are then computed at the nodes of this spherical isocontour with various implicit and explicit methods for comparisons. The first explicit method is classical and consists in fitting full quadric [12,22]. Discrete explicit methods implemented in the ParaView software [31] or the GTS library [29] are also tested. The mean curvature is computed in ParaView from the length of neighbor edges and dihedral angles between normals of neighbor facets [11] while it is deduced from the 'Mean Curvature Normal Operator' in

5a   Extracted isocontour from a spherical function ($\frac{R}{\Delta x} = 4$).

5b   Triangles generated with a marching cube method.

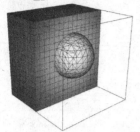

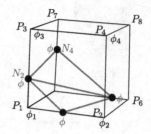

Fig. 5. Illustration of the marching cube method used to generate an isocontour representing the implicit interface contained in 3D data.

GTS [24]. The Gaussian curvature is computed with angle deficit methods both in ParaView and GTS [24] but with a more complex estimation of the area in the GTS library to deal with obtuse triangles. The implicit methods are based on diffuse approximation or Lagrange differentiation, as described in Sect. 2.

Figure 6a plots the maximum error on mean curvatures in log-scale as a function of the sphere resolution. Once again, relative errors on κ_K, κ_{min} and κ_{max} are very similar to mean curvature errors and are not shown for brevity. The explicit discrete methods of ParaView or GTS present both relative errors larger than 100% on curvatures. These discrete methods fail in estimating curvatures because the marching cube method generates a very irregular mesh of the sphere with non-uniform triangles (see Fig. 5a and Sect. 3.3). Fitting methods compute reasonable estimations of curvatures with relative errors about 10%

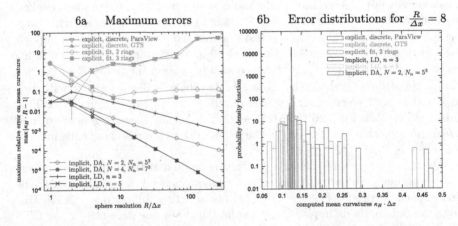

Fig. 6. Errors when computing mean curvatures of a spherical isocontour extracted from a 3D implicit function using a marching cube method.

with sufficiently refined spheres ($R/\Delta x \geq 8$). Note that increasing the number of rings of neighbor nodes to fit paraboloids slightly increased the accuracy with such irregular meshes. Implicit methods significantly improves the accuracy to estimate curvatures of an isocontour generated with a marching cube method. The relative error on curvatures is below 1% for sufficiently refined spheres and about 10% at low resolution ($\frac{R}{\Delta x}$ or $\frac{\delta}{\Delta x} \simeq 1$).

The error distributions of the computed curvatures are also plotted for $R/\Delta x = 8$ in Fig. 6b. The large width and flatness of distributions for discrete methods demonstrate that these methods are inaccurate on a large number of nodes and not only at an isolated node. The implicit methods with a second-order diffuse approximation and Lagrange differentiation with higher order ($n \geq 5$) present the narrowest distributions and therefore highest accuracies.

3.3   Accuracy Comparison Using a Regular Triangle Mesh

Compared to the previous test-case, the triangle mesh of the spherical isosurface is no more extracted by a marching cube method. It is '*artificially*' generated from the subdivision of an icosahedron constituted of $N_{\triangle,0} = 20$ equilateral triangles and 12 nodes. The ith refined icosahedron has $N_\triangle = 4^i \cdot N_{\triangle,0}$ equilateral triangles. In this test, curvatures are then computed at the nodes of a very regular triangle mesh of a sphere. This regular mesh, constituted of $N_\triangle$ equilateral triangles of side Δx, is artificially positioned at a radius $\frac{R}{\Delta x} = (N_\triangle \sqrt{3}/16\pi)^{1/2}$ since $4\pi R^2 \simeq N_\triangle \cdot \mathcal{A}_\triangle$, with $\mathcal{A}_\triangle = \sqrt{3}\Delta x^2/4$ being the area of an equilateral triangle. The implicit spherical function is then defined with a stiffness $\frac{\Delta x}{\delta} = \frac{\Delta x}{R}$ to allow comparisons between implicit/explicit methods.

The maximum errors on mean and Gaussian are plotted in log-scale as a function of the sphere resolution $R/\Delta x$ in Fig. 7a. All explicit methods exhibit a better accuracy with such artificially generated regular meshes than with previous irregular meshes. If the built-in discrete method of ParaView reaches about 10% of accuracy, the GTS discrete method reaches accuracies under 1% and even under 0.01% for the mean curvatures. The accuracy is also increased for fitting methods with such regular meshes even if less accurate than the GTS discrete method. Implicit methods still reach highest accuracies on curvatures even if nodes of the *artificial* isosurface are located any-where within the stencil where ϕ is discretized, which requires more interpolations than previous test-cases. Second-order numerical schemes (DA with $N = 2$ and LD with $n = 3$) are 1st-order accurate to compute curvatures whereas 4th-order (LD with $n = 5$ or DA with $N = 4$) are 3rd-order accurate. Implicit methods with high order schemes reach accuracies better than 1% for reasonable resolutions.

The error distributions at a moderate resolution $R/\Delta x \simeq 6.64$ are also shown in Fig. 7b. These distributions confirm that the errors are better bounded with regular meshes for explicit methods. The explicit discrete GTS methods and implicit methods are the most accurate since the error distribution is very narrow and centered around analytical solution. A shift from the analytical solution is present with the explicit fitting method and the computed curvatures are very dispersed with the discrete method implemented in ParaView.

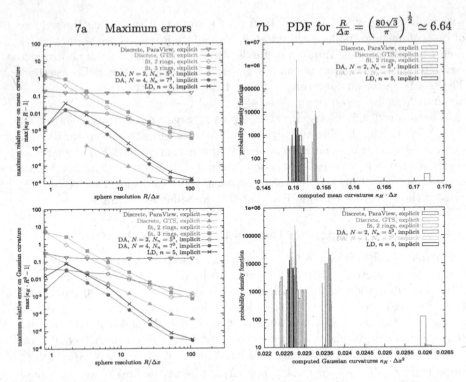

Fig. 7. Errors when computing mean and Gaussian curvatures of a regular triangle mesh (equilateral triangles artificially generated to avoid mesh irregularities of the isosurface).

The two previous test-cases show that explicit methods are really dependent on the regularity of the isosurface mesh. In opposite, the accuracy of implicit methods does not depend on the procedure used to extract the isocontour. It depends only on neighbor 3D scalar values and numerical differentiation schemes.

3.4   Accuracy Assessment to Measure Principal Directions

A donut is a nice geometry to test the proposed algorithm that computes principal directions. Indeed, this geometry presents elliptical points where the surface is convex ($\kappa_K > 0$), hyperbolic points where the surface is saddle shaped ($\kappa_K < 0$) and parabolic points ($\kappa_K = 0$) but does not contain umbillical points ($\kappa_{min} = \kappa_{max}$) where principal directions are undefined. An implicit function $\phi(x,y,z) = \tanh(\frac{\{((x^2+y^2)^{1/2}-R)^2+z^2\}^{1/2}-r}{\delta})$ is used to describe the torus with $r = \delta = \frac{R}{2}$. Figure 8a illustrates the implicit toric function as a cheese chunk and principal directions computed on an extracted isocontour with a marching cube method for a resolution $\frac{r}{\Delta x} = 4$.

An error between computed vectors and analytical solutions is then defined by $\|\boldsymbol{u} \wedge \boldsymbol{u}_{sol}\| = \sin(\widehat{\boldsymbol{u}, \boldsymbol{u}_{sol}})$ at each point of isocontours. Figure 8b then plots the

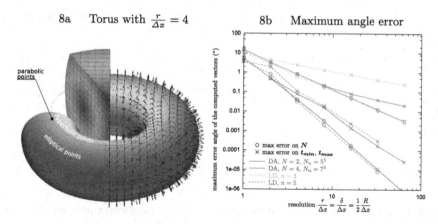

Fig. 8. Measurement of normals n and principal directions t_{min}, t_{max} of a torus with implicit methods. The maximum angle error of all computed vectors are plotted in Fig. b for various numerical schemes.

maximum angle error on computed normals and principal directions as a function of the torus resolution and for various differentiation schemes. The accuracy of implicit methods increases with the resolution and high order schemes. The largest angle errors are obtained with the 2nd-order LD implicit method that uses the smallest stencil. They are below $1°$ with a second order diffuse approximation and below $0.1°$ with 4th-order LD or DA schemes for sufficiently refined cases $\frac{\delta}{\Delta x} \geqslant 4$. The algorithm has been tested successfully through other cases that are not described here to be concise.

3.5 Speed Assessment of the Implicit Method

The CPU time of implicit methods depends on the used numerical scheme (LD, DA and N) and the stencil size $N_n = n^3$. Lagrange differentiation methods are faster than diffuse approximation ones for the same stencil size (see Fig. 9a). LD methods allow to reach high orders with less CPU time.

The speed of the implicit method is also compared to those of discrete and fitting methods in Fig. 9b. All programs are written in $C\#$ for this speed comparison. The speed increases linearly with the number of points on the discrete sphere for all methods. Implicit methods with Lagrange differentiation reach the highest speeds. The fourth-order LD ($n = 5$) and 2nd-order DA ($N_n = 3^3$) have comparable speeds to the built-in discrete method of the GTS library and the two-ring fitting method.

The curvature accuracy of these different methods is compared when computing Gaussian and mean curvatures of a same mesh extracted from a spherical implicit function with marching cubes like in Sect. 3.2. Figure 10a plots accuracy vs computation time for a coarse mesh and Fig. 10b for a finer mesh. These diagrams show that LD implicit methods reach the best compromises between speed and accuracy. Discrete method and the 1-ring fitting method fail in predicting

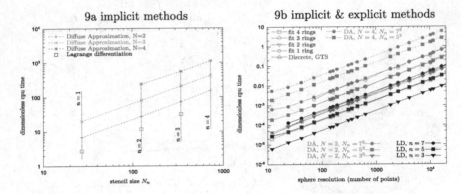

Fig. 9. Comparison of the computational speed to compute curvatures for different methods.

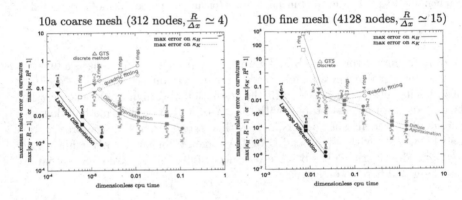

Fig. 10. Comparison diagram of computation time vs accuracy for various implicit and explicit methods. Curvatures are measured on a coarse and refined mesh generated from a spherical implicit function with marching cubes.

accurately curvatures at some nodes. DA implicit methods allow to reach higher accuracies than quadric fitting method but may require an additional cpu cost.

4   Applications

These methods are applied to determine curvatures of expanding and imploding flames. The flame images are obtained from Direct Numerical Simulations using the H-allegro in-house software [1] on a supercomputing infrastructure (PRACE).

4.1   Laminar Flames

An unburnt methane-air mixture at $T_u = 480\,\mathrm{K}$ is ignited at atmospheric pressure with a Gaussian profile in temperature and composition. The adiabatic

flame temperature of such a mixture is $T_b = 2260\,\mathrm{K}$ and the planar thermal
flame thickness is $\delta_L^0 = 202\,\mu\mathrm{m}$. The computational domain is a cube of $1.5\,\mathrm{cm}$
width discretized with a Cartesian grid of $(336)^3$ points, which implies a sufficient
refinement to solve the flame propagation ($\delta_L^0/\Delta x \simeq 4.5$).

Curvatures of the flame front are measured in the simulation results along
the flame propagation with different methods. To determine flame curvatures,
isocontours of a $569\,\mathrm{K}$ temperature (progress variable of 5%) are first extracted
with a marching cube method [19]. The ParaView software [31] is used to handle
the multiple-files of the parallel solution (512 files for one time-step) and the
complex topology of isocontours with a triangle mesh. Fifteen quasi-spherical
isocontours are extracted along the flame propagation; the flame evolves from a
small radius $R_S \simeq 12\delta_L^0$ to a larger radius $R_S \simeq 150\delta_L^0$.

Figure 11 shows the computed local flame radius $R_H = 1/\kappa_H$ based on mean
curvatures as a function of the area based flame radius $R_S = \sqrt{S/(4\pi)}$. Curves
for the Gaussian curvatures are not shown because too similar. At each flame
radius, the discrete methods of ParaView or GTS are inaccurate, because of
the irregularity of the mesh. The scattering of the local radius is so large that
we did not plot it out for comparisons. With fitting methods, the scattering is
reduced in particular when several rings to fit parabolas are used. Nevertheless,
the dispersion is still very high even with a 4 ring stencil. Implicit methods with
large stencils ($N_n \geqslant 5^3$ points) points significantly improves the accuracy when
predicting the local flame radius based on the mean or Gaussian curvatures,
which makes curvature analysis in flows possible. The small remaining scattering
at large radii is attributed to the effect of boundary conditions [3, 37] and not to
the inaccuracy of the proposed method.

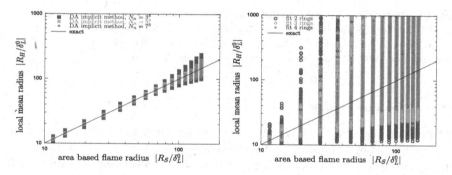

Fig. 11. Comparison of the measured local radii to the global radii based on the whole
surface when the flame propagates for DA implicit and fitting methods.

Probability density functions (PDF) of Gaussian and mean curvatures are
plotted in Fig. 12 at the middle time of the simulation ($t = 0.785\,\mathrm{ms}$). At this
stage, the flame is quasi-spherical with a radius $R \simeq -0.3\,\mathrm{cm}$. The probability
density function and cumulative function are respectively expected to be close

to a Dirac function and a Heaviside function. For the implicit method, the probability density is as expected a peak centered around $\kappa_H = 1/R$ or $\kappa_K = 1/R^2$ and the cumulative probability is the stiffest step compared to the fitting and discrete methods. The absence of peaks on the PDF for fitting and discrete methods shows that explicit methods are really inaccurate to estimate the curvatures of this stiff flame ($\delta_L^0/\Delta x \simeq 4.5$).

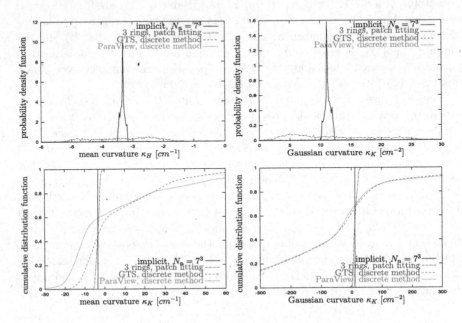

Fig. 12. Probability density function and cumulative distribution function of measured Gaussian, mean, minimal and maximal curvatures of a quasi-spherical flame ($t = 0.785\,\text{ms}$). Note that only the implicit method has a PDF close to a Dirac function and a cumulated probability close to a Heaviside function.

4.2   Turbulent Flames

A turbulent imploding hydrogen flame premixed with air diluted with 20% of steam is ignited by a Gaussian profile of temperature. After a transient, the flame propagates inward and becomes highly wrinkled. An isocontour of temperature (corresponding to a progress variable value of 50%) is extracted with a marching cube method at 1.5 ms after the ignition. The mean, Gaussian, minimum and maximum curvatures of this isocontour are then computed at every nodes of the mesh making use of (i) a 2nd-order DA implicit method with $N_n = 7^3$, (ii) a fitting method with a 3 ring patch, (iii) the explicit discrete method of GTS and (iv) the explicit discrete method of ParaView. Results are compared in Fig. 13.

If all methods seem to compute very similar curvatures, a careful look shows numerous oscillations for the discrete methods (compare zooms on κ_H in Fig. 13). Both the fitting method (with a large stencil) and the DA implicit method show

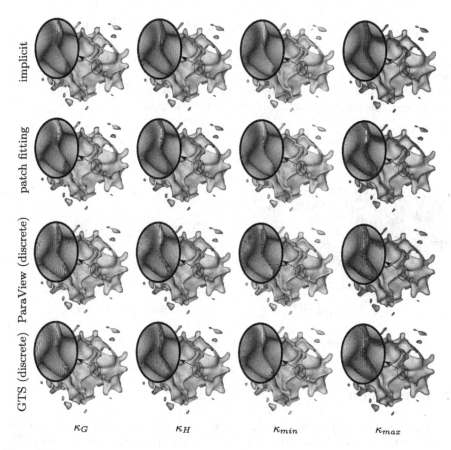

Fig. 13. Test of different methods to compute Gaussian, mean, minimal and maximal curvatures on a triangular unstructured mesh. Curvatures are plotted in colors from blue to red ($\kappa_H, \kappa_{min}, \kappa_{max} \in [-50; 50\,\mathrm{cm}^{-1}]$). (Color figure online)

very similar results. However, the fitting method fails at some node locations to fit a parabola; this method also generates visible oscillations near highly curved surface compared to the implicit method (cf. zooms). However, the similitude between results confirms the ability of the implicit method to compute mean, Gaussian, minimum and maximum curvatures, even for complex geometries. From previous test-cases, it may be expected that the implicit method give the most accurate estimates of curvatures. Explicit methods show a more acceptable prediction of curvatures compared to the inaccuracies observed on the laminar flame, this is attributed to the quality of the isosurface mesh that is better defined in the turbulent case with a different progress variable (50% vs. 5%) and a more refined flame ($\delta_L^0/\Delta x \simeq 7.8 > 4.5$).

Figure 14 shows some examples of local analyses that may be conducted to study turbulent flames with implicit methods. Some computed principal directions of the previous hydrogen imploding flame are shown in Fig. 14a.

The implicit method may be used to measure other flame properties from normals and curvatures. It may for instance be used to measure local consumption of the gaz mixture, local flame speeds, local flame stretches. Figure 14b shows a scatter plot of the local consumption speed as a function of mean and principal curvatures for a turbulent methane-air expanding flame, which highlights a link between gaz consumption and mean flame curvatures in this case.

14a   principal directions   14b   local mixture consumption

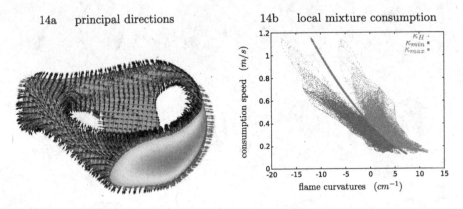

Fig. 14. Example of local flame analysis using implicit methods.

5   Conclusion

A computational method has been presented to compute curvatures and principal directions of implicit surfaces from 3D scalar data. Curvatures are computed at any vertex within the stencil where the implicit surface is discretized. This implicit method makes use of high-order Diffuse Approximation (DA) or Lagrange differentiation (LD) schemes to interpolate/differentiate the 3D scalar function. Implicit methods are compared to standard explicit methods (surface fitting, discrete methods) when computing curvatures along isocontours representing the implicit surface.

It is evidenced through numerous test cases that explicit methods fail in measuring accurately curvatures of surfaces with irregular meshes. However, the extraction of surfaces with complex topologies is not trivial and commonly used techniques like marching-cube methods do not always generate regular meshes. It is shown that implicit methods do not require regular meshes and may reach high accuracy, which makes these methods interesting to analyse curvatures of isosurfaces from 3D data. If the stiffness of the implicit function is known and not too high, the maximum error on curvatures with implicit methods may be estimated from Fig. 6a. The implicit method is also found to be competitive with discrete methods in terms of computational speed. A 2nd-order DA or LD schemes of higher order are recommended when computing curvatures to reduce the error without requiring much cpu overhead.

The proposed implicit method has been successfully applied to measure properties of flame surfaces based on normals and curvatures and shows very promising results to conduct accurate local 3D analysis of wrinkled flames (local consumption speed, displacement speeds, curvatures and stretches of the flame). A perspective of this work would be to extend this methodology to implicit surfaces defined by stiffer implicit functions using non-oscillatory schemes for interpolation/differentiation since discontinuous 3D scalar data may be encountered for instance in multiphase flows or medical images.

A   Derivation of Formulas for Principal Directions of Implicit Surfaces

The methodology of Lehmann et al. [18] is used to build the curvature tensor in the $(\boldsymbol{u}, \boldsymbol{v}, \boldsymbol{n})$ frame. In this frame, the orthogonal projector T onto the tangent space and the Hessian matrix of ϕ are respectively:

$$T = Id_3 - \boldsymbol{n} \cdot \boldsymbol{n}^t = \begin{pmatrix} 1 & 0 & 0 \\ 0 & 1 & 0 \\ 0 & 0 & 0 \end{pmatrix} \quad and \quad \nabla^2 \phi = \begin{pmatrix} \phi_{uu} & \phi_{uv} & \phi_{uN} \\ \phi_{vu} & \phi_{vv} & \phi_{vN} \\ \phi_{Nu} & \phi_{Nv} & \phi_{NN} \end{pmatrix}_{(\boldsymbol{u},\boldsymbol{v},\boldsymbol{n})}$$

The curvature tensor has then a simple expression in the normal frame:

$$E = \frac{T \cdot \nabla^2 \phi \cdot T}{|\phi_n|} = \begin{pmatrix} \frac{\phi_{uu}}{|\phi_n|} & \frac{\phi_{uv}}{|\phi_n|} & 0 \\ \frac{\phi_{uv}}{|\phi_n|} & \frac{\phi_{vv}}{|\phi_n|} & 0 \\ 0 & 0 & 0 \end{pmatrix}_{(\boldsymbol{u},\boldsymbol{v},\boldsymbol{n})} \tag{12}$$

Its expression is far more complicated in an arbitrary basis $(\boldsymbol{x}, \boldsymbol{y}, \boldsymbol{z})$ where we could not find eigenvectors. In the normal frame, main eigenvalues and corresponding eigenvectors of the curvature tensor are then found equal to:

$$\kappa_1 = K_h - \sqrt{|\kappa_h^2 - \kappa_k|} \cdot \zeta \qquad \kappa_2 = K_h + \sqrt{|\kappa_h^2 - \kappa_k|} \cdot \zeta \tag{13}$$

$$\boldsymbol{t_1} = \frac{1}{D_1} \begin{bmatrix} \phi_{uv} \\ \kappa_1 \phi_n - \phi_{uu} \\ 0 \end{bmatrix}_{\boldsymbol{u},\boldsymbol{v},\boldsymbol{n}} \qquad \boldsymbol{t_2} = \frac{1}{D_2} \begin{bmatrix} \kappa_2 \phi_n - \phi_{vv} \\ \phi_{uv} \\ 0 \end{bmatrix}_{\boldsymbol{u},\boldsymbol{v},\boldsymbol{n}} \tag{14}$$

with $\kappa_k = \frac{\phi_{uu}\phi_{vv} - \phi_{uv}^2}{\phi_n^2}$, $\kappa_h = \frac{\phi_{uu} + \phi_{vv}}{2|\phi_n|}$, $D_1 = D_2 = \sqrt{\phi_{uv}^2 + (\kappa_1 \phi_n - \phi_{uu})^2}$ and $\zeta = \pm 1$.

Since derivatives in the $(\boldsymbol{u}, \boldsymbol{v}, \boldsymbol{n})$ basis are linked to the derivatives in the default basis $(\boldsymbol{x}, \boldsymbol{y}, \boldsymbol{z})$ and coordinates of $\boldsymbol{u}, \boldsymbol{v}, \boldsymbol{n}$:

$$\phi_n = \boldsymbol{\nabla}\phi \cdot \boldsymbol{n} = \phi_x n_x + \phi_y n_y + \phi_z n_z = \|\boldsymbol{\nabla}\phi\| = |\phi_n| \tag{15a}$$

$$\phi_{uv} = \boldsymbol{u}^t \cdot \nabla^2 \phi \cdot \boldsymbol{v} = \phi_{xx} u_x v_x + \phi_{yy} u_y v_y + \phi_{zz} u_z v_z$$
$$+ \phi_{xy}(u_x v_y + u_y v_x) + \phi_{xz}(u_x v_z + u_z v_x) + \phi_{yz}(u_y v_z + u_z v_y) \tag{15b}$$

$$\phi_{uu} = \boldsymbol{u}^t \cdot \nabla^2 \phi \cdot \boldsymbol{u} \tag{15c}$$

$$\phi_{vv} = \boldsymbol{v}^t \cdot \nabla^2 \phi \cdot \boldsymbol{v} \tag{15d}$$

Replacing ϕ_n, ϕ_{uv}, ϕ_{uv} and ϕ_{uv} into curvature expressions (2b), (3b) and making some formal simplication, we recover the intrinsic expressions of curvatures (2a), (3a) that are independent on the choice of u, v. To sum up, principal curvatures do not depend on the choice of (u, v) whereas principal directions depend on their coordinates. A bad choice of (u, v) may then conduct to null vectors and then bad estimations. To circumvent bad choices, we then propose a criteria on ζ to ensure that t_1 and t_2 are non-null vectors in Sect. 1.

To better understand the role of the parameter ζ in avoiding degeneracies, it is advised to compute curvatures and principal directions by hand with $\zeta = \pm 1$ at $u = v = 0$ and $n = |R|$ for the following case studies:

- $\phi(u, v, n) = u^2 + n^2 - R^2 \Rightarrow$ cylinder of axis v that requires $\zeta = +1$
- $\phi(u, v, n) = R^2 - u^2 - n^2 \Rightarrow$ cylinder of axis v that requires $\zeta = -1$
- $\phi(u, v, n) = v^2 + n^2 - R^2 \Rightarrow$ cylinder of axis u that requires $\zeta = -1$
- $\phi(u, v, n) = R^2 - v^2 - n^2 \Rightarrow$ cylinder of axis u that requires $\zeta = +1$
- $\phi(u, v, n) = \pm u * v + n \Rightarrow$ saddle shaped surface ($\zeta = \pm 1$)

This work was granted access to the HPC resources of the RZG of the Max Planck Society made available within the Distributed European Computing Initiative by the PRACE-2IP, receiving funding from the European Community's Seventh Framework Programme (FP7/2007–2013) under grant agreement n° RI-283493. The research leading to these results has received funding from the European Research Council under the ERC grant agreement n° 247322, GREENEST. The authors thank Xiang He for his help in code development and Bruno Denet for his feedback in using the GTS library.

References

1. Albin, E., D'Angelo, Y.: Assessment of the evolution equation modelling approach for three-dimensional expanding wrinkled premixed flames. Combust. Flame **159**(5), 1932–1948 (2012)
2. Albin, E., D'Angelo, Y., Vervisch, L.: Using staggered grids with characteristic boundary conditions when solving compressible reactive Navier-Stokes equations. Int. J. Numer. Methods Fluids **68**(5), 546–563 (2010)
3. Albin, E., D'Angelo, Y., Vervisch, L.: Flow streamline based Navier-Stokes characteristic boundary conditions: modeling for transverse and corner outflows. Comput. Fluids **51**(1), 115–126 (2011)
4. Brackbill, J.U., Kothe, D.B., Zemach, C.: A continuum method for modeling surface tension. J. Comput. Phys. **100**(2), 335–354 (1992)
5. Cazals, F., Faugère, J.C., Pouget, M., Rouillier, F.: The implicit structure of ridges of a smooth parametric surface. Comput. Aided Geom. Des. **23**(7), 582–598 (2006)
6. Chang, Y.C., Hou, T.Y., Merriman, B., Osher, S.: Level set formulation of Eulerian interface capturing methods for incompressible fluid flows. J. Comput. Phys. **124**(2), 449–464 (1996)
7. Chauveau, C., Birouk, M., Gökalp, I.: An analysis of the d^2-law departure during droplet evaporation in microgravity. Int. J. Multiph. Flow **37**(3), 252–259 (2011)
8. D'Angelo, Y., Joulin, G., Boury, G.: On model evolution equations for the whole surface of three-dimensional expanding wrinkled premixed flames. Combust. Theor. Model. **4**(3), 317–338 (2000)

9. Denet, B.: Nonlinear model equation for three-dimensional Bunsen flames. Phys. Fluids **16**(4), 1149–1155 (2004)
10. Douros, I., Buxton, B.F.: Three-dimensional surface curvature estimation using quadric surface patches. In: Scanning (2002)
11. Dyn, N., Hormann, K., Kim, S.J., Levin, D.: Optimizing 3D triangulations using discrete curvature analysis. In: Mathematical Methods for Curves and Surfaces, pp. 135–146 (2001)
12. Gatzke, T., Grimm, C.M.: Estimating curvature on triangular meshes. Int. J. Shape Model. **12**(1), 1–28 (2006)
13. Giuliani, D.: Gaussian curvature: a growth parameter for biological structures. Math. Comput. Modell. **42**(11), 1375–1384 (2005)
14. Goldman, R.: Curvature formulas for implicit curves and surfaces. Comput. Aided Geom. Des. **22**(7), 632–658 (2005)
15. Hameiri, E., Shimshoni, I.: Estimating the principal curvatures and the Darboux frame from real 3-D range data. IEEE Trans. Syst. Man Cybern. B Cybern. **33**(4), 626–637 (2003)
16. Jeffreys, H., Jeffreys, B.S.: Methods of Mathematical Physics, §9.011 Lagrange's Interpolation Formula, p. 261. Cambridge University Press, Cambridge (1988)
17. Lebas, R., Menard, T., Beau, P.A., Berlemont, A., Demoulin, F.X.: Numerical simulation of primary break-up and atomization: DNS and modelling study. Int. J. Multiph. Flow **35**(3), 247–260 (2009)
18. Lehmann, N., Reif, U.: Notes on the curvature tensor. Graph. Models **74**, 321–325 (2012)
19. Lorensen, W.E., Cline, H.E.: Marching cubes: a high resolution 3D surface construction algorithm. ACM Siggraph Comput. Graph. **21**(4), 163–169 (1987)
20. Macklin, P., Lowengrub, J.: An improved geometry-aware curvature discretization for level set methods: application to tumor growth. J. Comput. Phys. **215**(2), 392–401 (2006)
21. Marchandise, E., Geuzaine, P., Chevaugeon, N., Remacle, J.F.: A stabilized finite element method using a discontinuous level set approach for the computation of bubble dynamics. J. Comput. Phys. **225**(1), 949–974 (2007)
22. McIvor, A.M., Valkenburg, R.J.: A comparison of local surface geometry estimation methods. Mach. Vis. Appl. **10**(1), 17–26 (1997)
23. Ménard, T., Tanguy, S., Berlemont, A.: Coupling level set/VOF/ghost fluid methods: validation and application to 3D simulation of the primary break-up of a liquid jet. Int. J. Multiph. Flow **33**(5), 510–524 (2007)
24. Meyer, M., Desbrun, M., Schröder, P., Barr, A.H.: Discrete differential-geometry operators for triangulated 2-manifolds. Visual. Math. **III**, 35–57 (2003)
25. Musuvathy, S., Cohen, E., Damon, J., Seong, J.K.: Principal curvature ridges and geometrically salient regions of parametric B-spline surfaces. Comput. Aided Des. **43**(7), 756–770 (2011)
26. Osher, S., Fedkiw, R.: Level Set Methods and Dynamic Implicit Surfaces. Springer, New York (2003). doi:10.1007/b98879
27. Page, D.L., Sun, Y., Koschan, A.F., Paik, J., Abidi, M.A.: Normal vector voting: crease detection and curvature estimation on large, noisy meshes. Graph. Models **64**(3), 199–229 (2002)
28. Peng, J., Li, Q., Kuo, C.C.J., Zhou, M.: Estimating Gaussian curvatures from 3D meshes. Electron. Imaging **5007**, 270–280 (2003)
29. Popinet, S.: The GNU triangulated surface library (2004)
30. Prax, C., Sadat, H., Dabboura, E.: Evaluation of high order versions of the diffuse approximate meshless method. Appl. Math. Comput. **186**(2), 1040–1053 (2007)

31. Squillacote, A.H.: The ParaView Guide: A Parallel Visualization Application. Kitware, New York (2007)
32. Sussman, M., Puckett, E.G.: A coupled level set and volume-of-fluid method for computing 3D and axisymmetric incompressible two-phase flows. J. Comput. Phys. **162**(2), 301–337 (2000)
33. Thirion, J.P., Gourdon, A.: The 3D marching lines algorithm and its application to crest lines extraction. Research report (1992)
34. Treece, G.M., Prager, R.W., Gee, A.H.: Regularised marching tetrahedra: improved iso-surface extraction. Comput. Graph. **23**(4), 583–598 (1999)
35. Trott, M.: The Mathematica Guidebook for Graphics. Springer, New York (2004). doi:10.1007/978-1-4419-8576-7
36. Yan, B., Li, B., Baudoin, E., Liu, C., Sun, Z.W., Li, Z.S., Bai, X.S., Aldén, M., Chen, G., Mansour, M.S.: Structures and stabilization of low calorific value gas turbulent partially premixed flames in a conical burner. Exp. Thermal Fluid Sci. **34**(3), 412–419 (2010)
37. Yoo, C.S., Im, H.G.: Characteristic boundary conditions for simulations of compressible reacting flows with multi-dimensional, viscous and reaction effects. Combust. Theor. Model. **11**(2), 259–286 (2007)
38. Yu, R., Bai, X.S.: Direct numerical simulation of lean hydrogen/air auto-ignition in a constant volume enclosure. Combust. Flame **160**(9), 1706–1716 (2013)

Coefficient–Based Spline Data Reduction by Hierarchical Spaces

Cesare Bracco, Carlotta Giannelli[✉], and Alessandra Sestini

Dipartimento di Matematica e Informatica "U. Dini", Università di Firenze,
Viale Morgagni 67/A, 50134 Firenze, Italy
{cesare.bracco,carlotta.giannelli,alessandra.sestini}@unifi.it

Abstract. We present a data reduction scheme for efficient surface storage, by introducing a coefficient–based least squares spline operator that does not require any pointwise evaluation to approximate (in a lower dimension spline space) a given bivariate B–spline function. In order to define an accurate approximation of the target spline with a significant reduction of the space dimension, this operator is subsequently combined with the hierarchical spline framework to design an adaptive method that exploits the capabilities of truncated hierarchical B–splines (THB–splines). The resulting THB–spline simplification approach is validated by several numerical tests. The target B–spline surfaces include approximations of functions whose analytical expression is available, reconstructions of geographic data and parametric surfaces.

Keywords: Data reduction · Quasi–interpolation · Hierarchical splines · THB–splines

1 Introduction

A general data reduction scheme indicates any process that enables to store a certain set of information by (strongly) decreasing the amount of data needed for its reliable reconstruction. For example, an image compression algorithm represents a data reduction approach for images. A natural choice in this context relies in considering a *reference spline* representation that has to be previously generated in a suitably large spline space in order to guarantee a certain accuracy of the approximation. The data reduction scheme can then be applied to reduce the dimension of the spline space while preserving the quality of the approximation. Examples that consider an initial reference spline in the univariate case may be found in [3,17,29]. In these schemes the dimension reduction of the spline space was obtained through *simplification* of the reference spline by placing/removing the knots according to the shape of interest.

We here consider the problem of data reduction for efficient surface representation, see e.g., [20], by assuming an initial description of the target surface in standard tensor–product B–spline form. We then look for a new *spline data reduction* approach for surfaces that can also allow us to deal with complex shapes when extended to multi–patch B–spline descriptions. Obviously,

M. Floater et al. (Eds.): MMCS 2016, LNCS 10521, pp. 23–41, 2017.
https://doi.org/10.1007/978-3-319-67885-6_2

when combined with a preliminary spline approximation phase, this kind of data reduction approach can also be applied to different surface representation formats, as for example gridded sets of space points that define geographic areas described by scanner acquisitions. In order to design a localized data reduction algorithm in a multivariate spline setting, different adaptive spline constructions may be considered. We mention T–splines [22], spline spaces over T–meshes [5] or locally refined (LR) box–partitions [7], as well as hierarchical splines [8]. In the bivariate context shape simplification with T–splines and polynomial splines over hierarchical T–meshes were discussed in [21] and [6], respectively. The use of LR B–splines for large data sets approximations was recently proposed [23].

Hierarchical B–splines were introduced as one of the first generalizations of tensor–product B–spline representations by considering a multilevel approach [8]. The idea of exploiting a multi–resolution spline scheme constitutes a powerful framework for data fitting with local refinements [9,14] and adaptive surface reconstruction [10,15]. The hierarchical levels are identified in terms of nested sequences of refined areas that define the domain hierarchy. A basis of hierarchical spline spaces may be easily constructed by selecting basis functions from different refinement levels according to the domain hierarchy [16]. By assuming mild conditions on the hierarchical mesh configuration, suitable choices of hierarchical B–spline bases span the entire space of piecewise polynomial functions of a certain degree and smoothness that are defined on the underlying grids, see e.g., [1,19]. A renewed interest in this kind of construction has been prompted by the introduction of the truncated basis for hierarchical splines [12]. Truncated hierarchical B–splines (THB–splines) slightly modify the selection mechanism for the hierarchical basis construction to recover the partition of unity property and reduce the influence of coarser basis functions in refined areas. Additional properties of the truncated basis have been derived by also considering a more general hierarchical setting, not necessarily restricted to the tensor–product B–spline model [13]. A relevant peculiarity of the truncated basis consists in facilitating the construction of hierarchical quasi–interpolants [25]. For example, a bivariate hierarchical Hermite quasi–interpolation scheme based on THB–splines was proposed in [2]. Additional results and examples within this approach were recently discussed [24].

By exploiting the truncated basis for hierarchical splines, we propose a data reduction approach by combining multilevel spline spaces with a coefficient–based operator applicable to spline functions. In particular our quasi–interpolant is based on a local least squares operator which uses only the de Boor coefficients of the target spline, and, consequently, no pointwise function evaluation is required. Its formulation in hierarchically refined spline spaces ensures a high level of data reduction, while simultaneously preserving the shape details of the given spline.

The structure of the paper is as follows. The coefficient–based spline operator is introduced in Sect. 2 while the construction and properties of (truncated) hierarchical B–splines are recalled in Sect. 3. Section 4 presents the THB–spline formulation of the new coefficient–based operator and the related spline simplification scheme. Finally, Sect. 5 provides several examples, including data reduction for functions whose analytical expression is available, geographic data approximation and geometric models, and Sect. 6 concludes the paper.

2   Coefficient–Based Data Reduction Operator

Let V be the multivariate tensor–product spline space of degree $\mathbf{d} = (d_1, d_2, ..., d_r)$, $r \in \mathbb{N}$ and $r \geq 1$, defined on a tensor–product mesh $\mathcal{G}$, with the associated basis of tensor–product B–splines

$$\mathcal{B}_{\mathbf{d}} := \{B_J, \, J \in \Gamma_{\mathbf{d}}\},$$

for the multi–index set $\Gamma_{\mathbf{d}}$.

Let us consider a spline $F \in V$ in the B–spline form

$$F = \sum_{J \in \Gamma_{\mathbf{d}}} c_J B_J ,$$

with each $c_J \in \mathbb{R}$.

Let $\bar{V} \subseteq V$ be another space of splines of degree $\mathbf{d}$ defined on a tensor–product mesh $\bar{\mathcal{G}}$, and let

$$\bar{\mathcal{B}}_{\mathbf{d}} := \{\bar{B}_J, J \in \bar{\Gamma}_{\mathbf{d}}\}$$

be the corresponding B–spline basis.

Since $\bar{V} \subseteq V$, we have a linear relation between the basis of the two spaces:

$$\bar{\mathbf{B}}^{(\mathbf{d})} = R^T \mathbf{B}^{(\mathbf{d})},$$

where

$$\mathbf{B}^{(\mathbf{d})} := [B_J]_{J \in \Gamma_{\mathbf{d}}} \quad \text{and} \quad \bar{\mathbf{B}}^{(\mathbf{d})} := [\bar{B}_J]_{J \in \bar{\Gamma}_{\mathbf{d}}}$$

are vectors of length $|\Gamma_{\mathbf{d}}|$ and $|\bar{\Gamma}_{\mathbf{d}}|$, respectively, while R is the matrix of size $|\Gamma_{\mathbf{d}}| \times |\bar{\Gamma}_{\mathbf{d}}|$ obtained by using the knot insertion formula to move from $\bar{V}$ to V (see, e.g., [4]). We define the operator $Q : V \to \bar{V}$ as follows,

$$Q(F) := \sum_{J \in \bar{\Gamma}_{\mathbf{d}}} \bar{c}_J \bar{B}_J , \tag{1}$$

with each coefficient $\bar{c}_J$ obtained by setting $\bar{c}_J = d_J^J$, where d_J^J is the component of index J of the set of coefficients $\{d_K^J\}_{K \in \bar{L}_J}$ solution of the local least squares problem

$$\min_{d_K^J : K \in \bar{L}_J} \sum_{H \in L_J} \left[\left(\sum_{K \in \bar{L}_J} r_{H,K} \, d_K^J \right) - c_H \right]^2 , \tag{2}$$

with $r_{H,K}$ denoting the element of R in the H-th row and K-th column, and

$$\bar{L}_J := K \in \bar{\Gamma}_{\mathbf{d}} : \operatorname{supp}(\bar{B}_K) \cap \operatorname{supp}(\bar{B}_J) \neq \emptyset, \tag{3}$$

$$L_J := H \in \Gamma_{\mathbf{d}} : \operatorname{supp}(B_H) \cap \operatorname{supp}(\bar{B}_J) \neq \emptyset.$$

Note that, considering (2) and (3), we can state that the coefficient $\bar{c}_J$ is the central coefficient of a local approximation of the restriction of F to the support of $\bar{B}_J$ defined on the analogous restriction of $\bar{V}$.

Since $\bar{V} \subset V$, we are approximating a spline surface with another spline surface belonging to a coarser space. Moreover, note that the computation of the coefficients of $Q(F)$ does not require any evaluation of the target spline F to be approximated. The next Proposition proves that Q is a projector into $\bar{V}$.

Proposition 1. *For any $F \in \bar{V}$, $Q(F) = F$.*

Proof. Since $F \in \bar{V}$, we have

$$F = \sum_{K \in \bar{\Gamma}_\mathbf{d}} \bar{a}_K \bar{B}_K \,,$$

which can also be written in the form

$$F = \sum_{H \in \Gamma_\mathbf{d}} c_H B_H \,,$$

with

$$c_H = \sum_{K : r_{H,K} > 0} r_{H,K} \bar{a}_K, \qquad H \in \Gamma_\mathbf{d}.$$

Note that for any $J \in \bar{\Gamma}_\mathbf{d}$, by the definitions of L_J and $\bar{L}_J$ in (3), if $H \in L_J$ it is

$$\{K \in \bar{\Gamma}_\mathbf{d} : r_{H,K} > 0\} = \{K \in \bar{L}_J : r_{H,K} > 0\}.$$

Therefore, we get

$$c_H = \sum_{K \in \bar{L}_J : r_{H,K} > 0} r_{H,K} \bar{a}_K = \sum_{K \in \bar{L}_J} r_{H,K} \bar{a}_K, \qquad H \in L_J.$$

This implies that, for any $J \in \bar{\Gamma}_\mathbf{d}$, it is

$$\sum_{H \in L_J} \left[\left(\sum_{K \in \bar{L}_J} r_{H,K} \bar{a}_K \right) - c_H \right]^2 = 0.$$

Since each coefficient $\bar{c}_J$ in (1) is obtained by solving the minimum problem (2), we must have $\bar{c}_J = \bar{a}_J$ for any $J \in \bar{\Gamma}_\mathbf{d}$, and, consequently, $Q(F) = F$.

3   Hierarchical Spline Spaces

This section briefly reviews (truncated) hierarchical B–spline—(T)HB–spline — construction and quasi–interpolation in hierarchical spline spaces. For a detailed introduction to (T)HB–splines and hierarchical quasi–interpolation, we refer to [12,13] and [2,24,25], respectively.

3.1   Hierarchical B–spline Bases

Let $V^{\ell-1} \subset V^\ell$ and $\Omega^{\ell-1} \supseteq \Omega^\ell$, $\ell = 1, \ldots, M$ be two nested sequences of multivariate tensor–product spline spaces and closed domains, respectively. By starting from an initial tensor–product configuration, each spline space V^ℓ is defined over a grid of level ℓ, obtained through h-refinement of the grid of level $\ell - 1$. The B–spline basis of degree $\mathbf{d}$ that spans the space V^ℓ is indicated as

$$\mathcal{B}_{\mathbf{d}}^\ell := \{B_J^\ell, J \in \Gamma_{\mathbf{d}}^\ell\},$$

for a certain multi–index set $\Gamma_{\mathbf{d}}^\ell$. We assume $\Omega^0 = \Omega$ and $\Omega^M = \emptyset$. Each Ω^ℓ is defined as a collection of cells with respect to the tensor–product grid of level $\ell - 1$.

At each level ℓ, the set of B–splines B_J^ℓ whose support is completely inside Ω^ℓ but not in successive refined domains is included in the hierarchical B–spline (HB–spline) basis [16, 28].

Definition 1. *The hierarchical B–spline basis $\mathcal{H}_{\mathbf{d}}(\mathcal{G}_\mathcal{H})$ of degree $\mathbf{d}$ with respect to the mesh $\mathcal{G}_\mathcal{H}$ is defined as*

$$\mathcal{H}_{\mathbf{d}}(\mathcal{G}_\mathcal{H}) := \{B_J^\ell \in \mathcal{B}_{\mathbf{d}}^\ell : J \in A_{\mathbf{d}}^\ell, \ \ell = 0, \ldots, M - 1\},$$

where

$$A_{\mathbf{d}}^\ell := \{J \in \Gamma_{\mathbf{d}}^\ell : \operatorname{supp} B_J^\ell \subseteq \Omega^\ell \wedge \operatorname{supp} B_J^\ell \not\subseteq \Omega^{\ell+1}\},$$

is the active set of multi–indices of level ℓ, $A_{\mathbf{d}}^\ell \subseteq \Gamma_{\mathbf{d}}^\ell$, and $\operatorname{supp} B_J^\ell$ denotes the intersection of the support of B_J^ℓ with Ω^0.

In view of the linear independence of hierarchical B–splines, they form a basis for the space $S_\mathcal{H} := \operatorname{span} \mathcal{H}_{\mathbf{d}}(\mathcal{G}_\mathcal{H})$ associated to the mesh $\mathcal{G}_\mathcal{H}$.

Definition 2. *Let*

$$s = \sum_{J \in \Gamma_{\mathbf{d}}^{\ell+1}} \sigma_J^{\ell+1} B_J^{\ell+1},$$

be the representation in the B–spline basis of $V^{\ell+1} \supset V^\ell$ of $s \in V^\ell$. The truncation operators

$$\operatorname{trunc}^{\ell+1} : V^\ell \to V^{\ell+1} \quad \text{and} \quad \operatorname{Trunc}^{\ell+1} : V^\ell \to S_\mathcal{H} \subseteq V^{M-1}$$

are defined as

$$\operatorname{trunc}^{\ell+1} s := \sum_{J \in \Gamma_{\mathbf{d}}^{\ell+1} \, : \, \operatorname{supp} B_J^{\ell+1} \not\subseteq \Omega^{\ell+1}} \sigma_J^{\ell+1} B_J^{\ell+1}, \quad \ell = 0, \ldots, M - 1,$$

and

$$\operatorname{Trunc}^{\ell+1} := \operatorname{trunc}^{M-1}(\operatorname{trunc}^{M-2}(\cdots(\operatorname{trunc}^{\ell+1}(s))\cdots)), \quad \ell = 0, \ldots, M - 1,$$

respectively.

The operators introduced in Definition 2 allow us to define an alternative basis for the hierarchical spline space $S_{\mathcal{H}}$, known as truncated hierarchical B–spline (THB–spline) basis [12].

Definition 3. *The truncated hierarchical B–spline basis* $\mathcal{T}_{\mathbf{d}}(\mathcal{G}_{\mathcal{H}})$ *of degree* $\mathbf{d}$ *with respect to the mesh* $\mathcal{G}_{\mathcal{H}}$ *is defined as*

$$\mathcal{T}_{\mathbf{d}}(\mathcal{G}_{\mathcal{H}}) := \left\{ T_J^{\ell} : J \in A_{\mathbf{d}}^{\ell},\ \ell = 0,...,M-1 \right\},\quad with\quad T_J^{\ell} := \mathrm{Trunc}^{\ell+1}(B_J^{\ell}).$$

In view of the B–spline refinement rule and the non–negativity of HB–splines, by subtracting from coarser THB–splines the values of B–splines inserted at subsequent hierarchical levels, the truncated basis forms a convex partition of unity [12]. The truncation also guarantees the property of coefficient preservation: THB–splines preserve the coefficients of functions represented with respect to one of the bases $\mathcal{B}_{\mathbf{d}}^{\ell}$. This property is stated in [13, Theorem 12] and can be summarized as follows. Let $s|_{D^{\ell}}$ be the restriction of $s \in \mathrm{span}\,\mathcal{T}_{\mathbf{d}}(\mathcal{G}_H)$ to $D^{\ell} = \Omega^{\ell} \setminus \Omega^{\ell+1}$ and consider its representation with respect to $\mathcal{T}_{\mathbf{d}}(\mathcal{G}_H)$ and $\mathcal{B}_{\mathbf{d}}^{\ell}$,

$$s|_{D^{\ell}} = \sum_{k=0}^{M-1} \sum_{I \in A_{\mathbf{d}}^k} d_I^k T_I^k = \sum_{J \in \Gamma_{\mathbf{d}}^{\ell}} c_J^{\ell} B_J^{\ell}.$$

The coefficient d_I^{ℓ} of each THB–spline T_I^{ℓ} of level ℓ is equal to the coefficient c_I^{ℓ} of the B–spline B_I^{ℓ} from which T_I^{ℓ} is originated via truncation, namely $d_I^{\ell} = c_I^{\ell}$, $I \in A_{\mathbf{d}}^{\ell}$. In addition, THB–splines form a strongly stable basis: the constants arising in the stability analysis of the basis do not depend on the number of refinement levels, see [13, Theorem 19].

3.2   THB–Spline Quasi–Interpolation

The property of coefficient preservation mentioned at the end of the previous section directly leads to the generalization of any quasi–interpolation operator to the hierarchical setting [25]. Let $f \in C(\Omega^0)$ and let

$$Q^{\ell}(f) := \sum_{J \in \Gamma_{\mathbf{d}}^{\ell}} \lambda_J^{\ell}(f)\, B_J^{\ell},\qquad \ell = 0,\ldots,M-1,$$

be a sequence of quasi–interpolants defined in terms of certain linear functionals $\lambda_J^{\ell}(f)$. Let also the B–spline B_J^{ℓ} related to the truncated basis function $T_J^{\ell} = \mathrm{Trunc}^{\ell+1}(B_J^{\ell})$ through Definition 3, be the *mother* B–spline of T_J^{ℓ}. Thanks to the preservation of coefficients, the hierarchical quasi–interpolant is simply defined by associating at each THB–spline the linear functional of its mother function, namely

$$Q_{\mathcal{H}}(f) := \sum_{\ell=0}^{M-1} \sum_{J \in A_{\mathbf{d}}^{\ell}} \lambda_J^{\ell}(f)\, T_J^{\ell}.$$

Note that the property of reproducing polynomials is preserved by the hierarchical construction:

$$Q(p) = p \quad \Rightarrow \quad Q_{\mathcal{H}}(p) = p, \quad \forall p \in \mathbb{P}^{\mathbf{d}},$$

where $\mathbb{P}^{\mathbf{d}}$ is the space of tensor–product polynomials of degree $\mathbf{d}$. While [25] introduced the general framework for hierarchical quasi–interpolation based on the truncated basis together with the related properties, the hierarchical Hermite BS quasi–interpolation scheme was presented in [2]. THB–spline quasi–interpolation was recently discussed also in [24].

4   THB–Spline Simplification

Given a tensor–product B–spline function, possibly obtained by approximation of a set of gridded data or by interactive modeling and processing, our data reduction scheme produces an accurate THB–spline approximation with a strongly reduced number of degrees of freedom. This result is obtained by locally applying to the original B–spline function the coefficient–based operator introduced in Sect. 2 to compute the coefficient associated with each truncated basis function. Note that, in the case of regular grids, the refinement matrices which express the relation between the coefficients on different levels of the hierarchy and are needed by the least–squares operator depend only on the spline degree. Consequently, they can be computed once and for all in the implementation of the method.

4.1   The Hierarchical Coefficient–Based Operator

Let $\mathcal{G}_{\mathcal{H}}$ be a hierarchical mesh with M levels, and let $V^0 \subset \cdots \subset V^{M-1}$ be the sequence of associated nested tensor–product spline spaces with $V^{M-1} \subseteq V$. We recall from the previous section that $\mathcal{B}_{\mathbf{d}}^{\ell}$ is the B–spline basis of V^{ℓ}, while $\mathcal{G}^{\ell}$ is the associated tensor-product mesh. For any $F \in V$, of the form

$$F = \sum_{H \in \Gamma_{\mathbf{d}}} c_H B_H, \tag{4}$$

we define the hierarchical operator

$$Q_{\mathcal{H}}(F) := \sum_{\ell=0}^{M-1} \sum_{J \in A_{\mathbf{d}}^{\ell}} c_J^{\ell} T_J^{\ell}, \tag{5}$$

where each c_J^{ℓ} is the coefficient of the corresponding tensor–product operator of type (1) defined in the space V^{ℓ} and expressed as

$$Q^{\ell}(F) := \sum_{J \in \Gamma_{\mathbf{d}}^{\ell}} c_J^{\ell} B_J^{\ell}.$$

Analogously to the tensor–product case, each coefficient c_J^ℓ is obtained by solving the local least squares problem

$$\min_{c_K^\ell : K \in \bar{L}_J^\ell} \sum_{H \in L_J^\ell} \left[\left(\sum_{K \in \bar{L}_J^\ell} r_{H,K}^\ell c_K^\ell \right) - c_H \right]^2, \tag{6}$$

where c_H are the coefficients in the tensor–product B-spline representation of F provided by (4),

$$\bar{L}_J^\ell := K \in \Gamma_{\mathbf{d}}^\ell : \operatorname{supp}(B_K^\ell) \cap \operatorname{supp}(B_J^\ell) \neq \emptyset,$$
$$L_J^\ell := H \in \Gamma_{\mathbf{d}} : \operatorname{supp}(B_H) \cap \operatorname{supp}(B_J^\ell) \neq \emptyset,$$

and $r_{H,K}^\ell$ is the element in the H-th row and K-th column of the matrix R^ℓ so that

$$\mathbf{B}^{(\mathbf{d},\ell)} = (R^\ell)^T \mathbf{B}^{(\mathbf{d})}, \tag{7}$$

with

$$\mathbf{B}^{(\mathbf{d},\ell)} := [B_J^\ell]_{J \in \Gamma_{\mathbf{d}}^\ell} \quad \text{and} \quad \mathbf{B}^{(\mathbf{d})}$$

representing the B–spline bases of V^ℓ and V, respectively. Note that, for given

$$0 \leq \ell \leq M - 1 \quad \text{and} \quad J \in A_{\mathbf{d}}^\ell,$$

only a submatrix of R^ℓ is employed for computing the solution of (6), namely

$$R_J^\ell := [r_{H,K}]_{H \in L_J^\ell, K \in \bar{L}_J^\ell}.$$

This matrix can be obtained as the Kronecker product of matrices expressing the relation between univariate B–splines:

$$R_J^\ell = R_{J,1}^\ell \otimes R_{J,2}^\ell \otimes \cdots \otimes R_{J,r}^\ell,$$

where

$$\mathbf{B}_{J,h}^{(\mathbf{d},\ell)} = (R_{J,h}^\ell)^T \mathbf{B}_{J,h}^{(\mathbf{d})}$$

with $\mathbf{B}_{J,h}^{(\mathbf{d})}$ and $\mathbf{B}_{J,h}^{(\mathbf{d},\ell)}$ being the vectors containing the univariate B–splines whose tensor–product gives the r-variate B–splines B_H, $H \in L_J^\ell$ and B_K, $K \in \bar{L}_J^\ell$, respectively.

Remark 1. We observe that, when we consider uniform meshes on each level and $V = V^{M-1}$, each matrix $R_{J,h}^\ell$ depends only on the degree $\mathbf{d}$, and on the number of dyadic refinements needed to pass from $\mathbf{B}_{J,h}^{(\mathbf{d},\ell)}$ to $\mathbf{B}_{J,h}^{(\mathbf{d})}$, that is, $M - 1 - \ell$. For example, in the case of only single knots at all levels, when $r = 2$, $\mathbf{d} = (2,2)$ and $M - 1 - \ell = 1$, for any J, we have

$$R_{J,h}^\ell = \begin{bmatrix} 3/4 & 1/4 & 0 & 0 & 0 \\ 1/4 & 3/4 & 0 & 0 & 0 \\ 0 & 3/4 & 1/4 & 0 & 0 \\ 0 & 1/4 & 3/4 & 0 & 0 \\ 0 & 0 & 3/4 & 1/4 & 0 \\ 0 & 0 & 1/4 & 3/4 & 0 \\ 0 & 0 & 0 & 3/4 & 1/4 \\ 0 & 0 & 0 & 1/4 & 3/4 \end{bmatrix}, \quad h = 1, \ldots, r.$$

The following proposition proves that $Q_{\mathcal{H}}$ reproduces the polynomial space $\mathbb{P}_{\mathbf{d}}$.

Proposition 2. *For any* $q \in \mathbb{P}_{\mathbf{d}}$, $Q_{\mathcal{H}}(q) = q$.

Proof. Note that, by Proposition 1 we have

$$Q^{\ell}(q) = q \quad \text{for any} \quad q \in \mathbb{P}_{\mathbf{d}}, \quad \ell = 0, ..., M - 1.$$

As a consequence, by applying Theorem 3 in [25], we obtain the thesis.

In addition, the hierarchical operator $Q_{\mathcal{H}}$ reproduces all splines of the subspace V^0, as it is proved in Proposition 3 below.

Proposition 3. *For any* $F \in V^0$, $Q_{\mathcal{H}}(F) = F$.

Proof. Let us consider c_J^{ℓ} in (5) determined by solving problem (6). Since $F \in V^0 \subseteq V^{\ell} \subseteq ... \subseteq V^{M-1} \subseteq V$, we have

$$F = \sum_{K \in \Gamma_{\mathbf{d}}^0} a_K^0 \, B_K^0 = \sum_{K \in \Gamma_{\mathbf{d}}^{\ell}} a_K^{\ell} \, B_K^{\ell} = \sum_{H \in \Gamma_{\mathbf{d}}} c_H B_H \,,$$

with

$$c_H = \sum_{K : r_{H,K}^{\ell} > 0} r_{H,K}^{\ell} a_K^{\ell}, \quad H \in \Gamma_{\mathbf{d}},$$

where each $r_{H,K}^{\ell}$ is the element in the H-th row and K-th column of the matrix R^{ℓ} in (7). Analogously to the proof of Proposition 1, this is enough to prove that $c_J^{\ell} = a_J^{\ell}$. This in turn, by using the THB–spline property of coefficient preservation [13], implies that $Q_{\mathcal{H}}(F) = F$.

It is clear that the accuracy of the hierarchical approximation $Q_{\mathcal{H}}(F)$ of $F \in V$ strongly depends on the choice of the hierarchical mesh $\mathcal{G}_{\mathcal{H}}$, and a strategy for its automatic generation is crucial.

4.2   The Adaptive Data Reduction Scheme

Let V be a $\mathbf{d}$–degree tensor–product spline space, $\mathbf{d} = (d_1, d_2, ..., d_r)$, $r \in \mathbb{N}$ and $r \geq 1$. For simplicity, we assume that V is defined on a grid $\mathcal{G}^{M_{max}-1}$ obtained from a coarser grid $\mathcal{G}^0$ by applying $M_{max} - 1$ successive dyadic refinements. Consequently, the mesh $\mathcal{G}^{\ell}$ is obtained by one dyadic refinement of the cells of $\mathcal{G}^{\ell-1}$, $\ell = 1, ..., M - 1$, with $M \leq M_{max}$. Let

$$\mathbf{T} := [T_K]_{K \in \Gamma_{\mathbf{d}}^T} \quad \text{with} \quad \Gamma_{\mathbf{d}}^T := \{(\ell, I) : I \in A_{\mathbf{d}}^{\ell}, 0 \leq \ell \leq M - 1\}$$

be the set of THB–splines defined by the spline hierarchy. We can then write

$$\mathbf{T} = P^T \mathbf{B}^{\mathbf{d}}$$

where the transpose of P is the matrix that expresses the linear relation between the basis of the hierarchical spline space and the basis of V. We denote by $p_{J,K}$

the element of P in the J-th row and K-th column. For subsequent use, we also rewrite (5) as

$$Q_{\mathcal{H}}(F) = \sum_{K \in \Gamma_{\mathbf{d}}^{\mathcal{T}}} c_K^{\mathcal{T}} T_K . \tag{8}$$

The following ascending algorithm summarizes the main steps to compute a THB–spline with $M \leq M_{max}$ levels which approximates $F \in V$ with knots in $\mathcal{G}_{M_{max}-1}$ within a given tolerance ϵ. As previously mentioned, for simplicity, we assume that V is a spline space whose mesh can be dyadically simplified $M_{max} - 1$ times.

Input:

- the set of coefficients $\{c_J, J \in \Gamma_{\mathbf{d}}\}$ defining $F \in V$ with knots in $\mathcal{G}_{M_{max}-1}$;
- a dyadic coarsening $\mathcal{G}_0$ of $\mathcal{G}_{M_{max}-1}$;
- a maximum number of hierarchical levels $M \leq M_{max}$;
- the tolerance $\epsilon > 0$.

1. initialize $\mathcal{G}_{\mathcal{H}} = \mathcal{G}_0$ and, consequently, $\Gamma_{\mathbf{d}}^{\mathcal{T}}$ and P;
2. compute the coefficients $c_K^{\mathcal{T}}$, $K \in \Gamma_{\mathbf{d}}^{\mathcal{T}}$, of $Q_{\mathcal{H}}(F)$ in (8) by solving for each of them the local least square system in (6);
3. while

$$\left| \sum_{K : p_{J,K} > 0} p_{J,K} c_K^{\mathcal{T}} - c_J \right| \leq \epsilon \cdot \left(\max_{H \in \Gamma_{\mathbf{d}}} c_H - \min_{H \in \Gamma_{\mathbf{d}}} c_H \right) \tag{9}$$

is not satisfied for all $J \in \Gamma_{\mathbf{d}}$ and the current number of levels is less than M, repeat the following steps:
   (a) for all $J \in \Gamma_{\mathbf{d}}$ which do not satisfy (9), mark the cells which belong to $\mathrm{supp}(B_J^{\ell})$ for all $K = (\ell, I) \in \Gamma_{\mathbf{d}}^{\mathcal{T}}$ such that $p_{J,K} > 0$;
   (b) obtain the new mesh $\mathcal{G}_{\mathcal{H}}$ by dyadically refining each marked cell belonging to $\mathcal{G}_\ell, \ell < M - 1$ and update $\Gamma_{\mathbf{d}}^{\mathcal{T}}$ and P;
   (c) compute the new coefficients $c_K^{\mathcal{T}}$, $K \in \Gamma_{\mathbf{d}}^{\mathcal{T}}$, of $Q_{\mathcal{H}}(F)$ in (8) by solving for each of them the local least square system in (6).

Output: THB–spline approximation $Q_{\mathcal{H}}(F)$ with $M \leq M_{max}$ levels of the form (8) approximating F within the given tolerance ϵ.

In the stopping criterion, the tolerance ϵ is compared with the error in the current hierarchical approximation of c_J, scaled with respect to the data, according to (9). The right–hand side of (9) vanishes if all the coefficients $c_H, H \in \Gamma_{\mathbf{d}}$, are equal to a constant. Even if this case is not of practical interest, we may note that it is still covered by the algorithm since the target spline is just a constant exactly represented already in V_0.

Note that, in step 3(a) of the algorithm, in order to avoid additional computations, instead of marking the cells in the support of the THB–splines associated to the coefficients that do not satisfy the desired tolerance, we simply consider the support of the corresponding B–splines. This is justified since the support of a THB–spline is contained in the one of its mother function, namely the B–spline from which the truncated basis functions is obtained by truncation.

Remark 1. *It is worth to mention that the whole algorithm can be naturally generalized for applying the data reduction scheme to tensor–product B–spline parametric surfaces where the coefficients are replaced by control points. Since c_H is now a vector and no more a scalar, the necessary changes consist in replacing the square brackets in (2) and (6) and the absolute value in (9) with the euclidean norm, and substituting the normalizing factor $(\max_{H \in \Gamma_d} c_H - \min_{H \in \Gamma_d} c_H)$ in (9) with $\max_{H,K \in \Gamma_d} \|c_H - c_K\|$.*

Remark 2. *Note that, when $M = M_{max}$, the algorithm always succeeds at meeting any tolerance, at most by producing a hierarchical mesh with M_{max} levels. This is due to the fact that, at each iteration of the algorithm, if (9) is not satisfied for a certain $J \in \Gamma_d$, all the cells belonging to the supports of the B–splines B_I^ℓ, $K = (I, \ell)$, such that $p_{J,K} > 0$ are refined. As a consequence, at each iteration the level ℓ of the indices $K = (\ell, I)$ such that $p_{J,K} > 0$ increases by 1. Eventually, in the worst case we will get $\ell = M_{max} - 1$ and $\{K = (\ell, I) : p_{J,K} > 0\} = \{J\}$, that is, the obtained hierarchical space is locally a tensor–product space. Therefore, $c_J^{M_{max}-1} = c_J$, which of course implies that (9) is satisfied for c_J.*

5 Numerical Experiments

For testing the proposed hierarchical data reduction scheme, we implemented the coefficient–based scheme in MATLAB and combined it with THB–splines by relying on the hierarchical B–spline implementation within the MATLAB package GeoPDEs, see [11,27]. Open knot vectors are considered for all the examples.

Example 1. *We first consider two test tensor-product B–spline surfaces, S_i, $i = 1, 2$, shown in Fig. 1(a) and (b). Each of them was obtained with a preliminary spline approximation of a corresponding set of 129×129 uniformly gridded functional data. More precisely, the tensor–product extension of the BS Hermite QI scheme introduced in [18] was adopted for this aim. The two discrete data sets used to generate S_1 and S_2 were defined by uniformly sampling the following two test functions,*

$$f_1(x, y) = \frac{\tanh(9y - 9x) + 1}{9}, \qquad (x, y) \in [-1, 1]^2,$$

$$f_2(x, y) = \frac{2}{3 \exp{(10x - 3)^2 + (10y + 4)^2}}, \qquad (x, y) \in [-1, 1]^2.$$

Example 2. *We applied the algorithm to a tensor–product B–spline surface S_3 approximating the set of geographic data available at [26] and describing the terrain elevation in a mountain region of the Hawaii Islands, see Fig. 1(c). The tensor–product surface was obtained with a modified version of the BS Hermite QI scheme (mentioned in [18]). Such variant, unlike the basic one, does not require the values of the first and second–order mixed partial derivatives of the approximated function on the rectangular mesh defining the spline knots.*

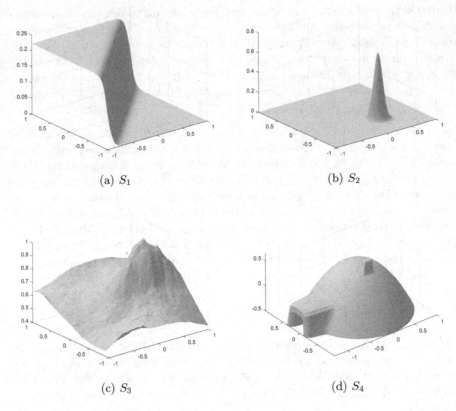

(a) S_1 (b) S_2

(c) S_3 (d) S_4

Fig. 1. The reference tensor–product spline surfaces $S_i, i = 1, 2, 3, 4$. The spline break-points are 129 uniformly spaced points in $[-1, 1]$ with respect to both directions.

Example 3. *In this example, we applied the data reduction algorithm to the "igloo" model S_4, defined in a tensor–product B–spline space of degree $(3, 3)$ on a 128×128 uniform grid, see Fig. 1(d). In this case the reference parametric surface is obtained through control point modification and the control points c_J belong to $\mathbb{R}^3$.*

In all the experiments we set $M = M_{max}$. For each test, we report the spline degree, the number M of levels, the tolerance ϵ used for generating the hierarchical mesh and the dimension of the spaces $S_{\mathcal{H}}$ and V. In addition, the last column of the table shows the discrete approximation of the infinity norm of the error e_i

$$e_i := Q_{\mathcal{H}}(S_i) - S_i, \qquad i = 1, 2, 3, \qquad e_4 := \|Q_{\mathcal{H}}(S_4) - S_4\|_2,$$

computed by sampling the error at the vertices of the original tensor-product grid. It is clear that the data reduction approximation error can be controlled by setting a suitable tolerance for the marking strategy considered in the algorithm.

Note that, for any considered test, there is a significant reduction of the number of degrees of freedom, thanks to the local refinement capabilities of hierarchical spline spaces.

Table 1 shows the results obtained by applying the hierarchical operator to the four reference surfaces with different tolerance values (ϵ=5e-2,1e-2,5e-3). The adaptive nature of the refinements obtained with the application of the algorithm is evident from the hierarchical meshes generated by the THB–spline simplification approach, see Figs. 2, 3 and 4 (right). The comparison between the approximated surfaces shown in Figs. 2, 3 and 4 (left) and the original surfaces $S_i, i = 1, \ldots, 4$ of Fig. 1 suggests that the shape of the data is also well reproduced. The corresponding contour plots are shown in Figs. 5 and 6 which confirm the good quality of the approximations (only very minor differences between the original and the approximated contour plot are present). Different experiments with periodic (rather than open) knot vectors suggest that this choice leads to more refined meshes near the boundary (and consequently more degrees of freedom).

Table 1. Numerical results obtained by applying the hierarchical quasi-interpolation operator $Q_{\mathcal{H}}$ to the tensor–product splines S_1, S_2, S_3, S_4.

S_1 ($d_1 = d_2 = 3$)				
M	ϵ	$\dim(S_{\mathcal{H}})$	$\dim(V)$	$\|e_1\|_\infty$
3	5e-2	361	17161	5.519e-3
4	1e-2	973	17161	2.546e-4
4	5e-3	1027	17161	2.546e-4
S_2 ($d_1 = d_2 = 3$)				
M	ϵ	$\dim(S_{\mathcal{H}})$	$\dim(V)$	$\|e_2\|_\infty$
4	5e-2	550	17161	9.251e-3
5	1e-2	820	17161	1.609e-3
5	5e-3	928	17161	3.456e-4
S_3 ($d_1 = d_2 = 2$)				
M	ϵ	$\dim(S_{\mathcal{H}})$	$\dim(V)$	$\|e_3\|_\infty$
3	5e-2	190	16900	2.125e-2
6	1e-2	2485	16900	4.733e-3
6	5e-3	5718	16900	2.504e-3
S_4 ($d_1 = d_2 = 3$)				
M	ϵ	$\dim(S_{\mathcal{H}})$	$\dim(V)$	$\|e_4\|_\infty$
6	5e-2	2848	17161	1.373e-2
6	1e-2	3739	17161	3.835e-3
6	5e-3	3952	17161	3.334e-4

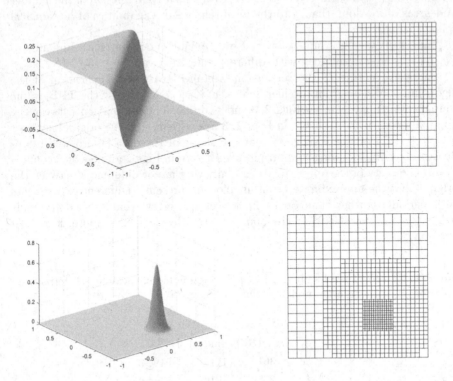

Fig. 2. THB–spline approximations (left) and corresponding hierarchical meshes (right) obtained by applying $Q_{\mathcal{H}}$ to the tensor–product splines of Example 1 with $\epsilon=1e\text{-}2$.

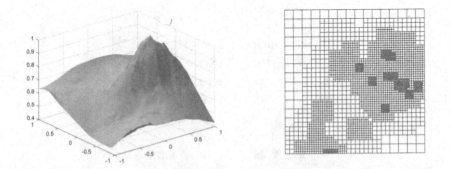

Fig. 3. THB–spline approximation (left) and corresponding hierarchical mesh (right) obtained by applying $Q_{\mathcal{H}}$ to the tensor–product spline of Example 2 with $\epsilon=1e\text{-}2$.

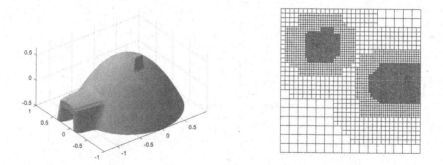

Fig. 4. THB–spline approximation (left) and corresponding hierarchical mesh (right) obtained by applying $Q_{\mathcal{H}}$ to the tensor–product spline of Example 3 with ϵ=1e-2.

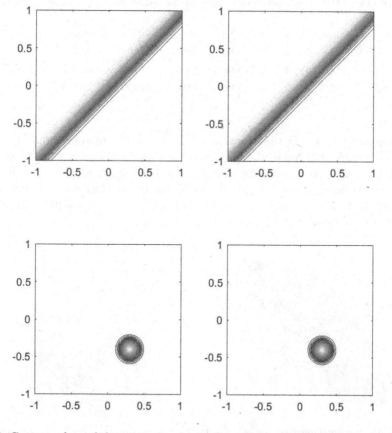

Fig. 5. Contour plots of the tensor–product splines (left) of Example 1: S_1 (top) and S_2 (bottom) and of their THB–spline approximations (right) obtained with ϵ=1e-2.

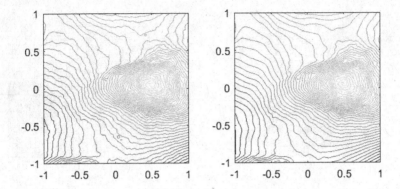

Fig. 6. Contour plots of the tensor–product spline of Example 2 (left) and of its THB–spline approximation (right) obtained with ϵ=1e-2.

Table 2. Numerical results obtained by applying the hierarchical quasi-interpolation operator $Q_{\mathcal{H}}$ to the tensor–product splines S_1, S_2, S_3, shown in Fig. 1(a), (b) and (c).

test	M	$d_1 = d_2$	ϵ	$\dim(S_{\mathcal{H}})$	$\dim(V)$	$\|e_i\|_\infty$
S_1	6	3	5.0e-6	7597	17161	1.702e-7
S_2	6	3	5.0e-6	3142	17161	5.157e-7
S_3	6	2	5.0e-3	5718	16900	2.504e-3

In order to show that the adaptive scheme can generate approximations with the same accuracy of the tensor–product case with a reduced number of degrees of freedom, we also present the results in Table 2. In this case the tolerance values were chosen of the same order of the error obtained by approximating the original data with tensor–product B–splines. The corresponding meshes are shown in Figs. 7 and 8.

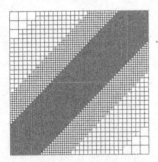 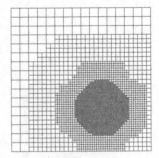

Fig. 7. Hierarchical meshes (right) obtained by applying $Q_{\mathcal{H}}$ to the tensor–product splines of Example 1 with ϵ=5e-6.

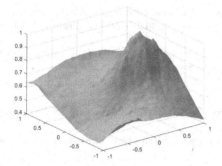 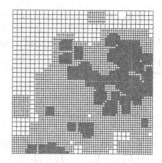

Fig. 8. THB–spline approximation (left) and corresponding hierarchical mesh (right) obtained by applying $Q_{\mathcal{H}}$ to the tensor–product spline of Example 2 with ϵ=5e-3.

6   Conclusions

In order to reduce the computational costs connected with the reconstruction of large data sets, we introduced a data reduction operator that does not require any pointwise functional evaluation and its THB–spline generalization. Such operator can be applied to any initial (highly refined) standard bivariate spline, preliminarily constructed by suitable classical spline approximation, or alternatively obtained either by control point modification of an initial spline configuration, or as the result of modeling techniques. The THB–spline simplification algorithm ensures accurate spline representations with a strongly reduced number of degrees of freedom. The algorithm can also be exploited for interactive design and model simplification. In principle, the data reduction scheme can also be applied to other kind of Bernstein/B–spline-type representations, assuming to start with a target function represented in this alternative form. The analysis of the influence of the chosen representation on the final approximation is beyond the scope of this paper.

Acknowledgements. The support by MIUR "Futuro in Ricerca" programme through the project DREAMS (RBFR13FBI3) and by the Istituto Nazionale di Alta Matematica (INdAM) through Gruppo Nazionale per il Calcolo Scientifico (GNCS)—"Finanziamento Giovani Ricercatori" and "Progetti di ricerca" programmes—and Finanziamenti Premiali SUNRISE are gratefully acknowledged.

References

1. Berdinsky, D., Kim, T.-W., Bracco, C., Cho, D., Mourrain, B., Oh, M.-J., Kiatpanichgij, S.: Dimensions and bases of hierarchical tensor-product splines. J. Comput. Appl. Math. **257**, 86–104 (2014)
2. Bracco, C., Giannelli, C., Mazzia, F., Sestini, A.: Bivariate hierarchical Hermite spline quasi-interpolation. BIT **56**, 1165–1188 (2016)
3. Conti, C., Morandi, R., Rabut, C., Sestini, A.: Cubic spline data reduction choosing the knots from a third derivative criterion. Numer. Algorithms **28**, 45–61 (2001)

4. de Boor, C.: A Practical Guide to Splines. Springer, New York (2001). Revised ed
5. Deng, J., Chen, F., Feng, Y.: Dimensions of spline spaces over T-meshes. J. Comput. Appl. Math. **194**, 267–283 (2006)
6. Deng, J., Chen, F., Li, X., Hu, C., Tong, W., Yang, Z., Feng, Y.: Polynomial splines over hierarchical T-meshes. Graph. Models **70**, 76–86 (2008)
7. Dokken, T., Lyche, T., Pettersen, K.F.: Polynomial splines over locally refined box-partitions. Comput. Aided Geom. Des. **30**, 331–356 (2013)
8. Forsey, D.R., Bartels, R.H.: Hierarchical B-spline refinement. Comput. Graphics **22**, 205–212 (1988)
9. Forsey, D.R., Bartels, R.H.: Surface fitting with hierarchical splines. ACM Trans. Graphics **14**, 134–161 (1995)
10. Forsey, D.R., Wong, D.: Multiresolution surface reconstruction for hierarchical B-splines. In: Graphics, Interface, pp. 57–64 (1998)
11. Garau, E.M., Vázquez, R.: Algorithms for the implementation of adaptive isogeometric methods using hierarchical splines. Appl. Numer. Math. **123**, 58–87 (2018)
12. Giannelli, C., Jüttler, B., Speleers, H.: THB-splines: the truncated basis for hierarchical splines. Comput. Aided Geom. Des. **29**, 485–498 (2012)
13. Giannelli, C., Jüttler, B., Speleers, H.: Strongly stable bases for adaptively refined multilevel spline spaces. Adv. Comput. Math. **40**, 459–490 (2014)
14. Greiner, G., Hormann, K.: Interpolating and approximating scattered 3D-data with hierarchical tensor product B-splines. In: Le Méhauté, A., Rabut, C., Schumaker, L.L. (eds.) Surface Fitting and Multiresolution Methods, pp. 163–172. Vanderbilt University Press, Nashville (1997)
15. Kiss, G., Giannelli, C., Zore, U., Jüttler, B., Großmann, D., Barner, J.: Adaptive CAD model (re-)construction with THB-splines. Graph. Models **76**, 273–288 (2014)
16. Kraft, R.: Adaptive and linearly independent multilevel B-splines. In: Le Méhauté, A., Rabut, C., Schumaker, L.L. (eds.) Surface Fitting and Multiresolution Methods, pp. 209–218. Vanderbilt University Press, Nashville (1997)
17. Lyche, T., Mørken, K.: A data-reduction strategy for splines with applications to the approximation of functions and data. IMA J. Numer. Anal. **8**, 185–208 (1988)
18. Mazzia, F., Sestini, A.: The BS class of Hermite spline quasi-interpolants on nonuniform knot distribution. BIT **49**, 611–628 (2009)
19. Mokriš, D., Jüttler, B., Giannelli, C.: On the completeness of hierarchical tensor-product B-splines. J. Comput. Appl. Math. **271**, 53–70 (2014)
20. Morandi, R., Sestini, A.: Data reduction in surface approximation. In: Lyche, T., Schumaker, L. (eds.) Mathematical Methods for Curves and Surfaces: Oslo 2000, pp. 315–324. Vanderbilt University Press, Nashville (2001)
21. Sederberg, T.W., Cardon, D.L., Finnigan, G.T., North, N.S., Zheng, J., Lyche, T.: T-spline simplification and local refinement. ACM Trans. Graphics **23**, 276–283 (2004)
22. Sederberg, T.W., Zheng, J., Bakenov, A., Nasri, A.: T-splines and T-NURCCS. ACM Trans. Graphics **22**, 477–484 (2003)
23. Skytt, V., Barrowclough, O., Dokken, T.: Locally refined spline surfaces for representation of terrain data. Comput. Graphics **49**, 58–68 (2015)
24. Speleers, H.: Hierarchical spline spaces: quasi-interpolants and local approximation estimates. Adv. Comput. Math. **43**, 235–255 (2017)
25. Speleers, H., Manni, C.: Effortless quasi-interpolation in hierarchical spaces. Numer. Math. **132**, 155–184 (2016)
26. U.S. Geological Survey. https://www.usgs.gov/, http://dds.cr.usgs.gov/pub/data/nationalatlas/el_usa_hawaii.bil_nt00924.tar.gz

27. Vázquez, R.: A new design for the implementation of isogeometric analysis in Octave and Matlab: GeoPDEs 3.0. Comput. Math. Appl. **72**, 523–554 (2016)
28. Vuong, A.-V., Giannelli, C., Jüttler, B., Simeon, B.: A hierarchical approach to adaptive local refinement in isogeometric analysis. Comput. Methods Appl. Mech. Eng. **200**, 3554–3567 (2011)
29. Wever, U.A.: Global and local data reduction strategies for cubic splines. Comput. Aided Des. **23**, 127–132 (1991)

A Versatile Strategy for the Implementation of Adaptive Splines

Andrea Bressan[1]([⊠]) and Dominik Mokriš[2]

[1] Department of Mathematics, University of Oslo, Oslo, Norway
andbres@math.uio.no
[2] Institute of Applied Geometry, Johannes Kepler University Linz, Linz, Austria
dominik.mokris@jku.at

Abstract. This paper presents an implementation framework for spline spaces over T-meshes (and their d-dimensional analogs). The aim is to share code between the implementations of several spline spaces. This is achieved by reducing evaluation to a generalized Bézier extraction.

The approach was tested by implementing hierarchical B-splines, truncated hierarchical B-splines, decoupled hierarchical B-splines (a novel variation presented here), truncated B-splines for partially nested refinement and hierarchical LR-splines.

Keywords: Implementation · Bézier extraction · THB-splines · LR-splines

1 Introduction

A common method to represent shapes in Computer-Aided Design (CAD), Computer-Aided Engineering (CAE) and Computer-Aided Manufacturing (CAM) is to parametrize the desired geometry (or its boundary) with Non-Uniform Rational B-Splines (NURBS). B-splines have a global tensor-product structure, where each d-variate basis function is a product of d univariate basis functions. This means that changes in spatial resolution cannot be confined to a small region; they necessarily spread to a union of stripes of the domain (Fig. 1).

Different constructions that allow for local refinement were proposed during the last two decades and gained support with the introduction of IsoGeometric Analysis (IGA) [24]. Indeed, IGA pushed the use of splines in numerical simulation where local refinement is a prerequisite of adaptive methods. The following list includes the best known constructions:

– Hierarchical B-splines (HB) [16]. This is a multiscale approach: each scale is associated to a different tensor-product B-spline space. Functions from each scale are *selected* depending on the locally required resolution and together they form the hierarchical B-spline basis. There are many variations of HB, among them: the Truncated Hierarchical B-splines (THB) [17], the Truncated Decoupled Hierarchical B-splines (TDHB) [32], the Truncated B-splines for partially nested refinement (TBPN) [43] and Decoupled Hierarchical B-splines (DHB) introduced here for the first time.

© Springer International Publishing AG 2017
M. Floater et al. (Eds.): MMCS 2016, LNCS 10521, pp. 42–73, 2017.
https://doi.org/10.1007/978-3-319-67885-6_3

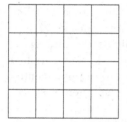 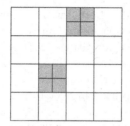 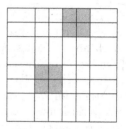

Fig. 1. Limitation of the tensor-product construction. Left: the coarse grid; Middle: the desired refinement; Right: the coarsest tensor-product grid refined on the grey area.

- T-splines (T) [38,39]. The central notion is the T-mesh: a planar graph with lengths. A B-spline corresponds to each vertex of the graph and its knot vectors depend on the length of the neighboring edges. These B-splines generate the space. Unfortunately, they can be linearly dependent. *Analysis Suitable T-splines* (AST) avoid linear dependencies by restricting the class of allowed T-meshes [11,31]. AST spaces can be constructed in 2D [35] and also defined for 3D domains [34].
- Locally Refined splines (LR) [14]. Their definition is given in terms of *minimally supported* B-splines contained in a space of piecewise polynomials. The generators are not always linearly independent. A bivariate construction that avoids linear dependencies is the hierarchical LR-splines (HLR) [5].

Several other spaces and alternative bases exist, e.g., [6,8,13,27,36]. On the one hand, the mentioned spaces contain piecewise polynomials over box-shaped subdomains and allow for smooth functions. On the other hand, each construction was defined for a specific application and, as a consequence, described and analyzed with its own set of tools. Thus it is difficult to make a comparison involving more than a few spaces and having criteria that are not application-specific. A comparison of HB, THB and LR based on the conditioning of the mass matrix is presented in [25]. A similar approach was used in [23].

Our aim is to describe a software framework allowing to implement various spline spaces in a systematic way. The main criterion is the versatility of the code, that is, the possibility to include further spline spaces to this framework with relative ease. In this way we hope to facilitate both the comparison of different spline spaces and experimenting with alternative definitions. The proposed method is a generalization of Bézier extraction [2,15,37], which is a well-established tool in IGA. As a proof of concept, three spline spaces available in the literature and a newly designed space were implemented. The choice of the spaces has been based on authors' personal research interests and contains only spaces with multilevel structure. Less structured spaces such as LR-splines or T-splines could be implemented as well, but they would probably require more effort due to their intrinsic complexity, particularly so when non-dyadic refinement and knot lines with multiplicities would be considered.

The framework is presented in Sect. 2 without any reference to specific spline spaces. Section 3 discusses the space and time complexity of the proposed approach and presents possible optimizations. Section 4 highlights the differences from Bézier extraction, while Sect. 5 describes how the framework can be applied to HB, THB, DHB, TBPN and HLR splines. These spaces were implemented and their implementations are used in Sect. 6 to show how the different spaces behave in a few selected cases.

The following notation conventions are used throughout the paper.

Style	Example	Used for
Lowercase Latin letters	$a, b, \ldots$	Real numbers
Bold lowercase Latin letters	$\mathbf{a}, \mathbf{b}, \ldots$	Vectors of real numbers
Lowercase Greek letters	$\alpha, \beta, \ldots$	Functions
Bold lowercase Greek letters	$\boldsymbol{\alpha}, \boldsymbol{\beta}, \ldots$	Vectors of functions
Uppercase Greek letters	$\Omega, \Delta, \ldots$	Subsets of $\mathbb{R}^d$ or $\mathbb{R}^{d-1}$
Uppercase Latin letters	$A, B, \ldots$	Sets
Bold uppercase Latin letters	$\mathbf{A}, \mathbf{B}, \ldots$	Matrices and operators
Calligraphic uppercase Latin letters	$\mathcal{A}, \mathcal{B}, \ldots$	Function spaces

2 Implementation Method

The aim of an implementation is to evaluate the generators of a spline space at a set X of points contained in the domain $\Omega \subseteq \mathbb{R}^d$. This is sufficient for the application to interpolation problems and for the implementation of Galerkin methods based on numerical quadrature.

The spline spaces of interest are generated by piecewise polynomials on a partition of Ω into axis-aligned boxes called *elements*. Thus their restriction to an element can be expressed in terms of tensor-product Bernstein polynomials. By doing so it is possible to repurpose Finite Element Method (FEM) codebases to IGA. This approach was proposed for NURBS in [2] under the name *Bézier extraction* and later extended to other spaces [15,37].

The main idea of this paper is to replace the elements with more general subdomains and the Bernstein basis with an arbitrary local basis, possibly a different one for each subdomain. This allows the implementations to be closer to the mathematical definitions of the spline spaces, which are typically described in terms of B-splines and not of Bernstein polynomials. A detailed comparison with Bézier extraction is provided in Sect. 4.

2.1 Description

Let $G = \{\gamma_1, \ldots, \gamma_n\}$ be the generating set of a spline space and assume that there exists a partition $T = \{\Delta_1, \ldots, \Delta_s\}$ of the domain Ω and a corresponding set of local bases[1] $B = \{B_1, \ldots, B_s\}$ such that the restriction of each $\gamma \in G$ to any Δ_i admits a representation in span B_i. More precisely,

$$\forall \gamma \in G, \ \forall_{i=1}^s, \ \forall \mathbf{x} \in \Delta_i : \quad \gamma(\mathbf{x}) = \sum_{\beta \in B_i} m_{\beta,\gamma} \beta(\mathbf{x}), \tag{1}$$

and thus

$$\boldsymbol{\gamma}(\mathbf{x}) = \boldsymbol{\beta}_i(\mathbf{x}) \mathbf{M}_i, \tag{2}$$

where $\boldsymbol{\gamma}(\mathbf{x})$ and $\boldsymbol{\beta}_i(\mathbf{x})$ are the row vectors

$$\begin{aligned}
\boldsymbol{\gamma}(\mathbf{x}) &= (\gamma_1(\mathbf{x}), \ldots, \gamma_n(\mathbf{x})) = \big(\gamma(\mathbf{x})\big)_{\gamma \in G}, \\
\boldsymbol{\beta}_i(\mathbf{x}) &= \big(\beta(\mathbf{x})\big)_{\beta \in B_i}
\end{aligned} \tag{3}$$

and $\mathbf{M}_i = \big(m_{\beta,\gamma}\big)_{\beta \in B_i, \gamma \in G}$ is the matrix containing the coefficients from (1). The matrices $\mathbf{M}_i$ can be collected as blocks of the matrix $\mathbf{M}$ with $\sum_{i=1}^s \#B_i$ rows and $\#G$ columns as depicted in Fig. 2.

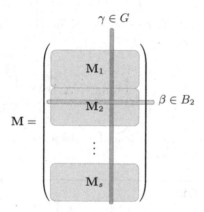

Fig. 2. Structure of the representation matrix.

The generating set G is uniquely determined by T, B and $\mathbf{M}$ through (2). This suggests an implementation in which T, B and $\mathbf{M}$ are provided by the space-specific code and the evaluation of $\boldsymbol{\gamma}(\mathbf{x})$ is performed using (2). Note that different choices of T, B and $\mathbf{M}$ can describe the same G and thus there is a certain freedom to optimize for different scenarios (see Sect. 5).

For such implementation T requires a method findSubdomain that given a point $\mathbf{x} \in \Omega$ returns the index i of the corresponding subdomain $\Delta_i \ni \mathbf{x}$. The implementation of the local bases B_i should contain a method eval that returns a matrix containing the values of the basis functions $\beta \in B_i$ in a given set of

---

[1] (Or, more generally, generating sets).

points $X \subset \Omega$. For consistency the same interface should be implemented by the resulting spline space.

Before describing a suggested implementation of the three components T, B and $\mathbf{M}$ it is worth describing the eval interface in more detail. For the expected applications it is necessary to compute both function values and function derivatives at a set of points $X \subset \Omega$. As shown in (2) the values (and also the derivatives) can be transformed by a matrix multiplication. This suggest a format that allows the transformation of all the data with a single operation.

Let $W = \{\mathbf{v}_1, \ldots, \mathbf{v}_w\}$ be the list of the multiindices corresponding to the desired derivatives. For instance, in two dimensions, the value corresponds to $(0, 0)$, the first partial derivative with respect to the first direction to $(1, 0)$ and the second mixed derivative to $(1, 1)$ and $W = \{(0, 0), (1, 0), (1, 1)\}$ means that all these three are computed. Then for a set of functions $F = \{\varphi_1, \ldots, \varphi_f\}$ the base format can be

$$\mathbf{E}_F(\mathbf{x}, W) = \begin{pmatrix} \partial^{\mathbf{v}_1}\varphi_1(\mathbf{x}) & \cdots & \partial^{\mathbf{v}_1}\varphi_f(\mathbf{x}) \\ \vdots & \ddots & \vdots \\ \partial^{\mathbf{v}_w}\varphi_1(\mathbf{x}) & \cdots & \partial^{\mathbf{v}_w}\varphi_f(\mathbf{x}) \end{pmatrix}.$$

The values at multiple points can be stored by collecting similar blocks. In particular, for $X = \{\mathbf{x}_1, \ldots, \mathbf{x}_r\}$ let

$$\mathbf{B}_i(X, W) = \begin{pmatrix} \mathbf{E}_{B_i}(\mathbf{x}_1, W) \\ \vdots \\ \mathbf{E}_{B_i}(\mathbf{x}_r, W) \end{pmatrix} \quad \text{and} \quad \mathbf{G}(X, W) = \begin{pmatrix} \mathbf{E}_G(\mathbf{x}_1, W) \\ \vdots \\ \mathbf{E}_G(\mathbf{x}_r, W) \end{pmatrix}.$$

Assuming the above format, a general implementation of eval for G is given by the following procedure.

Procedure: eval(X, W)
    Input: the set of points $X = \{\mathbf{x}_1, \ldots, \mathbf{x}_r\} \subset \Omega$
    Input: the requested data $W = \{\mathbf{v}_1, \ldots, \mathbf{v}_w\}$
    Output: $\mathbf{G}(X, W)$
    foreach $\mathbf{x} \in X$ **do**
        $i = \text{findSubdomain}(\mathbf{x})$
        $\mathbf{E}_{B_i}(\mathbf{x}, W) = B_i.\text{eval}(\mathbf{x}, W)$
        $\mathbf{E}_G(\mathbf{x}, W) = \mathbf{E}_{B_i}(\mathbf{x}, W)\mathbf{M}_i$
        /* $\mathbf{E}_G(\mathbf{x}, W)$ is directly written into $\mathbf{G}(X, W)$ */
    end

For the common case when X is contained in one Bézier element of the space it can be useful to provide the following optimized procedure that accepts the containing subdomain as an input parameter.

Procedure: `evalSubdomain(`X, W, i`)`
    Input: the set of points $X = \{\mathbf{x}_1, \ldots, \mathbf{x}_r\} \subset \Omega$
    Input: the requested data $W = \{\mathbf{v}_1, \ldots, \mathbf{v}_w\}$
    Input: the index i of the subdomains containing X
    Output: $\mathbf{G}(X, W)$
    $\mathbf{B}_i(X, W) = B_i.\text{eval}(X, W)$
    $\mathbf{G}(X, W) = \mathbf{B}_i(X, W)\mathbf{M}_i$

Now T, B and $\mathbf{M}$ will be described in more detail.

A suitable implementation of T is a binary decision tree (more precisely a binary space partition, cf. [40, 41]). For the spaces of interest it is possible to assume that the subdomains Δ_i are polytopes with axis-aligned faces. Each fork in the tree corresponds to a spatial split along an axis-aligned affine hyperspace, i.e., to a comparison for a specific coordinate. Each branch corresponds to taking one of the corresponding half-spaces. Each leaf of the tree corresponds to the intersection of the taken half-spaces and Ω. Thus T can be represented by a tree storing in each leaf the index of the subdomain containing the corresponding box. Figure 3 depicts a partition and a corresponding tree.

Binary space partitions not only provide an efficient implementation of the method `findSubdomain` but also offer useful representations of piecewise constant maps $\Omega \to \mathbb{N}$. They enable efficient computation of binary operations (see the references above for union and intersection) that can be employed both for the geometrical computation required by the construction of the different spaces and by refinement strategies.

For instance, given two trees that assign to each point a refinement level, it is easy to compute the coarsest common refinement by the pointwise-max operation. Similarly, the finest common submesh can be computed with a pointwise-min operation.

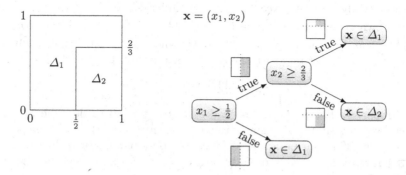

Fig. 3. A partition of Ω in Δ_1 and Δ_2 and a decision tree describing it. The darkened area next to each branch highlights the region corresponding to the branch.

The collection of local bases $B = \{B_1, \ldots, B_s\}$ is simply a list of polymorphic objects implementing the `eval` interface. This allows for arbitrary local bases and thus, for example, Bernstein polynomials as in Bézier extraction, B-splines as in all the implementations presented here, or enriched spaces such as generalized B-splines [3] with piecewise trigonometric or exponential functions.

Finally, $\mathbf{M}$ is a sparse matrix. However, the initialization of the matrix for a particular spline space usually requires most of the space-specific code.

2.2   Subspaces and Functions

Consider a subspace span $G' \subset$ span G generated by $G' = \{\gamma_1', \ldots, \gamma_k'\}$. If $\mathbf{N} = (n_{\gamma,\gamma'})_{\gamma \in G, \gamma' \in G'}$ is a matrix that contains in the i-th column the expansion of γ_i' in $\gamma_1, \ldots, \gamma_n$, i.e.,

$$\forall \mathbf{x} \in \Omega, \ \boldsymbol{\gamma}'(\mathbf{x}) = \boldsymbol{\gamma}(\mathbf{x})\mathbf{N},$$

then G' can be implemented by $T, B, \mathbf{M}'$ with

$$\mathbf{M}' = \mathbf{MN}.$$

As a consequence, `eval` is not only a suitable implementation of G, but also of a single function $\varphi \in$ span G. Indeed, this corresponds to $\mathbf{M}'$ having a single column and $\mathbf{N}$ being the column vector of the coefficients of φ. Another application of this method is enforcing homogeneous linear constraints on the space, such as boundary condition or smoothness constraints.

2.3   Multipatch Domains

The proposed framework can be extended to allow for multipatch domains. Multipatch domains are used to describe geometries Ω with nontrivial topology and for which no regular parametrization $G : \widehat{\Omega} \to \Omega$ with a box $\widehat{\Omega} \subset \mathbb{R}^d$ exists. A simple example is the unit sphere in 3D for which there exists no regular parametrization defined on a rectangle. However, it is possible to partition such a domain Ω into mutually disjoint (except at their boundaries) *patches* $\Omega_1, \ldots, \Omega_w$, each with its own regular parametrization $G_p : \widehat{\Omega}_p \subset \mathbb{R}^d \to \Omega_p$. Then Ω can be thought of as the image of $G : \widehat{\Omega} \to \Omega$, where $\widehat{\Omega}$ is the disjoint union of $\widehat{\Omega}_1, \ldots, \widehat{\Omega}_w$ and the points with the same image have been identified, i.e.,

$$\widehat{\Omega} = \coprod_{p=1}^{w} \widehat{\Omega}_p /_\sim$$

for a proper $\sim$. The proposed method can be extended to describe functions defined on $\widehat{\Omega}$ by simply changing `findSubdomain` to take the different patches into account. This can be achieved by an optional parameter p. In particular, `evalSubdomain` does not need modifications as long as Δ_i is contained in $\widehat{\Omega}_j$ if B_i is defined on $\widehat{\Omega}_j$.

The construction of C^k function spaces on multipatch domains is an active research topic in IGA [7,10,28]. This corresponds, by the isoparametric approach, to the construction of subspaces of patchwise C^k functions on $\widehat{\Omega}$. The relations that define the subspace depend on G and its derivatives and do not necessarily correspond to smoothness conditions on $\widehat{\Omega}$ after simple point identification.

The space of patchwise $\mathcal{C}^k$ functions can be described in the proposed framework by a block-diagonal matrix $\mathbf{M}$, where each block represents the space of $\mathcal{C}^k$ functions on each patch. As described in the previous subsection, the representation of a subspace is obtained by multiplying $\mathbf{M}$ by an appropriate $\mathbf{N}$. This strategy was used in [7], where, due to a different implementation of the patch spaces, the multiplication by $\mathbf{N}$ is done at a post-processing stage and thus incurs in an additional cost.

3 Complexity

Delegating the evaluation to a local basis and computing the linear combination incurs in an additional computational cost. Moreover, storing the coefficients of $\mathbf{M}$ can require a substantial amount of memory.

3.1 Space Complexity

The tested implementation uses a row-compressed format: only the nonzero coefficients are stored in lexicographic order of their indices. The column position of the nonzero entries is stored in a second vector. The row position is deduced by storing a pointer to the first nonzero of each row. This means that the total required memory is proportional to the sum of the number of rows plus the number of nonzero entries of $\mathbf{M}$.

The number of rows of $\mathbf{M}$ equals $\sum_{i=1}^{s} \#B_i$ and thus there is a memory cost associated to functions of the local bases even if they are not used in any Δ_i to represent G.

The number of nonzero coefficients in $\mathbf{M}$ depends on the complexity of the mesh and on the shape of the generators. The number of nonzero coefficients in the column corresponding to $\gamma \in G$ is

$$\sum_{i:\gamma|_{\Delta_i} \neq 0} \#\{\beta : m_{\beta,\gamma} \neq 0\}. \tag{4}$$

Thus it is minimized if γ is supported in a single Δ_i and if $\gamma = \beta$ for some $\beta \in B_i$. In contrast, the generators γ whose supports intersect many subdomains or whose shape requires many coefficients to be represented in a subdomain require more memory.

3.2 Time Complexity

The time cost of the `eval` procedure is proportional to the cardinality of X and depends on the cost of the local basis evaluation, which is unknown. Denoting $\mathfrak{C}(matrix)$ the cost of the computation of $\mathbf{B}_i(\mathbf{x}, W)\mathbf{M}_i$, it can be written as

$$\mathfrak{C}(\texttt{eval}) = \#X\Big(\mathfrak{C}(B_i.\texttt{eval}) + \mathfrak{C}(matrix) + \mathfrak{C}(\texttt{findSubdomain})\Big).$$

Remembering that $w = \#W$ is the number of rows in the blocks $\mathbf{E}_{\square,W}$ it is possible to describe each term in more detail.

The complexity of B_i.eval depends on the specific local basis used and is clearly bounded from below by the output size $w\#B_i$. For $d \geq 2$ tensor-product B-splines can be implemented in such a way that the cost is quasi-optimal, i.e., proportional to the output size with a factor that does not depend on their degree but which depends on the dimension:

$$\mathfrak{C}(B_i.\text{eval}) \cong dw\#B_i.$$

The cost of the matrix-matrix product $\mathbf{B}_i(\mathbf{x})\mathbf{M}_i$ using standard algorithms is proportional to the product of the three dimensions of the two matrices:

$$\mathfrak{C}(matrix) \cong w\#B_i\#G.$$

The cost of findSubdomain depends on the tree structure and on the complexity of the mesh. For a balanced tree this would be proportional to $\log_2 \ell$, where ℓ is the number of leaves in T. However, a balanced tree is not necessarily optimal, as the tree should take the usage pattern into account. For instance, if we assume a uniform sampling of the domain, then the optimal tree will have leaves of depth inversely proportional to the measure of the corresponding region. Already when avoiding unnecessary splits (without any balancing), the cost of findSubdomain was negligible in the profiling tests.

The total evaluation cost is thus of the following magnitude

$$\mathfrak{C}(\text{eval}) \cong \#X\mathfrak{C}(matrix) \cong w\#X\#B_i\#G. \tag{5}$$

The same result is obtained for evalSubdomain with the difference that $w\#X$ then means the number of rows in $\mathbf{B}_i(X)$.

Comparing this with the output size $w\#X\#G$ shows that the method is rather expensive if $\#B_i$ is big. The next sections show how this cost can be reduced.

3.3 Local Basis and Compression

If the functions in G and B_i have small supports, then the number of nonzero columns in $\mathbf{B}_i(X, W)$ and in $\mathbf{G}(X, W)$ is small compared to $\#B_i$ and $\#G$, respectively. This suggests the use of a compressed format for $\mathbf{B}_i(X, W)$ and $\mathbf{G}(X, W)$, where only the nonzero values and their positions are stored. This is standard in FEM as well as in other numerical methods and it is used in our implementation too.

Assume that $X \subset \Delta_i$. Let L be the set of functions in B_i such that the corresponding columns in $\mathbf{B}_i(X, W)$ are not zero and let A be the corresponding set of functions in G defined by

$$A = \{\gamma \in G : \exists \beta \in L : m_{\beta,\gamma} \neq 0\}.$$

A function φ is called *active* on X if $\varphi \in L$ or $\varphi \in A$.

Let $\mathbf{L}(X,W)$ and $\mathbf{A}(X,W)$ be the corresponding submatrices of $\mathbf{B}_i(X,W)$ and $\mathbf{G}(X,W)$,

$$\mathbf{L}(X,W) = \begin{pmatrix} \mathbf{E}_L(\mathbf{x}_1,W) \\ \vdots \\ \mathbf{E}_L(\mathbf{x}_r,W) \end{pmatrix}, \qquad \mathbf{A}(X,W) = \begin{pmatrix} \mathbf{E}_A(\mathbf{x}_1,W) \\ \vdots \\ \mathbf{E}_A(\mathbf{x}_r,W) \end{pmatrix}.$$

Then $\mathbf{B}_i(X,W)$ can be implemented with the pair $(L, \mathbf{L}(X,W))$ where the set L is implemented as a list of indices. This reduces the lower bound on $\mathfrak{C}(B_i.\texttt{eval})$ to

$$\mathfrak{C}(B_i.\texttt{eval}) \cong w\#X\#L.$$

Similarly, $\mathbf{G}(X,W)$ can be implemented by the pair $(A, \mathbf{A}(X,W))$. The coefficients in $\mathbf{A}$ are computed by

$$\mathbf{A}(X,W) = \mathbf{L}(X,W)\mathbf{M}_{L,A},$$

where $\mathbf{M}_{L,A}$ is the submatrix of $\mathbf{M}$ containing the $m_{\beta,\gamma}$, $\beta \in L$ and $\gamma \in A$. This reduces the cost of the linear combination to

$$\mathfrak{C}(matrix) \cong w\#X\#L\#A,$$

improving (5) by the factor

$$\frac{\#L\#A}{\#B_i\#G}. \tag{6}$$

The described compression can be applied both to $\texttt{eval}$ and $\texttt{evalSubdomain}$. In $\texttt{eval}$ it is applied to the evaluation at single points. Then either the list containing the (per point) compressed matrices is returned or all of the matrices are merged into one matrix. The first approach is faster and simpler, the second returns a standard matrix.

The application of compression to $\texttt{evalSubdomain}$ is straightforward but it is important to limit X to points contained in a small region so that (6) is minimized.

When using compression, the cost of evaluation is proportional to the output size $w\#X\#A$ multiplied by $\#L$. For polynomial splines and with X contained in a single element, $\#L$ depends only on the polynomial degree. As a consequence the cost of the evaluation per unit of output data does not depend on the mesh (h-refinement) but depends on the degree (p-refinement).

3.4   Tensor Factorization

The tensor-product structure allows to reduce d-variate computations to computations on univariate objects. In our case it allows to replace the computation of the linear combination of d-variate functions with d linear combinations of univariate functions. This is advantageous because the cost of the matrix-matrix product is roughly proportional to the product of the three involved dimensions. In the optimal case this optimization reduces one of the dimensions to its d-th root.

The above strategy can be applied under the weaker assumption that each $\gamma \in G$ and each $\beta \in B_i$ can be factored into products of univariate functions:

$$\gamma(\mathbf{x}) = \prod_{c=1}^{d} \gamma^{(c)}(x_c), \qquad \beta(\mathbf{x}) = \prod_{c=1}^{d} \beta^{(c)}(x_c), \tag{7}$$

where $\square^{(c)}$ means the *factor* of $\square$ corresponding to the c-th coordinate. Of the implemented spaces only HB and HLR satisfy (7), thus this optimization was not implemented and the following is only a theoretical analysis.

In analogy with (7), the same notation is used to denote the factors corresponding to the coordinate directions of tensors $\square = \bigotimes_{c=1}^{d} \square^{(c)}$ and of Cartesian grids of points $\square = \bigtimes_{c=1}^{d} \square^{(c)}$. This should not be confused with the *components* of vectors and tensors that are denoted by subscripts as in $\mathbf{x} = (x_1, x_2)$. In particular, the following factors are defined:

$$B_i^{(c)} = \{\beta^{(c)} : \beta \in B_i\};$$
$$G^{(c)} = \{\gamma^{(c)} : \gamma \in G\};$$
$$X^{(c)} = \{x_c : \mathbf{x} = (x_1, \ldots, x_d) \in X\};$$
$$W^{(c)} = \{v_c : \mathbf{v} = (v_1, \ldots, v_d) \in W\}.$$

In the following $w^{(c)}$ denotes $\#W^{(c)}$, i.e., the number of derivatives of $\gamma^{(c)}$ that are required to compute all requested partial derivatives in W.

Necessarily $G^{(c)} \subseteq \text{span } B_i^{(c)}$, which means that there exists a matrix $\mathbf{M}_i^{(c)}$ such that

$$\forall \mathbf{x} \in \Delta_i : \quad \boldsymbol{\gamma}^{(c)}(x_c) = \boldsymbol{\beta}_i^{(c)}(x_c)\mathbf{M}_i^{(c)}.$$

Here, analogously to (3), $\boldsymbol{\gamma}^{(c)}(x_c)$ and $\boldsymbol{\beta}_i^{(c)}(x_c)$ denote the vectors having components indexed by $G^{(c)}$ and $B_i^{(c)}$, respectively, i.e.,

$$\boldsymbol{\gamma}^{(c)}(x_c) = \left(\gamma^{(c)}(x_c)\right)_{\gamma^{(c)} \in G^{(c)}},$$
$$\boldsymbol{\beta}_i^{(c)}(x_c) = \left(\beta^{(c)}(x_c)\right)_{\beta^{(c)} \in B_i^{(c)}}.$$

Let S be the set of the multiindices that define G as a subset of $\bigotimes_{c=1}^{d} G^{(c)}$:

$$G = \left\{ \prod_{c=1}^{d} \gamma_{\mathbf{s}_c}^{(c)} : (\mathbf{s}_1, \ldots, \mathbf{s}_d) \in S, \gamma_{\mathbf{s}_c}^{(c)} \in G^{(c)} \right\} \subseteq \bigotimes_{c=1}^{d} G^{(c)}.$$

For simplicity it is assumed that $B_i = \bigotimes_{c=1}^{d} B_i^{(c)}$ and $X = \bigtimes_{c=1}^{d} X^{(c)}$, but a proper subset (similarly as for G) can be considered at the expense of a more involved notation and implementation.

The tensor-product structure propagates to the set of active functions. Here L contains the multiindices of the functions of B_i corresponding to nonzero columns of $\mathbf{B}_i$. Similarly, A contains the subset of the multiindices in S, i.e., the multiindices of functions in G that correspond to nonzero columns in $\mathbf{G}$. Analogously to the other symbols, $L^{(c)}$ and $A^{(c)}$ denote the collection of the entries relative to the c-th coordinate in L and A respectively.

The procedure compose assembles the matrix $\square$ out of the matrices of its factors $\square^{(c)}$ and a list of the necessary products P as in the description of S above.

Procedure: compose$(\square^{(1)}, \ldots, \square^{(d)}, X, W, P)$
   Input: the tensor components $\square^{(c)}$
   Input: the list of points X
   Input: the list of derivatives W
   Input: the list of required products P
   Output: $\square$
   foreach $\mathbf{p} = (p_1, \ldots, p_d) \in P$ **do**
      foreach $\mathbf{x} = (x_1, \ldots, x_d) \in X$ **do**
         /* write row block of the derivatives of $\square_{\mathbf{p}}$ at $\mathbf{x}$ */
         foreach $\mathbf{v} = (v_1, \ldots, v_d) \in W$ **do**
            $\partial^{\mathbf{v}} \square_{\mathbf{p}}(\mathbf{x}) = \prod_{c=1}^{d} \partial^{v_c} \square_{p_c}^{(c)}(x_c)$
            /* each value is written in $\square$: the row
            corresponds to the pair $(\mathbf{x}, \mathbf{v})$, the column to $\mathbf{p}$
            */
         end
      end
   end

If P is omitted, it is assumed that $\square = \bigotimes_{c=1}^{d} \square^{(c)}$ and thus that P contains all the Cartesian multiindices.

For a given domain dimension d the cost of the procedure compose is proportional to the size of its output with factor d,

$$\mathfrak{C}(\text{compose}) = dw\#P\#X.$$

Only an application of this optimization to evalSubdomain is presented, but it can also be applied to eval. By using compose, the evaluation of the local basis can be split in two steps: evaluation of the components of the local basis and their composition. The original evalSubdomain can be equivalently rewritten as:

Procedure: `evalSubdomain(X, W, i)`
    Input: the set of points $X = \{\mathbf{x}_1, \ldots, \mathbf{x}_r\} \subset \Omega$
    Input: the requested data $W = \{\mathbf{v}_1, \ldots, \mathbf{v}_w\}$
    Input: the index i of the subdomains containing X
    Output: $\mathbf{G}(X, W)$
    for $c = 1, \ldots, d$ **do**
        $\mathbf{B}^{(c)} = B_i^{(c)}.\mathtt{eval}(X^{(c)}, W^{(c)})$ `/* local evaluation */`
    end
    $\mathbf{B}_i(X) = \mathtt{compose}(\mathbf{B}^{(1)}, \ldots, \mathbf{B}^{(d)}, X, W)$ `/* composition */`
    $\mathbf{G}(X, W) = \mathbf{B}_i(X, W)\mathbf{M}_i$ `/* linear combination */`

As described, the computational cost of the above is determined by the computation of the matrix-matrix product. This can be reduced by leaving `compose` as the last operation and computing d products of smaller matrices as follows:

Procedure: `evalSubdomain(X, W, i)`
    Input: the set of points $X = \{\mathbf{x}_1, \ldots, \mathbf{x}_r\} \subset \Omega$
    Input: the requested data $W = \{\mathbf{v}_1, \ldots, \mathbf{v}_w\}$
    Input: the index i of the subdomains containing X
    Output: $\mathbf{G}(X, W)$
    for $c = 1, \ldots, d$ **do**
        $\mathbf{B}^{(c)} = B_i^{(c)}.\mathtt{eval}(X^{(c)}, W^{(c)})$ `/* local evaluation */`
        $\mathbf{G}^{(c)} = \mathbf{B}^{(c)}\mathbf{M}_i^{(c)}$ `/* linear combination */`
    end
    $A = S \cap \bigcap_{c=1}^{d} A^{(c)}$ `/* actives */`
    $\mathbf{G}(X, W) = \mathtt{compose}(\mathbf{G}^{(1)}, \ldots, \mathbf{G}^{(d)}, X, W, A)$ `/* composition */`

The cost estimates of each step for the two different versions are reported in Table 1.

If there exists m such that for $c = 1, \ldots, d$

$$\#L^{(c)} \#A^{(c)} \leq m \#A, \tag{8}$$

then the evaluation cost of the optimized version is proportional to the output size $w \#X \#A$. This result holds independently of the mesh size (h-refinement) and the polynomial degree (p-refinement). Note however that the output size depends on the number of active functions $\#A$ and on the size of the requested data, thus on the polynomial degree, on the set of points X and on w.

The assumption in (8) holds in situations that are of interest for the applications. In particular, it holds for splines of degree p and for points X contained in a single element if $\#A^{(c)} \leq m(p+1)^{d-1}$ for some m independent of the degree. This is the case for m-admissible HB meshes [9], where $\#A^{(c)} \leq m(p+1)$, and for the HLR basis described in [5], for which $\#A^{(c)} \leq 2(p+1)$.

Table 1. Comparison of the cost for the standard and optimized evaluation for spaces with tensor-product structure.

	Standard	Optimized
Linear combination	$w\#X\#A\#L$	$\sum_{c=1}^{d} w^{(c)}\#X^{(c)}\#A^{(c)}\#L^{(c)}$
Composition	$dw\#X\#L$	$dw\#X\#A$

4 Comparison with Bézier Extraction

Bézier extraction was proposed for NURBS [2] and extended to T-splines [37] and THB [22]. It is an implementation technique aimed at reusing standard FEM codebases in IGA by representing the basis functions as linear combinations of Bernstein polynomials on each element. The *extraction operator*, i.e., the linear transformation of the basis, is stored in a combination of a per element matrix and a global index of the per element active functions called IEN.

The proposed framework reduces to Bézier extraction when the following choices are made:

T is the partition of the domain into elements;
B_i is the Bernstein basis remapped to the element Δ_i;
$\mathbf{M}_i$ contains the expansion of γ_j in the Bernstein basis in column j.

Thus all the spaces that can be implemented with Bézier extraction – such as T-splines or LR-splines – can be also incorporated to the proposed framework.

Even if the underlying concepts are the same, the implementation differs as the two approaches are optimized for different scenarios. The main differences are collected in Table 2 and their consequences are described below.

Table 2. Qualitative comparison of the two frameworks.

	Bézier extraction	Proposed framework
Mesh description	List	Tree
Expansion in local basis	Per element	Global matrix
Local basis	Bernstein polynomials	Any

4.1 Mesh Description

Bézier extraction is based on per element data structures. This reflects the original aim: IGA with per element quadrature integration. For such application it is both simple and efficient. The drawback is that it does not provide a feature-rich description of regions that can be used in the implementation of different spaces. Every space must implement its own strategy both for identifying the element containing a point and for the description of the mesh used in its construction.

The tree-based description of the subdomains in the proposed framework provides an efficient tool to describe "arbitrary" regions contained in the domain, to compute intersections, unions and for testing whether a region contains a point (see findSubdomain in Sect. 2). This common code is shared by all the implemented spaces, thus decreasing the per-space code requirements.

Iteration on the elements in this framework is done by nested iteration: the outer iteration is on the tree leaves and the inner on the elements provided by the local basis and contained in the region corresponding to the current leaf. The nested loop is a part of the shared code.

4.2   Expansion in Local Basis

Both approaches store the expansion of the generators with respect to a local basis. In Bézier extraction the linear operator is represented by submatrices (the *extraction operators*) together with their indices (the IEN), whereas in the proposed framework it is represented by a sparse matrix.

From this point of view Bézier extraction can be seen as a specialized matrix format. However, avoiding the specialized format makes the implementation of subspaces and multipatch straightforward, for they correspond to matrix multiplication as has been described in Subsects. 2.2 and 2.3 and the code is already provided and optimized by the linear algebra library. It is true that products involving submatrices of a sparse matrix are less efficient then products involving full matrices, but by allowing different generators the total memory requirement can be lowered as it is discussed in the next subsection.

4.3   Local Basis

While Bézier extraction was only described for spaces of piecewise tensor-product polynomials with Bernstein polynomials as local generators, there is no practical issue to extend it to other local bases such as enriched splines spaces as those from [3]. This means that both techniques are roughly equally applicable. Nevertheless, there are two advantages of the proposed framework compared to Bézier extraction.

The first is that most of the spaces of interest are defined in terms of collections of B-splines functions and not Bernstein polynomials. Thus the proposed code stays closer to the definition of the space and it is easier to write.

The second is that by using B-splines as local generators the coefficients are shared between many elements and thus the memory requirement is decreased. Applying (4) shows that the amount of coefficients stored per a generator $\gamma \in G$ in Bézier extraction equals

$$\mathbf{e}_\gamma \dim \mathbb{T},$$

where $\mathbf{e}_\gamma$ is the number of elements contained in support γ and $\mathbb{T}$ is the space of tensor-product polynomials. This is an upper bound for the amount of coefficients stored using B-splines as local generators. Increasing the number of coefficients does not only increase the memory requirements, but it also increases the cost of space initialization.

In favor of Bézier extraction stands the fact that in IGA applications it is not necessary to evaluate the Bernstein polynomials on the quadrature nodes: it is enough to scale the derivatives computed on a reference element according to the current element.

Summarizing, the proposed framework permits testing of various definitions with a reduced development time. Bézier extraction optimizes the matrix assembling in IGA applications for a specific space.

5   Implemented Spaces

The proposed strategy was tested by implementing HB, THB, TBPN, DHB and HLR splines spaces in the $G+Smo$ object oriented library [26]. The spaces were coded as templated C++03 classes. They all derive from a common base class that is the realization of the described approach.

The interested reader can compare with other implementations that are either available or described in the literature. (T)HB are implemented in the $G+Smo$ open-source library [19]. The code, as of 2014, is described in [29]. Another implementation of (T)HB tailored for IGA research is described in [42]. The source code of bivariate LR-splines is available as a part of the goTools library [20], but no technical description is available.

The first subsection describes the shared code. The following subsections specialize to various spaces. The last subsection reports the size of the implementation measured in lines of code.

5.1   Shared Code

The shared code contains the implementation of the ideas described in Sect. 2 as well as common utilities such as input-output, tensor-product B-splines and debugging functions.

Part of the required code was already present in the $G+Smo$ library, in particular, vectors, matrices, sparse matrices (all of them based on the Eigen library [21]) and tensor-product B-splines. Some parts were coded anew such as a specialized version of Boehm's knot-insertion algorithm, the binary tree, functions for transforming between flat-indices and multiindices of multivariate B-splines and others to export data in the *ParaView* format [1].

The implementation of T uses the binary tree as described in Sect. 2 and includes an interface for performing arbitrary unary and binary operations, possibly by restricting the operation to a box contained in the domain. This mechanism is used in the construction of the spaces: for instance, the Kraft selection mechanism of (T)HB corresponds to finding the minimum of the indices of the subdomains intersecting the support of a function, the decoupling procedure requires methods to compute intersections and unions of polytopes with axis-aligned faces. The implementation of T automatically removes unnecessary branches at construction by collapsing equal subtrees.

A component of $G+Smo$ that was developed for smooth multipatch spaces [7] was reused for $\mathbf{M}$. At its core it is a sparse matrix with additional methods for computing A from L and extracting $\mathbf{M}_{L,A}$, see Subsect. 3.3. It also contains conversion functions to and from other data types related to the implementation of multipatch geometries and boundary conditions in *G+Smo*.

The base class of all the implemented spaces contains the reference to T, $\mathbf{M}$ and B, the evaluation procedure, constructors allowing for multipatch domains and utilities to obtain functions and subspaces as described in Subsect. 2.2.

5.2   (Truncated) Hierarchical B-splines

The hierarchical basis is defined from a sequence $P_1, \ldots, P_s$ of tensor-product B-spline bases and a corresponding sequence $\Omega = \Omega_1 \supseteq \cdots \supseteq \Omega_s = \emptyset$ of closed domains. For simplicity it is assumed here that Ω is a box in $\mathbb{R}^d$ and that the bases have clamped knots on its boundary. It is required that the tensor-product spaces form a *hierarchy*, i.e.:

$$i < j \Rightarrow \operatorname{span} P_i \subset \operatorname{span} P_j. \tag{9}$$

The *hierarchical basis* (HB-splines) is defined by Kraft's *selection* criteria [30]:

$$H = \bigcup_{i=1}^{s} \{\psi \in P_i \; : \; \operatorname{support} \psi \subseteq \Omega_i \text{ and } \operatorname{support} \psi \cap (\Omega_i \setminus \Omega_{i+1}) \neq \emptyset\}. \tag{10}$$

The *truncated hierarchical basis* (THB-splines) described in [17] is defined by recursive truncation

$$H' = \{\mathbf{T}_s \cdots \mathbf{T}_{i+1} \psi \; : \; \psi \in H \cap P_i\}.$$

The truncation operator $\mathbf{T}_i : \operatorname{span} P_i \to \operatorname{span} P_i$ is defined by

$$\mathbf{T}_i(\varphi) = \sum_{\psi \in P_i \; : \; \psi|_{\Omega \setminus \Omega_i} \neq 0} c_{\varphi,\psi} \psi,$$

where the coefficients $c_{\varphi,\psi}$ are taken from the expansion of φ in P_i:

$$\varphi = \sum_{\psi \in P_i} c_{\varphi,\psi} \psi.$$

The matrix representation of $\mathbf{T}_i$ with respect to the basis P_i is thus diagonal with entries

$$t_{\psi,\psi} = \begin{cases} 1 & \text{if } \psi|_{\Omega \setminus \Omega_i} \neq 0, \\ 0 & \text{otherwise.} \end{cases}$$

The truncation procedure improves the locality of the resulting basis, guarantees that H' forms a convex partition of unity and preserves the same coefficients as P_i for polynomial expansion [18]. The drawback is that it breaks the tensor-product structure, i.e., the functions $\psi' \in H'$ are not tensor-product B-splines. Thus the optimization described in Sect. 3.4 cannot be applied for H'.

Note that the composition of the truncation operators differs from the truncation by the finest level: in general if $\psi \in P_i$ then for any $k \geq i$

$$\mathbf{T}_k \cdots \mathbf{T}_{i+1} \psi|_{\Omega_k} \neq \mathbf{T}_k \psi|_{\Omega_k}. \tag{11}$$

The equality in (11) holds if the mesh is sufficiently graded.

Two implementations are described. Both assume that the bases $P_1, \ldots, P_s$ have the same degree (i.e., only h-refinement is allowed) and that the subdomains Ω_i are unions of elements of span P_i. The first implementation is closer to the definition and has actually been coded. The second is described in order to show that memory requirements can be lowered with more complex code.

Implementation 1. The simplest implementation defines T by:

$$\Delta_i = \Omega_i \setminus \Omega_{i+1}, \quad i = 1, \ldots, s,$$

and B by $B_i = P_i, i = 1, \ldots, s$.

Most of the construction of $\mathbf{M}$ is common for HB and THB. In the coded implementation truncation is controlled by a construction option in order to decrease code duplication. The procedure `constructTHB` describes the constructor.

```
Procedure: constructTHB(Ω₁,...,Ωₛ, P₁,...,Pₛ, t)
    Input: Subdomains Ω₁,...,Ωₛ
    Input: Bases P₁,...,Pₛ
    Input: Option t: switch between HB and THB
    for ℓ = s to 1 do
        Lₗ = {ψ ∈ Pₗ : ψ|Ωₗ ≠ 0}
        foreach τ ∈ Lₗ do
            ℓₘ = minLevelIntersecting(support τ)
            if ℓₘ == ℓ then                              /* τ ∈ H due to (10) */
                γ = addGenerator()
                m_{τ,γ} = 1                    /* New column of M with exactly one 1. */
                d = (d_ψ)_{ψ∈Pₗ} = (0,...,0,d_τ = 1,0,...,0)
                for j = ℓ+1 to ℓₘ do
                    d = refine(d, j)                        /* Now d = (d_ψ)_{ψ∈P_j}. */
                    if t then                                          /* THB */
                        foreach ψ ∈ P_j do
                            if ψ ∈ L_j \ H then             /* Save the coeff. */
                                m_{ψ,τ} = d_ψ
                            else                            /* Truncate the coeff. */
                                d_ψ = 0
                        end
                    end
                    /* Now Σ_{ψ∈P_j} d_ψ ψ|Ω_j = T_j ... T_i τ|Ω_j = γ|Ω_j. */
                    else                                               /* HB */
                        foreach ψ ∈ L_j do             /* Save the coeff. */
                            m_{ψ,γ} = d_ψ
                        end
                    end
                end
            end
        end
    end
end
```

The lists L_ℓ are constructed by traversing the leaves of T and using the implementation of tensor-product B-splines. The procedure `minLevelIntersecting(box)` returns the minimum of $\{i : \varDelta_i \cap \mathtt{box} \neq \emptyset\}$ and it is provided by the shared code. The procedure `refine(d, j)` uses Boehm's algorithm to compute the expansion of the following function σ in terms of level j,

$$\sum_{\psi \in P_{j-1}} d_\psi \psi = \sigma = \sum_{\varphi \in P_j} c_{\varphi,\sigma} \varphi,$$

and returns $(c_{\varphi,\sigma})_{\varphi \in P_j}$.

The levels are iterated from the finest to the coarsest. In this way the difference $L_j \setminus H$ can be computed because $H \cap P_j$ is already known. Consequently the full construction of the space can be performed in one loop over the levels. A different solution (used for instance in $G+Smo$) is to delay the computation of the expansion after determining the selected functions from all levels.

The fact that only the coefficients $m_{\psi,\tau}$ with $\psi \in L_j$ are saved is a memory optimization, the same code runs with P_j in place of L_j except for the test for $\psi \in L_j \setminus H$ that would be modified accordingly.

This strategy was tested against the reference implementation in $G+Smo$. The comparison showed both faster evaluation and smaller memory consumption for selected 2D examples.

Implementation 2. The choices above are the simplest, but they can cause a very high memory consumption. According to Subsect. 3.1 the memory usage depends on the total number of rows in $\mathbf{M}$. For dyadic refinement of the P_i the number of rows grows as $2^{d(s+1)}$, where d is the domain dimension and s is the number of levels. Since each row requires a memory pointer, this means that an empty $\mathbf{M}$ for a 3D example with 10 levels exceeds 10 GB in size.

The problem can be solved with slightly more complex code. The main idea is to remove the rows of $\mathbf{M}$ containing only zeros, that is, to define B_i and $\varDelta_i$ so that $\psi|_{\varDelta_i} = 0$ does not happen for any $\beta \in B_i$.

Denoting $\widetilde{T}_i$ the set of leaves of the binary partition tree representing the domains $\widetilde{\varDelta}_i = \Omega_i \setminus \Omega_{i+1}$, define

$$T = \bigcup_{i=1}^{s} \widetilde{T}_i.$$

For each $\varDelta_k \in T$ there is exactly one j such that $\varDelta_k \in \widetilde{T}_j$; define

$$B_k = \{\beta \in P_j : \beta|_{\varDelta_k} \neq 0\}.$$

Since $\varDelta_k$ is a box (a Cartesian product of intervals), B_k is a tensor-product basis. Note that typically $\#T > s$ but all B_k are quite small.

The construction of $\mathbf{M}$ is done as in the previous implementation except for the necessary shifts of indices. This solution is not coded for (T)HB, but the required machinery was implemented for DHB.

5.3   Truncated B-splines for Partially Nested Refinement

This is a generalization of (T)HB-splines that was proposed in [43]. It allows for independent refinement in different parts of the domain (see Fig. 4) and can help for multipatch geometries as shown in Example 3.

The requirement (9) is dropped and the sequence of nested domains is replaced by a partition of Ω into *patches* $\Lambda_1, \ldots, \Lambda_s$. Note that here the word "patch" has a different meaning from the context of multipatch domains, cf. Subsect. 2.3. The construction requires the following compatibility condition: if Λ_i and Λ_j share a $(d-1)$-dimensional interface $\Gamma_{i,j} = \partial\Lambda_i \cap \partial\Lambda_j$, then

$$\text{span } P_i \subset \text{span } P_j \quad \text{or} \quad \text{span } P_j \supset \text{span } P_i.$$

This means that $\{\text{span } P_i : i = 1, \ldots, s\}$ is not totally ordered anymore, only *partially ordered*. In particular, if the boundaries are disjoint or their intersection is not $(d-1)$-dimensional, the spaces span P_i and span P_j do not have to be comparable for inclusion. Note that the construction requires "sufficient separation" of the patches associated to two incomparable spaces. See [43] for details.

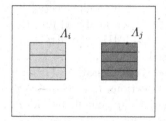 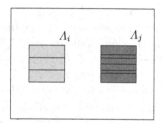

Fig. 4. Left: TBPN-splines allow to refine the subdomains Λ_a and Λ_b independently. Right: THB-splines require nested knot vectors for any pair of subdomains.

Basis functions are again a subset of $\bigcup_{i=1}^{s} P_i$ and are selected using a modification of Kraft's procedure based on *slave functions*. A function $\psi \in P_i$ is called a *slave* if it is active on an $(n-1)$-dimensional interface $\Gamma_{i,j}$ with span $P_j \subset$ span P_i. The set of slaves of level i can be written as

$$S_i = \{\psi \in P_i : \exists j : \psi|_{\Gamma_{i,j}} \neq 0, \text{ span } P_j \subset \text{span } P_i, \dim \Gamma_{i,j} = d-1\}.$$

The above can be explained as follow. Slave functions are the generators in P_i whose coefficient is uniquely determined by the restriction of the function and its derivatives (up to the smoothness) on the interfaces $\Gamma_{i,j}$ with span $P_j \subset$ span P_i. This means that their coefficients are determined by the coefficients of the functions of coarser bases together by the smoothness on the interfaces.

The selected functions are defined by

$$M = \bigcup_{i=1}^{s} M_i,$$

where M_i contains the *master functions* of level i, i.e., the functions of P_i that are active on Λ_i and that are not slaves:

$$M_i = \{\psi \in P_i : \psi|_{\Lambda_i} \neq 0, \ \psi \notin S_i\}. \tag{12}$$

Truncation is defined in the same way as in the case of THB-splines. The resulting basis is called *truncated B-splines for partially nested refinement* (TBPN). The set M forms a non-negative partition of unity, it is a basis and, similarly to THB, it preserves the coefficients of polynomial representation. Moreover, if (9) holds, then TBPN reduces to THB with the same bases and appropriate subdomains. See [43] for details.

Implementation. Only the truncated version of the construction was implemented. The partition T can be defined as

$$\Delta_i = \Lambda_i, \quad i = 1, \dots, s$$

and B by

$$B_i = P_i, \quad i = 1, \dots, s.$$

The matrix $\mathbf{M}$ is built iteratively while discovering the functions selected by the modified Kraft procedure. First the bases $P_1, \dots, P_s$ are analyzed and the nesting relations are stored in a matrix $\mathbf{Z}$. Then, as for (T)HB, the lists L_i of the functions in P_i that are active on Δ_i are computed.

For each function $\psi \in L_i$, $i = s, \dots, 1$, the modified Kraft conditions (12) are tested. The test requires the computation of the intersections $\Gamma_{i,j} \cap$ support ψ that is achieved by computing support $\psi \cap \Lambda_i$ and then decomposing its boundary into segments. If $\dim(\Gamma_{i,j} \cap$ support $\psi) = d - 1$ for some j with span $P_j \subset$ span P_i then ψ is a slave and it is saved in the list S_i. Otherwise the conditions (12) are satisfied and a new column is added to $\mathbf{M}$. The coefficients $m_{\beta,\gamma}$ are computed using a recursive algorithm. For all j such that span $P_i \subset$ span P_j and $\dim(\Gamma_{i,j} \cap$ support $\psi) = d - 1$ the expansion of ψ with respect of P_j is computed by knot insertion. Then for each functions in S_j with a nonzero coefficient the procedure is repeated, giving the coefficients of slaves of finer levels. It is possible that the same $\beta \in B_k$ appears during different recursions while computing the representation of the same generator γ. In this case the sum of the coefficients computed from functions of the same coarsest level must be saved in $\mathbf{M}$.

The implementation described has the same problem as the first implementation of (T)HB: unreasonable memory consumption for the 3D case. This can be solved by using the same strategy described for (T)HB.

5.4   Decoupled Hierarchical B-splines

Contrarily to tensor-product B-splines, (T)HB do not always span the full space of piecewise polynomials on their mesh [33]. This observation was the starting point of the development of TDHB [32]. There *decoupling* is used in conjunction with truncation in order to enlarge the space and span the full piecewise polynomial space for a broader class of meshes. A modification of TDHB called decoupled hierarchical B-splines was coded and it is presented here for the first time. The novelty is that truncation is abandoned in favor of recursive decoupling. By doing so the spanned space can be further enlarged as showed in Example 1.

First, decoupling will be introduced in a slightly more general version compared to [32]. Let $\varphi \in$ span P be a function, let $c_{\varphi,\psi}$ be the coefficients of its expansion with respect to P,

$$\varphi = \sum_{\psi \in P} c_{\varphi,\psi} \psi,$$

and let $\Theta \subseteq \Omega$ be a domain. The *decoupling graph* $R(\varphi, P, \Theta)$ is the graph whose vertices are

$$R_V(\varphi, P, \Theta) = \{\psi \in P : c_{\varphi,\psi} \neq 0\} \tag{13}$$

and the edges are

$$R_E(\varphi, P, \Theta) = \{(\psi, \psi') : \text{support } \psi \cap \text{support } \psi' \cap \overline{\Theta} \neq \emptyset\}. \tag{14}$$

The *decoupling operator* $\mathbf{D}_{P,\Theta}$ associates to function $\varphi \in$ span P one or more *decoupled functions* in span P:

$$\mathbf{D}_{P,\Theta} : \varphi \mapsto \left\{ \sum_{\psi \in K} c_{\varphi,\psi} \psi : K \text{ is a connected component of } R(\varphi, P, \Theta) \right\}.$$

Let $P_1 \subset \cdots \subset P_s$ and $\Omega_1 \supseteq \cdots \supseteq \Omega_s \supset \Omega_{s+1} = \emptyset$ be as in (T)HB. Denote $D_s = P_s$ and

$$D_i = \bigcup_{\psi \in P_i} \mathbf{D}_{D_{i+1}, \Omega \setminus \Omega_{i+1}}(\psi). \tag{15}$$

Then using Kraft's method the *decoupled hierarchical basis* (DHB-splines) D is defined as

$$D = \bigcup_{i=1}^{s} \{\varphi \in D_i : \text{support } \varphi \subseteq \Omega_i \text{ and support } \varphi \cap (\Omega_i \setminus \Omega_{i+i}) \neq \emptyset\}.$$

Given $P_1, \ldots, P_s$ and $\Omega_1, \ldots, \Omega_s$, all of the HB, THB, TDHB and DHB bases are defined. Denoting Z the TDHB basis from [32], the following inclusions hold

$$\text{span } H = \text{span } H' \subseteq \text{span } Z \subseteq \text{span } D.$$

Recall that H, H' and D are the HB, THB and DHB bases, respectively, cf. Subsect. 5.2. See also Example 1.

Implementation. For DHB it is not possible to identify the local bases B_i with the defining bases P_i. This is because a function $\psi \in P_i$ can be decoupled in multiple functions that must be distinguished.

The definitions of T and B are the same as in Implementation 2 in Subsect. 5.2. The construction of $\mathbf{M}$ follows the definition of the space. First, each D_i is constructed: for each function in D_i its expansion with respect of D_{i+1} and its originating function in P_i are stored. Then the Kraft selection mechanism is employed and for each selected function a column is inserted in $\mathbf{M}$. Computing the coefficient $m_{\beta,\gamma}$ for $\gamma \in D_i$ and $\beta \in P_j$ $(j > i)$ is performed by first composing the precomputed change of bases from D_i to D_j and then storing the obtained coefficients according to the subdomain and the originating function.

5.5   Hierarchical Locally Refined Splines

HLR-splines are a special case of LR-splines. Let K denote a partition of Ω into boxes and let μ denote a function that assigns each interface between two boxes a nonnegative integer. To the triplet (K, μ, p) corresponds the spline space $\mathcal{S}$ of the piecewise polynomials of degree p in each variable on K and such that their smoothness across each interface Γ between two boxes is greater than or equal to $p - \mu(\Gamma)$.

A B-spline β is *nested* in a B-spline β' relatively to $\mathcal{S}$, written as $\beta \prec \beta'$, if there exists a sequence of B-splines $\beta = \beta_1, \ldots, \beta_n = \beta'$ such that each $\beta_i \in \mathcal{S}$ and such that each β_{i+1} is obtained from β_i by knot insertion.

LR-splines are the set of minimal elements for the ordering $\prec$ that are comparable with at least one Bernstein polynomial on Ω. They can be linearly dependent and do not necessarily span the entire space of piecewise polynomials satisfying the smoothness conditions [5,14]. However, many properties of the generators are linked together, in particular local linear independence and partition of unity property are equivalent [4].

HLR-splines are a class of LR-splines enjoying local linear independence and thus also the partition of unity property. This is achieved by mimicking the HB approach in constructing K. Take a sequence of tensor-product B-spline spaces $\mathcal{V}_1 \subset \ldots \subset \mathcal{V}_s$ with $\mathcal{V}_i = \text{span } P_i$ and a corresponding sequence of subdomains $\Omega_1 \supseteq \ldots \supseteq \Omega_s \supset \Omega_{s+1} = \emptyset$. Then define

$$K = \bigcup_{i=1}^{s} \{\Theta \text{ element of the partition corr. to } \mathcal{V}_i \; : \; \Theta \subseteq \Omega_i \setminus \Omega_{i+1}\} \qquad (16)$$

and μ that describes the smoothness of the space $\mathcal{V}_i$ on $\Omega_i \setminus \Omega_{i+1}$. Then assuming that each $\mathcal{V}_i$ is obtained from $\mathcal{V}_{i-1}$ by refining a single tensor-component of $\mathcal{V}_{i-1}$ (i.e., h-refinement in a single direction denoted by u_i) and that the borders of the subdomains Ω_i are sufficiently separated, the generators form a partition of unity and they are locally linearly independent [5].

Implementation. A simple choice is to define T by:

$$\Delta_i = \Omega_i \setminus \Omega_{i+1}, \quad i = 1, \ldots, s$$

and B by

$$B_i = P_i, \quad i = 1, \ldots, s.$$

The generators in HLR-splines are either function from P_i for some i or are obtained from a function of P_i by inserting a sequence of knots in u_i-th component of its knot vector. Both types of functions must be active on Ω_i. To find these functions, P_i is projected to a $(d-1)$-variate spaces $\bar{P}_i$ by ignoring its u_i-th tensor factor. For each function $\bar{\psi} \in \bar{P}_i$ the description of the subdomains T is restricted to support $\bar{\psi} \times \mathbb{R} \cap \Omega$, where $\mathbb{R}$ is in direction u_i. This information is used to build a sequence of knot-vectors for the u_i-th direction that describes all the functions in P_i or refinement of functions of P_i that are in LR and have the same knot vector as $\bar{\psi}$ in the directions different from u_i. For each such function a column is added to $\mathbf{M}$ and filled with the appropriate coefficients.

5.6 Implementation Size

This subsection reports the amount of code that implements the spaces described above. While this is a debatable metric, it is the standard approximation of the coding effort and it highlights the amount of code shared between the different spaces. The data reported is the result of the *cloc* tool [12] run on appropriate subsets of the files.

The code is in C++03 standard and contains verbose template parts that can be avoided. Only the (T)HB space is a complete implementation with initialization, refinement (with possible coefficient update) and serialization to the *G+Smo* xml format. For the other spaces only the initialization is provided.

The line count for the different components can be read in Tables 3 and 4. As shown, most of the code implements the shared functionality and the space-specific code amounts to roughly 500 lines per space. The numbers compare favourably with the size of the reference implementation of HB and THB in *G+Smo* that together amount to roughly 6000 lines of code.

Table 3. Lines of the shared code

	Files	Blank	Comment	Code
Utils	5	198	96	1008
Domain code (T)	11	352	302	1998
Matrix code ($\mathbf{M}$)	4	133	248	483
Base-class (`eval`)	2	105	59	443
Total	22	788	705	3932

Table 4. Lines of code of specific spaces.

	Files	Blank	Comment	Code
(T)HB	1	88	73	421
TBPN	1	80	22	396
DHB	1	92	31	548
HLR	1	127	33	550
Total	4	387	159	1915

6   Examples

This section contains selected examples that can be useful to grasp the similarities and the differences between the implemented spline spaces. The basis functions have been plotted with *ParaView* [1] using the data produced with the implementations described in the previous section. The only exception is Example 1 that has been created in Mathematica, because it involves TDHB [32], which we did not implement.

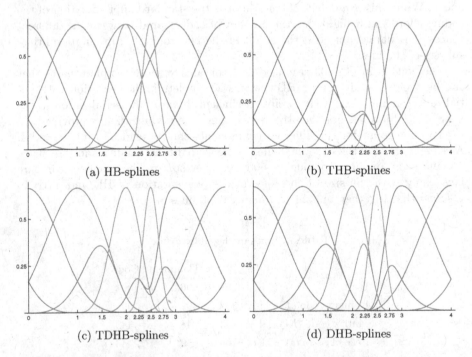

Fig. 5. Comparison of various hierarchical bases. Generators originating from functions of level 0 in blue, from level 1 in orange and from level 2 in green. (Color figure online)

Example 1. Compare the cubic univariate HB, THB, TDHB and DHB with dyadic refinement, the level 0 knot vector $(\ldots, -1, 0, 1, \ldots)$ and $\Omega_0 = [0, 4]$, $\Omega_1 = [1, 3]$ and $\Omega_2 = [2, 3]$. Figure 5 shows all the basis functions of each spline space and highlights the level of the B-spline from which they where derived by truncation or decoupling. Table 5 lists the number of generators according to the level of the original B-spline. Note that DHB is the only space that spans the entire space of $\mathcal{C}^2$ piecewise cubic polynomials on the mesh.

Table 5. Number of generators by the level of the originating function for Example 1.

	HB	THB	TDHB	DHB
Level 0	7	7	8	8
Level 1	1	1	1	2
Level 2	1	1	1	1
Total	9	9	10	11

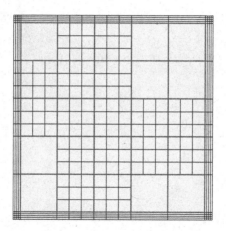

Fig. 6. Hierarchical mesh and a support of a function from Example 2.

Example 2. Consider bivariate hierarchical splines of bi-degree $(4, 4)$ on a mesh shown in Fig. 6. The function with the knot lines indicated in red is selected in the hierarchical basis (Fig. 7 left), truncated in the truncated hierarchical basis (Fig. 7 right) and decoupled into four different functions (that are selected) in the decoupled hierarchical basis (Fig. 8).

Note that due to properties of DHB the sum of the four functions in Fig. 8 equals the truncated function in Fig. 7 right.

Example 3. The design process often involves several patches. To achieve continuity between the patches without losing accuracy, it is necessary that the restrictions of the two spaces are compatible on the interface. That means that one space has to be a subspace of the other.

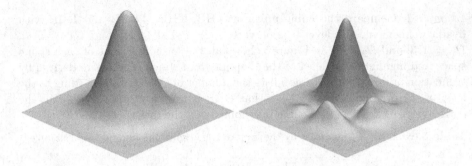

Fig. 7. Function with the support in Fig. 6 as selected into the hierarchical basis (left) and truncated in the truncated basis (right).

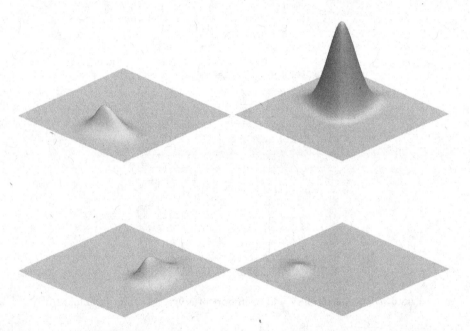

Fig. 8. Decoupled functions stemming from the B-spline with the support indicated in Fig. 6.

Sometimes a new patch must be introduced to bridge between two given patches that should not be modified. Thus the restriction of the space of the bridge patch to each boundary must be a superspace of the restrictions of the other space. If the two given patches have different knot vectors, THB-splines would lead to significant refinement. On the other hand, the TBPN space can achieve interface compatibility without adding unnecessary degrees of freedom.

The bicubic THB basis on the mesh depicted in Fig. 9 has 72 degrees of freedom, whereas the TBPN basis on the same mesh has only 60.

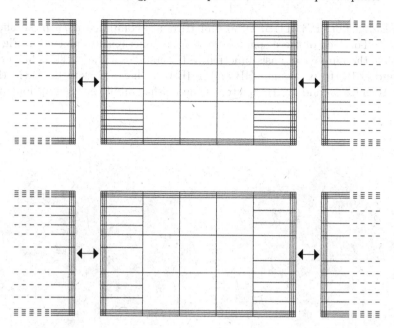

Fig. 9. Meshes from Example 3. Top: THB mesh; bottom: TBPN mesh.

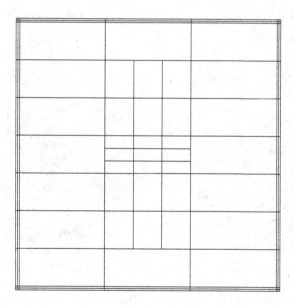

Fig. 10. Mesh from Example 4.

Example 4. Cubic HB, THB, DHB and HLR are compared on a mesh shown in Fig. 10. For each of these spaces all the basis functions are plotted in Fig. 11. Note that the number of basis function in the middle of the patch is higher for HLR and DHB. In particular, HB and THB basis have 49 elements each; HLR and DHB have 53 and are complete, as the meshes fulfill the assumptions from [5, 32].

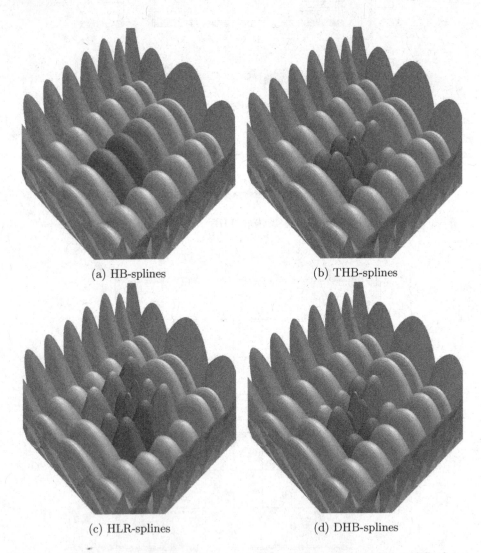

(a) HB-splines   (b) THB-splines

(c) HLR-splines   (d) DHB-splines

Fig. 11. Details of the bases from Example 4. Only the basis functions that differ have been marked in colour. Note that not all the HB basis functions are visible: three are hidden in the central area. (Color figure online)

7   Conclusions

The effectiveness of the proposed implementation framework is demonstrated by the implementation of various spline spaces that share the same evaluation code. The space-specific code is reduced to the initialization of the required data structures as demonstrated by the implementations of HB, THB, TBPN, DHB and HLR. Moreover, the proposed approach grants the following advantages:

1. code reduction both by sharing evaluation between different spaces and between spaces and functions;
2. arbitrary local bases that, in principle, open the way to experimentation with hierarchical constructions based on generalized splines [3], or to the use of ad-hoc functions near a priori known singularities;
3. transparent handling of multipatch domains.

Acknowledgments. The authors have been supported by the Austrian Science Fund (FWF, NFN S117 "Geometry + Simulation") and by the Seventh Framework Programme of the EU (project EXAMPLE, GA No. 324340). This support is gratefully acknowledged. The authors would also like to thank Dr. Rafael Vázquez for commenting on an earlier version of this paper and to the reviewers for their valuable suggestions.

References

1. Ayachit, U.: The paraview guide: a parallel visualization application (2015)
2. Borden, M.J., Scott, M.A., Evans, J.A., Hughes, T.J.R.: Isogeometric finite element data structures based on Bézier extraction of NURBS. Int. J. Numer. Methods Eng. **87**(1–5), 15–47 (2011)
3. Bracco, C., Lyche, T., Manni, C., Roman, F., Speleers, H.: Generalized spline spaces over T-meshes: dimension formula and locally refined generalized B-splines. Appl. Math. Comput. **272**(part 1), 187–198 (2016)
4. Bressan, A.: Some properties of LR-splines. Comput. Aided. Geom. Des. **30**(8), 778–794 (2013)
5. Bressan, A., Jüttler, B.: A hierarchical construction of LR meshes in 2D. Comput. Aided Geom. Des. **37**, 9–24 (2015)
6. Brovka, M., López, J., Escobar, J., Montenegro, R., Cascón, J.: A simple strategy for defining polynomial spline spaces over hierarchical T-meshes. Comput. Aided Des. **72**, 140–156 (2016)
7. Buchegger, F., Jüttler, B., Mantzaflaris, A.: Adaptively refined multi-patch B-splines with enhanced smoothness. Appl. Math. Comput. **272**(part 1), 159–172 (2016)
8. Buffa, A., Garau, E.M.: Refinable spaces and local approximation estimates for hierarchical splines. IMA J. Numer. Anal. **37**(3), 1125–1149 (2017)
9. Buffa, A., Giannelli, C.: Adaptive isogeometric methods with hierarchical splines: error estimator and convergence. Math. Models Methods Appl. Sci. **26**(1), 1–25 (2016)
10. Collin, A., Sangalli, G., Takacs, T.: Analysis-suitable G^1 multi-patch parametrizations for C^1 isogeometric spaces. Comput. Aided Geom. Des. **47**, 93–113 (2016)

11. Da Veiga, L.B., Buffa, A., Sangalli, G.; Vázquez, R.: Analysis-suitable T-splines of arbitrary degree: definition, linear independence and approximation properties. Math. Models Methods Appl. Sci. **23**(11), 1979–2003 (2013)

12. Danial, A.: CLOC: Count Lines of Code (2006–2017). https://github.com/AlDanial/cloc

13. Deng, J., Chen, F., Li, X., Hu, C., Tong, W., Yang, Z., Feng, Y.: Polynomial splines over hierarchical T-meshes. Graph. Models **70**(4), 76–86 (2008)

14. Dokken, T., Lyche, T., Pettersen, K.F.: Polynomial splines over locally refined box-partitions. Comput. Aided. Geom. Des. **30**(3), 331–356 (2013)

15. Evans, E.J., Scott, M.A., Li, X., Thomas, D.C.: Hierarchical T-splines: analysis-suitability, Bézier extraction, and application as an adaptive basis for isogeometric analysis. Comput. Methods Appl. Mech. Eng. **284**, 1–20 (2015)

16. Forsey, D.R., Bartels, R.H.: Hierarchical B-spline refinement. In: SIGGRAPH Computer Graphics, vol. 22, no. 4, pp. 205–212 (1988)

17. Giannelli, C., Jüttler, B., Speleers, H.: THB-splines: the truncated basis for hierarchical splines. Comput. Aided Geom. Des. **29**(7), 485–498 (2012)

18. Giannelli, C., Jüttler, B., Speleers, H.: Strongly stable bases for adaptively refined multilevel spline spaces. Adv. Comput. Math. **40**(2), 459–490 (2014)

19. Geometry + simulation modules (G+Smo): Open source C++ library for isogeometric analysis (2016). http://www.gs.jku.at/gismo

20. GoTools: Collection of C++ libraries connected to geometry (2016). https://github.com/SINTEF-Geometry/GoTools

21. Guennebaud, G., Jacob, B., et al.: Eigen v3 (2010). http://eigen.tuxfamily.org

22. Hennig, P., Müller, S., Kästner, M.: Bézier extraction and adaptive refinement of truncated hierarchical NURBS. Comput. Methods Appl. Mech. Eng. **305**, 316–339 (2016)

23. Hennig, P., Kästner, M., Morgenstern, P., Peterseim, D.: Adaptive mesh refinement strategies in isogeometric analysis-a computational comparison. Comput. Methods Appl. Mech. Eng. **316**, 424–448 (2016)

24. Hughes, T.J.R., Cottrell, J.A., Bazilevs, Y.: Isogeometric analysis: CAD, finite elements, NURBS, exact geometry and mesh refinement. Comput. Methods Appl. Mech. Eng. **194**(39–41), 4135–4195 (2005)

25. Johannessen, K.A., Remonato, F., Kvamsdal, T.: On the similarities and differences between classical hierarchical, truncated hierarchical and LR B-splines. Comput. Methods Appl. Mech. Eng. **291**, 64–101 (2015)

26. Jüttler, B., Langer, U., Mantzaflaris, A., Moore, S.E., Zulehner, W.: Geometry + simulation modules: Implementing isogeometric analysis. PAMM **14**(1), 961–962 (2014)

27. Kang, H., Xu, J., Chen, F., Deng, J.: A new basis for PHT-splines. Graph. Models **82**, 149–159 (2015)

28. Kapl, M., Vitrih, V., Jüttler, B., Birner, K.: Isogeometric analysis with geometrically continuous functions on two-patch geometries. Comput. Math. Appl. **70**(7), 1518–1538 (2015)

29. Kiss, G., Giannelli, C., Jüttler, B.: Algorithms and data structures for truncated hierarchical B-splines. In: Floater, M., Lyche, T., Mazure, M.-L., Mørken, K., Schumaker, L.L. (eds.) MMCS 2012. LNCS, vol. 8177, pp. 304–323. Springer, Heidelberg (2014). doi:10.1007/978-3-642-54382-1_18

30. Kraft, R.: Adaptive and linearly independent multilevel B-splines. In: Le Méhauté, A., Rabut, C., Schumaker, L.L. (eds.) Surface Fitting and Multiresolution Methods, pp. 209–218. Vanderbilt University Press, Nashville (1997)

31. Li, X., Zheng, J., Sederberg, T.W., Hughes, T.J.R., Scott, M.A.: On linear independence of T-spline blending functions. Comput. Aided Geom. Des. **29**(1), 63–76 (2012)

32. Mokriš, D., Jüttler, B.: TDHB-splines: the truncated decoupled basis of hierarchical tensor-product splines. Comput. Aided Geom. Des. **31**(7–8), 531–544 (2014)

33. Mokriš, D., Jüttler, B., Giannelli, C.: On the completeness of hierarchical tensor product B-splines. J. Comput. Appl. Math. **271**, 53–70 (2014)

34. Morgenstern, P.: Globally structured three-dimensional analysis-suitable T-splines: definition, linear independence and m-graded local refinement. SIAM J. Numer. Anal. **54**(4), 2163–2186 (2016)

35. Morgenstern, P., Peterseim, D.: Analysis-suitable adaptive T-mesh refinement with linear complexity. Comput. Aided. Geom. Des. **34**, 50–66 (2015)

36. Rabut, C.: Locally tensor product functions. Numer. Algorithms **39**(1–3), 329–348 (2005)

37. Scott, M.A., Borden, M.J., Verhoosel, C.V., Sederberg, T.W., Hughes, T.J.R.: Isogeometric finite element data structures based on Bézier extraction of T-splines. Int. J. Numer. Methods Eng. **88**(2), 126–156 (2011)

38. Sederberg, T.W., Cardon, D.L., Finnigan, G.T., North, N.S., Zheng, J., Lyche, T.: T-spline simplification and local refinement. ACM Trans. Graph. **23**(3), 276–283 (2004)

39. Sederberg, T.W., Zheng, J., Bakenov, A., Nasri, A.: T-splines and T-NURCCs. ACM Trans. Graph. **22**(3), 477–484 (2003)

40. Thibault, W.C., Naylor, B.F.: Set operations on polyhedra using binary space partitioning trees. In: Proceedings of the 14th Annual Conference on Computer Graphics and Interactive Techniques, SIGGRAPH 1987, pp. 153–162. ACM, New York (1987)

41. Toth, C.D., O'Rourke, J., Goodman, J.E.: Handbook of Discrete and Computational Geometry. CRC Press, Boca Raton (2004)

42. Vázquez, R., Garau, E.: Algorithms for the implementation of adaptive isogeometric methods using hierarchical splines. Tech. report 16–08, IMATI-CNR, Pavia, July 2016

43. Zore, U.: Constructions and properties of adaptively refined multilevel spline spaces. Dissertation, Johannes Kepler University Linz (2016). http://epub.jku.at/obvulihs/download/pdf/1273941

Machinability of Surfaces via Motion Analysis

Robert J. Cripps[1]([✉]), Ben Cross[1], Glen Mullineux[2], and Mat Hunt[2]

[1] Department of Mechanical Engineering, University of Birmingham,
Edgbaston B15 2TT, UK
r.cripps@bham.ac.uk
[2] Department of Mechanical Engineering, University of Bath, Bath BA2 7AY, UK

Abstract. The machinability of a surface describes its ability to be machined and the factors which affect this. These are independent of any material properties or cutting parameters but instead reflect an ability to replicate a desired tool path motion with sufficient control of the material removal process. Without this control there is a potential for surface defects and costly finishing stages.

Five-axis CNC milling machines are commonly used for machining complex free-form shapes. The processes required to obtain CNC instructions for a machine tool, starting from a target surface, are presented. An overview is first given and later formalised with mathematical methods. Specifically, a moving cutting tool is characterised by a tool path motion. Interpreting the moving cutter in terms of moving machine axes provides a diagnostic tool for detecting machining errors.

Examination of two case studies reveals different types of errors, machine-dependent and machine-independent. The contribution of geometry to machine-independent errors is discussed and related back to the machinability of a surface.

Keywords: Machinability · Five-axis machine tool · Tool path motion

1 Introduction

Computer Aided Design and Manufacture (CAD/CAM) provides a highly-automated process for the machining of components. The term machining is used to describe a process that begins with some raw material which is gradually removed via cutting until a desired shape is achieved. The *machinability* of a part thus refers to its ability to be machined and the factors which affect this [1]. Five-axis CNC milling machines are commonly used for machining parts with complex free-form shape due to their ability to control the position and orientation of a cutter.

The manufacturing process begins with the design of a part geometry inside a CAD environment [13]. The end goal is to machine this part geometry as efficiently as possible and for it to be of a sufficiently high quality. An immediate measure of quality is the dimensional accuracy of the part to that of the CAD model. Another consideration, which is decisive in whether or not a part is

M. Floater et al. (Eds.): MMCS 2016, LNCS 10521, pp. 74–95, 2017.
https://doi.org/10.1007/978-3-319-67885-6_4

deemed acceptable, lies in its visual appearance. Any aesthetic irregularities on the part surface require costly finishing stages even if the part is within dimensional tolerance.

An important factor in machining a satisfactory part is sufficient control of the cutting tool and thus the cutting conditions. This control is governed by the Computer Numerical Control (CNC) unit of the machine tool. The CNC controller does not exactly replicate the desired motion due to physical restraints. The difference between desired and actual motion affects the cutting conditions and can therefore be used to predict surface defects. Since the tool path motion is defined from the surface geometry the effect of geometry in causing these surface defects is important to understand. The machinability of a surface is used here to describe its ability to be machined independent of any material properties. In particular it is used here to describe the properties of a tool path motion generated from the surface and the ability to replicate this motion with sufficient control of the material removal process.

The CNC controller is able to manipulate the position and orientation of the cutter with respect to the workpiece by moving the machine axes. These movements must obey physical constraints in reality, such as limited speed and acceleration, in accordance with the laws of physics. Constraints on axes speed and acceleration are hard-coded into the CNC controller. Thus when a motion does not abide by these constraints an alternative to the desired motion is experienced. Positional tolerance of the cutter is of high priority in the CNC controller and the most likely effect is a slow down in cutting feed. This affects the surface integrity of the part, possibly producing dwell marks (Fig. 7). Since these affect the visual appearance of the part, analysis of the machine axes kinematics can be used to predict the occurrence of machining flaws.

In this paper two simplified case studies are examined which are based on real life examples where surface defects have occurred. Analysing irregularities in the motion of a cutting tool provides an explanation for the occurrence of these defects. Furthermore, a relationship between part geometry and potential sources of defects is established via analysis of the tool path motion.

The structure of the paper is described as follows. Section 2 presents an overview of the machining process with regards to forming tool paths from a part geometry [2]. Section 3 formalises these processes with the relevant mathematical preliminaries. Cutter poses are defined with the introduction of the *workpiece coordinate system*. The connection to machine axes is presented through the kinematic chain of the machine tool leading to the *machine coordinate system*. A relationship between these coordinate systems for a tool path motion is then presented via the Jacobian matrix.

Section 4 presents the first case study. A tool path is constructed to demonstrate machining errors identified as a feature of the machining singularity. The singularity is shown to correspond to a degenerate Jacobian matrix. Section 5 presents the second case study. The example considers a single tool path used for machining a simplified turbine blade. The machining errors here are shown to be linked to the geometry of the tool path motion. Section 6 discusses the link between the geometry and the machine axes' behaviour. The paper finishes with some concluding remarks on the two different types of machining errors.

2 Background to the Machining Process and Tool Path Motion Analysis

This section presents an overview of the procedures involved with the machining process, beginning from the CAD geometry and finishing with movements of the machine axes. It is summarised by the flow diagram given in Fig. 1. The mathematical formalisation of these procedures is presented in Sect. 3.

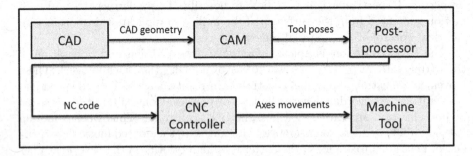

Fig. 1. Overview of the machining process.

In order to transform raw material (workpiece) into a desired shape, a machining strategy has to be developed. The first stage involves removing the bulk of the excess material in what is known as the roughing stage. The final machining stage traces over the surface geometry of the part removing the excess material remaining.

This finishing stage can be achieved in a variety of ways but all abide by the same principle. Given a target geometry to machine, a cutting tool is traced over the surface forming a sequence of connected tool paths. The envelope of the moving cutting tool must not intersect with the part geometry else over cutting (gouging) occurs. Furthermore, the volume represented by the envelope must remove all excess material leaving only the material coinciding with the part geometry. Thus the volumetric intersection of the tool path envelope with the CAD model should theoretically be the boundary surfaces of the part. Note that in practice the target surface geometry does not exactly coincide with the tool path envelope but is instead designed to be within a strict dimensional tolerance of the target shape. The formation of these tool paths is usually computed with the aid of Computer Aided Manufacturing (CAM) software [3].

The position and orientation of the cutter is referenced with respect to a coordinate system fixed relative to the workpiece. The combined position and orientation information is here referred to as a *pose*. When the cutter removes material in the finishing stage it touches the target surface at a particular point on the cutting tool edge. This point is referred to as the cutter contact (CC) point. However, the position of the cutter must be defined by a fixed point on the cutter called the cutter location (CL) point. This is usually defined at the tip of the cutter. Note that the location of the CC is not generally at a fixed displacement from the CL. Thus CL data is derived from varying offsets of the target surface which depend upon the orientation and geometry of the cutting tool (see Fig. 2). This offsetting can change the geometry of the CL data from that derived from the surface. However efforts are usually made to preserve the angle of the cutter to the surface normal and tangent direction in order to maintain a consistent chip pattern on the surface [4]. This geometric property proves insightful when considering the derivatives of the tool path motion.

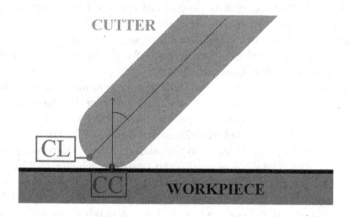

Fig. 2. The Cutter Contact (CC) and Cutter Location (CL) points.

The next stage in the manufacturing process involves converting the tool paths, in the form of positional CL and orientation data, into machine instructions for a CNC controller. These instructions, commonly known as NC code, depend upon what type of CNC machine tool is to be used. Information regarding the machine tool structure is required to determine the effect each axis has on the pose of the tool with respect to the workpiece. This information is contained within the *kinematic chain* which connects frames of reference across each axis of the machine as well as the frames of reference connecting the workpiece and cutter (more detail is given to the kinematic chain in Sect. 3). The kinematic chain can thus be used to define a locally one-to-one function between the tool poses of the CAM model and the five machine axis values that replicate each pose.

The tool path from the CAM software is discretized into a sequence of tool poses which are then converted into five corresponding machine axis values in the post-processing stage. This list of machine configurations forms part of the CNC instructions for the machine tool along with other machining parameters such as spindle speed. The CNC machine can then move the machine axes to interpolate each sequential configuration, machining the desired geometry (within tolerance). The amount of time desired between each pose is also stored in the NC code in an attempt to control the cutting speed (or feed rate) and thus cutting conditions. Alternatively an explicit feed rate can be given.

The cutting speed represents the relative speed between the cutter and the surface of the part. It is usually constant for each individual tool path to preserve cutting conditions. Consider the collection of CC points for such a tool path in the finishing stage. These derive from a curve on the target surface and are parameterised by some constant factor of the arc-length parameterisation for this curve. This provides enough information to fully define a tool path motion. It is a function that outputs the pose of the cutter for a given input of time. This concept is formalised in more detail in Sect. 3.

Given a tool path motion it is possible to convert the poses into machine axes values and analyse the changes in them with respect to time. These represent kinematical characteristics of the machine and therefore must abide by the laws of physics. For example, the acceleration for each axis is bounded by the amount of force (or power) that can be transmitted in accordance with Newton's second law. The maximum speed must also be bounded not least for safety but also for the sustainability of the machine. In order to smooth the movement of a machine these characteristics are controlled by the CNC controller. The exact details of the control algorithms used are not in the public domain as the intellectual property belongs to the manufacturer. However, most abide by similar principles. Data corresponding to position, velocity and acceleration is measured at an instance. Decisions regarding how much power to deliver to the axes are then made by the controller from this data in a closed feedback loop [5]. A fundamental component of this decision making is limiting the maximum speed and acceleration of the machine axes. These bounds are hard-coded inside the controller. If the desired tool path motion exceeds these bounds then the controller must make adjustments. Even if the controller is able to maintain positional accuracy, the cutting speed cannot be maintained. These dwells cause irregular chip patterns and rubbing of the cutting tool. This in turn affects the surface finish of the part and may cause it to be rejected (Fig. 7).

Therefore, given a tool path, it is possible to detect these types of machining issues (here referred to as machining errors). The process involves generating the machine axes values at each instance of time and checking that they lie within predetermined bounds. Analysis of two case studies leads to distinct causes of machining errors. In the first case the excessive speeds and accelerations can be eliminated by changing the kinematic chain. This is akin to having a different machine tool perform the same tool path motion. For this reason they are classified as machine-dependent errors. In the second case however it can be shown that no kinematic chain can resolve the issue and the cause of the error lies in the geometry of the tool path motion. These are referred to as machine-independent errors. Since the tool path motion is defined from the surface of the part, the link with surface geometry and machining errors is discussed in Sect. 6.

The next section presents the mathematical processes required to form the tool path motion. Then using knowledge of the CNC machine tool structure, in the form of a kinematic chain, the relationship between tool poses and machine axes values is given. These form two separate coordinate systems. The Jacobian between these coordinate systems is then presented which proves useful for analysing how the coordinate systems change with respect to each other.

3   Mathematical Preliminaries

3.1   Coordinate Systems and the Kinematic Chain

A workpiece coordinate system is defined with respect to some fixed coordinate frame rigidly attached to the workpiece. The position of the cutter is described with a triplet of $\mathbf{V}_W = (x, y, z)$ values. More precisely this is the CL point of the cutter which is normally chosen to be the tip of the cutter. The cutting tool rotates in the spindle about a line. The direction of this line is described with a unit vector, $\mathbf{O}_W = (i, j, k)$, which represents the orientation of the cutter. The combined position and orientation coordinates, $\mathbf{P}_W = (x, y, z, i, j, k)$ form the pose of the cutter. In the post-processing stage, the pose of the cutter must be converted into machine coordinates, $\mathbf{P}_M = (M_1, M_2, M_3, M_4, M_5)$, representing the values of each machine axis. This requires information about the CNC machine tool regarding the arrangement of the axes relative to the workpiece and cutter.

A kinematic diagram is a graph that represents the connectivity of links and joints in a machine (see Fig. 3). Nodes of the graph represent parts of the machine and the edges between them represent joints. Each joint can be represented as a rigid-body transform between coordinate systems fixed at each link. A joint is considered an actuator if its movements can be controlled by the machine. The kinematic graph of a five-axis CNC machine therefore comprises five actuator joints connected to the workpiece and the cutting tool. Only serially connected machines are considered here. This implies that each actuator is connected in series and the kinematic diagram is a tree with world space as the root. The five actuators combined with the workpiece and cutter form a closed kinematic loop [6]. That is to say if one begins at a node and applies the rigid-body transforms of each sequential edge as one loops round the graph one arrives back at the original node with the same frame of reference. The machine body is fixed to the ground (world space) and remains stationary whilst the machine is moving. This is signified with a connection to ground and helps to define a fixed reference frame for each set of links and joints.

The pose of the cutter with respect to the workpiece is represented in the kinematic diagram by the edge connecting them. Thus given values of each actuator the pose of the cutter can be determined by looping round the kinematic chain from the workpiece to the cutter thorough the actuator joints. Since each edge represents a rigid-body transform, applying these transformations sequentially results in the rigid-body transform corresponding to the pose of the cutter.

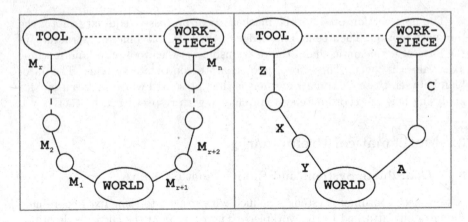

Fig. 3. Kinematic diagram of [left] a generic 5-axis machine and [right] the Hermle C600U.

Take the Hermle C600U machine tool for example. A schematic of the machine tool is given in Fig. 4. This machine consists of 3 translational axes (X, Y, Z) controlling the spindle position and two rotary axes (A, C) controlling the orientation of the workpiece mounted on the machine bed. For simplicity a reference (or world space) coordinate frame is chosen such that the origin corresponds to the intersection of the axes of rotation for the A and C rotary axes. The (x, y, z) directions align with the (X, Y, Z) translational movements of the spindle. The origin of the cutter's reference frame is chosen at the tip of the cutter. The orientation of the cutter's frame is aligned to the reference frame. The coordinate frame for the workpiece is chosen so that when the A and C rotary axes are set to zero it coincides with the reference frame. The kinematic diagram is given in Fig. 3.

The pose of the cutter can thus be inferred from the kinematic chain. Note the rigid-body transform from the reference frame to the cutting tool is simply a translation of (X, Y, Z). The rigid-body transform from the reference frame to the workpiece is a rotation of angle C in the Z direction followed by a rotation of angle A in the X direction. Thus the CL location is given by

$$\mathbf{V}_W = \begin{pmatrix} x \\ y \\ z \end{pmatrix} = \begin{pmatrix} \cos(C) & \sin(C) & 0 \\ -\sin(C) & \cos(C) & 0 \\ 0 & 0 & 1 \end{pmatrix} \begin{pmatrix} 1 & 0 & 0 \\ 0 & \cos(A) & \sin(A) \\ 0 & -\sin(A) & \cos(A) \end{pmatrix} \begin{pmatrix} X \\ Y \\ Z \end{pmatrix}$$
$$= \mathbf{R}_z(C)\mathbf{R}_x(A)\mathbf{V}_M. \tag{1}$$

Here A and C are the angles of the rotary axes and $\mathbf{R}_i(j)$ is the rotation matrix about the i axis of angle j. Note that the subscript of W or M represents the coordinate frames chosen to determine the CL position, where M is the machine coordinate frame and W is the workpiece coordinate frame. The orientation of the cutter can also be determined by reorienting through the kinematic chain, starting with it aligned in the Z direction.

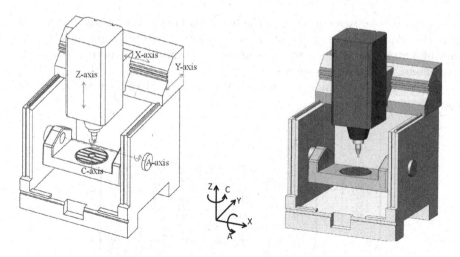

Fig. 4. Schematic of the Hermle C600U machine tool.

$$\mathbf{O}_W = \begin{pmatrix} i \\ j \\ k \end{pmatrix} = \begin{pmatrix} \cos(C) & \sin(C) & 0 \\ -\sin(C) & \cos(C) & 0 \\ 0 & 0 & 1 \end{pmatrix} \begin{pmatrix} 1 & 0 & 0 \\ 0 & \cos(A) & \sin(A) \\ 0 & -\sin(A) & \cos(A) \end{pmatrix} \begin{pmatrix} 0 \\ 0 \\ 1 \end{pmatrix}$$

$$= \mathbf{R}_z(C)\mathbf{R}_x(A)\mathbf{O}_M. \tag{2}$$

Given a cutter pose, $\mathbf{P}_W = (x, y, z, i, j, k)$, the corresponding machine axes values can be calculated. The first task is to find the A and C values that produce the orientation $\mathbf{O}_W = (i, j, k)$. This can be found by solving Eq. (2). Table 1 shows possible solutions [7]. The coordinates (X, Y, Z) can then be found from Eq. (1) by premultiplying by the appropriate rotation matrices.

Table 1. A and C angles for a given orientation $\mathbf{O}_W = (i, j, k)$.

$k = +1$	$A = 0$	C - undefined
$k = -1$	$A = \pi$	C - undefined
$-1 < k < 1$	$A = \tan^{-1}(\pm\sqrt{i^2 + j^2}, k)$	$C = \tan^{-1}(\pm i, \pm j)$

Either the workpiece coordinates or machine axes values (machine coordinates) can be used to describe a pose. The workpiece coordinates characterise the geometry of the tool path whereas the machine coordinates characterise the machine behaviour. A useful tool for describing the link between coordinate systems is the Jacobian matrix.

3.2   The Jacobian Matrix

The change in machine coordinates over time represents the kinematic properties of the individual axes. If the speed or acceleration of these axes becomes too large then surface defects can occur. The change in workpiece coordinates is described with the tool path motion. Thus, in order to gain a better insight into how the moving cutter corresponds to movements in machine axes, derivatives of the coordinates should be considered.

The velocity of machine axes movements can be inferred from the velocity of the pose via the chain rule.

$$\left[\frac{\partial \mathbf{P}_M}{\partial t}\right] = \left[\frac{\partial \mathbf{P}_M}{\partial \mathbf{P}_W}\right]\left[\frac{\partial \mathbf{P}_W}{\partial t}\right]$$

$$\begin{pmatrix} \frac{\partial M_1}{\partial t} \\ \frac{\partial M_2}{\partial t} \\ \frac{\partial M_3}{\partial t} \\ \frac{\partial M_4}{\partial t} \\ \frac{\partial M_5}{\partial t} \end{pmatrix} = \begin{pmatrix} \frac{\partial M_1}{\partial x} & \frac{\partial M_1}{\partial y} & \frac{\partial M_1}{\partial z} & \frac{\partial M_1}{\partial i} & \frac{\partial M_1}{\partial j} & \frac{\partial M_1}{\partial k} \\ \frac{\partial M_2}{\partial x} & \frac{\partial M_2}{\partial y} & \frac{\partial M_2}{\partial z} & \frac{\partial M_2}{\partial i} & \frac{\partial M_2}{\partial j} & \frac{\partial M_2}{\partial k} \\ \frac{\partial M_3}{\partial x} & \frac{\partial M_3}{\partial y} & \frac{\partial M_3}{\partial z} & \frac{\partial M_3}{\partial i} & \frac{\partial M_3}{\partial j} & \frac{\partial M_3}{\partial k} \\ \frac{\partial M_4}{\partial x} & \frac{\partial M_4}{\partial y} & \frac{\partial M_4}{\partial z} & \frac{\partial M_4}{\partial i} & \frac{\partial M_4}{\partial j} & \frac{\partial M_4}{\partial k} \\ \frac{\partial M_5}{\partial x} & \frac{\partial M_5}{\partial y} & \frac{\partial M_5}{\partial z} & \frac{\partial M_5}{\partial i} & \frac{\partial M_5}{\partial j} & \frac{\partial M_5}{\partial k} \end{pmatrix} \begin{pmatrix} \frac{\partial x}{\partial t} \\ \frac{\partial y}{\partial t} \\ \frac{\partial z}{\partial t} \\ \frac{\partial i}{\partial t} \\ \frac{\partial j}{\partial t} \\ \frac{\partial k}{\partial t} \end{pmatrix},$$

$$\dot{\mathbf{P}}_M = \mathbf{J}\dot{\mathbf{P}}_W. \tag{3}$$

The **J** term, referred to as the Jacobian matrix, describes the relationship between velocities in the two coordinates systems as a matrix transformation. The matrix is non-square because in workpiece coordinates the unit normal $\mathbf{O}_W = (i, j, k)$ has only two degrees of freedom. Nonetheless, an upper bound on this matrix transformation can be obtained from analysis of the *spectral norm* defined as:

$$||\mathbf{J}|| = \max_{||\mathbf{x}||=1} ||\mathbf{J}\mathbf{x}||.$$

This equals the square root of the largest eigenvalue of $\mathbf{J}^T\mathbf{J}$ [8] and is referred to here as the *bound of the Jacobian*.

3.3   Machine Singularities

The Jacobian, **J**, represents the change in machine coordinates with respect to workpiece coordinates. A similar matrix, **K**, for the change in workpiece coordinates with respect to machine coordinates can be derived as:

$$\left[\frac{d\mathbf{P}_W}{dt}\right] = \left[\frac{\partial \mathbf{P}_W}{\partial \mathbf{P}_M}\right]\left[\frac{d\mathbf{P}_M}{dt}\right],$$

$$\dot{\mathbf{P}}_W = \mathbf{K}\dot{\mathbf{P}}_M.$$

It can be calculated by differentiating the rigid-body transformation between cutter and workpiece as described in the kinematic chain.

If the rank of the matrix **K** is less than 5 then some machine axes movements become redundant in that they do not affect the pose of the cutter relative to the workpiece. This occurs when $\det(\mathbf{K}^T\mathbf{K}) = 0$ with the redundant movements corresponding to the eigenvector with zero eigenvalue. Furthermore the Jacobian, **J**, is undefined here. In this scenario there are not enough degrees of freedom in the system to accommodate all possible pose changes, this is identified as a machining singularity [9].

As $\det(\mathbf{K}^T\mathbf{K}) \to 0$ the spectral norm of the Jacobian $\|\mathbf{J}\| \to \infty$ since it becomes possible to change the machine axes values with a diminishing effect on the pose. Therefore local to a singularity a machine may require relatively large speeds to attain a constant feed rate [7].

Consider the Hermle C600U with its kinematic chain. Certain orientations correspond to a singularity, these occur when $\mathbf{O}_W = (0,0,1)^T$ (Table 1). This is due to the fact that when the cutter is oriented at $(0,0,1)^T$ it is possible to spin the C-axis and follow a circle in the XY plane centered on the C-axis of rotation without affecting the pose. This can be demonstrated with calculation of $\det(\mathbf{K}^T\mathbf{K})$. For simplicity consider only the sub-matrix $(\mathbf{K}_O)$ corresponding to orientation changes and rotary axes movements. Then

$$\left[\frac{d\mathbf{O}_W}{dt}\right] = \left[\frac{\partial\mathbf{O}_W}{\partial\mathbf{O}_M}\right]\left[\frac{d\mathbf{O}_M}{dt}\right] = [\mathbf{K}_O]\left[\frac{d\mathbf{O}_M}{dt}\right],$$

$$\begin{pmatrix} \dfrac{\partial i}{\partial t} \\[2mm] \dfrac{\partial j}{\partial t} \\[2mm] \dfrac{\partial k}{\partial t} \end{pmatrix} = \begin{pmatrix} \dfrac{\partial i}{\partial A} & \dfrac{\partial i}{\partial C} \\[2mm] \dfrac{\partial j}{\partial A} & \dfrac{\partial j}{\partial C} \\[2mm] \dfrac{\partial k}{\partial A} & \dfrac{\partial k}{\partial C} \end{pmatrix} \begin{pmatrix} \dfrac{\partial A}{\partial t} \\[2mm] \dfrac{\partial C}{\partial t} \end{pmatrix}$$

From (2)

$$\frac{\partial i}{\partial A} = \sin C \cos A, \quad \frac{\partial j}{\partial A} = \cos C \cos A, \quad \frac{\partial k}{\partial A} = -\sin A,$$

$$\frac{\partial i}{\partial C} = \cos C \sin A, \quad \frac{\partial j}{\partial C} = -\sin C \sin A, \quad \frac{\partial k}{\partial C} = 0.$$

Therefore

$$\begin{pmatrix} \dfrac{\partial i}{\partial A} & \dfrac{\partial i}{\partial C} \\[2mm] \dfrac{\partial j}{\partial A} & \dfrac{\partial j}{\partial C} \\[2mm] \dfrac{\partial k}{\partial A} & \dfrac{\partial k}{\partial C} \end{pmatrix}^T \begin{pmatrix} \dfrac{\partial i}{\partial A} & \dfrac{\partial i}{\partial C} \\[2mm] \dfrac{\partial j}{\partial A} & \dfrac{\partial j}{\partial C} \\[2mm] \dfrac{\partial k}{\partial A} & \dfrac{\partial k}{\partial C} \end{pmatrix} = \begin{pmatrix} 1 & 0 \\ 0 & \sin^2 A \end{pmatrix}$$

which gives

$$\|\mathbf{K}_O^T\mathbf{K}_O\| = \sin^2 A.$$

When $\mathbf{O}_W = (0,0,1)^T$ then $A = 0$ and thus $\|\mathbf{K}_0^T\mathbf{K}_O\| = \|\mathbf{K}^T\mathbf{K}\| = 0$.

The case study presented in the following section analyses machine axes movements as $\det(\mathbf{K}^T\mathbf{K}) \to 0$ and the spectral norm of the Jacobian, $\|\mathbf{J}\|$, increases.

4   Case Study One: Machine-Dependent Sources of Error

In this section an example of a tool path that produces machining errors (as defined in Sect. 2) with the Hermle C600U machine tool is examined. It is taken from a recent publication [7] wherein further details can be found.

The tool path motion is defined as

$$\mathbf{V}_W(t) = \begin{pmatrix} x \\ y \\ z \end{pmatrix} = \begin{pmatrix} \frac{100}{3}t \\ 0 \\ 0 \end{pmatrix}, \quad t \in [0, \tfrac{3}{2}]$$

$$\mathbf{O}_W(t) = \mathbf{R}_x(11°)\mathbf{R}_z(120°t)\mathbf{R}_x(-10°)\hat{\mathbf{z}}, \quad t \in [0, \tfrac{3}{2}]$$

The cutter tip ($\mathbf{V}_W$) moves in a straight line whilst the cutter orientation ($\mathbf{O}_W$) rotates at a uniform speed. A zenith (orientation closest to singularity) is reached halfway along the tool path at angle of $1°$. A graphical representation of this tool path is given in Fig. 5 which also illustrates the orientation of the cutter as well as a simulation for the expected shape of the part after cutting.

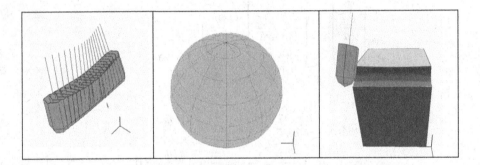

Fig. 5. [Left] Visualisation of the linear tool path. [Centre] The tool orientations ($\mathbf{O}_W$) visualized on the unit sphere. [Right] Simulated part shape after machining.

Applying the inverse kinematics (see Sect. 3) machine axes values can be determined at each moment of the motion and hence the speed of axes movements can be calculated. Figure 6 illustrates the machine kinematics of the rotary axes. Halfway through the tool path it is noted that the C-axis is required to spin at a maximum speed of 199.0 rpm which has been calculated from (3) as

$$\left.\frac{\partial C}{\partial t}\right|_{t=\frac{3}{4}} = \frac{20\sin(10°)}{\sin(1°)} \approx 199.0.$$

However the maximum speed of the rotary axes for the Hermle C600U is around 25 rpm, as stated by the machine tool manufacturers [10]. Therefore the CNC controller has to make a compromise on the desired tool path motion. From this analysis a surface defect is predicted between $x = 23$ mm and $x = 27$ mm.

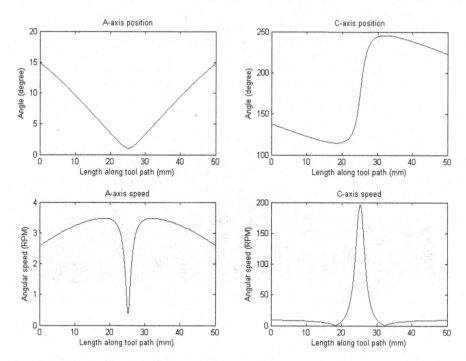

Fig. 6. Kinematics of the rotary axes with a feed rate 2000 mm/min. [Left] A-axis [Right] C-axis [Top] Axis angle [Bottom] Axis speed.

After machining, the part was inspected with a 3D micro coordinate measurement machine and surface roughness measurement device [11]. The images taken (Fig. 7) illustrate the presence of a surface defect in the form of a discolouration of the material at the center of the part. Furthermore, a roughness profile measurement taken across the part indicates an increase in surface roughness local to the predicted affected region between $x = 23$ mm and $x = 27$ mm across the part (Fig. 8). The surface defect is thought to be melting of aluminium from dwelling as a consequence of singular behaviour.

The singular behaviour can be eliminated by changing the Jacobian, which is defined from the kinematic chain of the machine tool. This is achieved by either choosing a machine tool with a different kinematic chain or changing the existing kinematic chain. The latter can be achieved by reorientating the workpiece on the machine bed (with the use of a jig [7]) which effectively inserts an extra node into the kinematic diagram (Fig. 3). Consequently the tool path has to be reorientated.

Therefore a second tool path, based upon the previous tool path, is constructed through reorientation of 10° in the direction away from the singularity (so its zenith is 11° away). This tool path motion is defined as

$$\mathbf{P}_W(t) = \begin{pmatrix} x \\ y \\ z \end{pmatrix} = \begin{pmatrix} \frac{100}{3}t \\ 0 \\ 0 \end{pmatrix}, \quad t \in [0, \tfrac{3}{2}]$$

$$\mathbf{O}_W(t) = \mathbf{R}_x(21°)\mathbf{R}_z(120°t)\mathbf{R}_x(-10°)\hat{\mathbf{z}}, \quad t \in [0, \tfrac{3}{2}].$$

Fig. 7. Image showing surface defect occurring near to singularity.

A graphical representation of this tool path is given in Fig. 9 which also illustrates the orientation of the cutter as well as a simulation for the expected

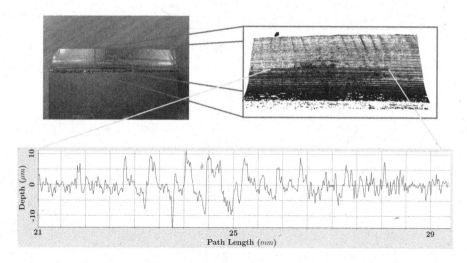

Depth (μm)

Path Length (mm)

Fig. 8. Surface roughness profile of the machined surface.

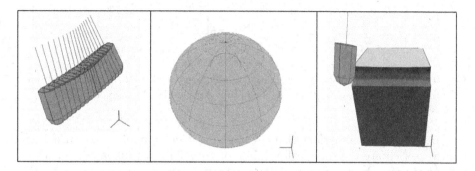

Fig. 9. [Left] Visualisation of the second tool path. [Center] The modified tool orientations ($\mathbf{O}_W$) visualized on the unit sphere. [Right] Simulated part shape after machining.

shape of the part after cutting. The corresponding machine kinematics of the rotary axes are given in Fig. 10.

Applying the reorientation has had the effect of reducing the C-axis speed from around 199.0 rpm to around 18.2 rpm, which has been calculated from (3) as

$$\frac{\partial C}{\partial t}\bigg|_{t=\frac{3}{4}} = \frac{20\sin(10°)}{\sin(11°)} \approx 18.2.$$

This value is below the C-axis maximum speed (Fig. 10) and is therefore not expected to cause the same issues as in the first tool path. The image in Fig. 11 confirms that the surface defect from the original tool path, explained as a consequence of singular behaviour, has been successfully removed through reorientation.

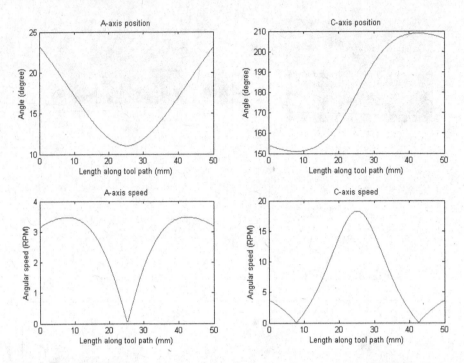

Fig. 10. Kinematics of the rotary axes (reorientated tool path) with a feed rate 2000 mm/min. [Left] A-axis [Right] C-axis [Top] Axis angle [Bottom] Axis speed.

This case study illustrates that, by applying a shape-preserving transformation to reorientate the tool path, sources of error be can eliminated. Therefore the error cannot be attributed to the geometry of the tool path but rather the Jacobian and thus the machine tool kinematic chain. For this reason these errors are referred to as machine-dependent. In the next case study an example of a machining flaw that cannot be resolved with a manipulation of the kinematic chain is presented.

5   Case Study Two: Machine-Independent Sources of Error

Consider the task of machining a turbine blade as in Fig. 12 [3]. Machining flaws often occur on the turbine blade between the flatter sections and the more rounded edges (as highlighted in Fig. 13). To gain insight into what might be causing these flaws a simplified 2D tool path is considered. This turbine blade boundary consists of a flat section followed by a semi-circular tip leading back to another flat section. A machining strategy that preserves the angle between the surface normal and tangential direction (described in Sect. 2) is chosen for use with the Hermle C600U machine tool that maintains a uniform cutting feed rate.

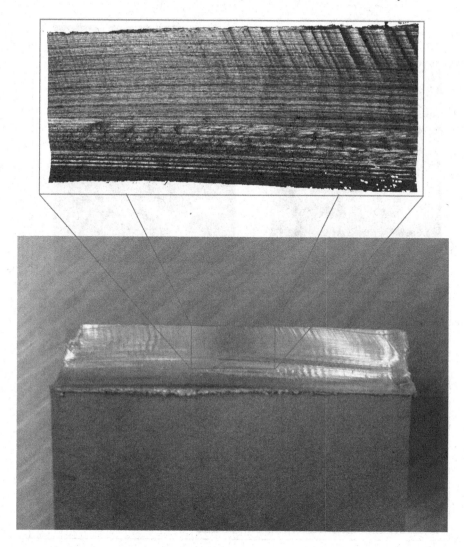

Fig. 11. Image showing improved surface finish due to singularity avoidance (compared to Fig. 7).

The tool path is chosen to lie in the xy-plane. To ensure the cutter orientation remains in this plane the A-axis value is fixed at $90°$. The angle of orientation in the xy-plane corresponds to the angle of the C-axis. Along the flat sections the orientation of the cutter is preserved. This requires the rotary axes to be stationary. In the circular section the orientation of the cutter rotates through $180°$ at a uniform speed. This requires the C-rotary axis to be moving at a constant speed.

At the joins between flat sections and circular sections, the rotary axes have to transition from being stationary to moving at a constant speed in an instance.

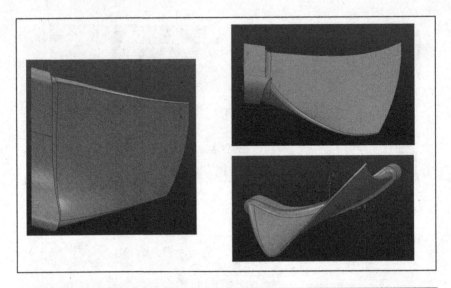

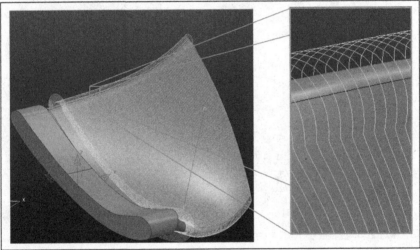

Fig. 12. [Top] CAD model of the blade and [Bottom] tool paths generated with the CAM software.

This corresponds to an infinite amount of acceleration/deceleration, which is physically impossible. The CNC controller must therefore make alterations to the tool path motion resulting in a different set of cutting conditions possibly leading to machining flaws.

Recall Eq. (3): $\dot{\mathbf{P}}_M = \mathbf{J}\dot{\mathbf{P}}_W$. Along the flat sections the last three components of $\dot{\mathbf{P}}_W$ are zero since these represent the change in orientation. Along circular sections the magnitude of the vector formed from the last three coordinates is a non-zero constant signifying the orientation speed. Thus there is a discontinuity

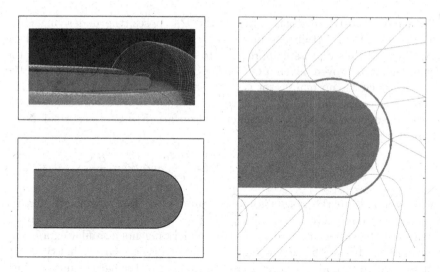

Fig. 13. [Top Left] Area of blade where machining flaws predicted. [Bottom Left] Simplified model of the turbine blade. [Right] Tool paths generated from the simple model.

in $\dot{\mathbf{P}}_W$ where the sections join. Since the Jacobian is locally constant at the join between the flat and circular sections, the machine coordinates $\dot{\mathbf{P}}_M$ must also experience a discontinuity and $\ddot{\mathbf{P}}_M$ becomes unbounded.

An infinite acceleration of the machine axes therefore occurs independent of the choice for the kinematic chain of the machine tool. For this reason the error is referred to as machine-independent. Equation (3) can be used to distinguish between a machine dependent error and a machine-independent error. Machine-dependent errors relate to the Jacobian matrix whereas machine-independent errors relate to the $\dot{\mathbf{P}}_W$ data. Thus a solution to eliminate machine-independent errors requires a reformulation of the pose data $\mathbf{P}_W$. This can be achieved by either altering the machining strategy, such as no longer preserving the angles with normal and tangent, or by altering the CAD geometry.

6   Discussion: Geometric Contribution to Machine Errors

Case study one showed that singularity errors are machine-dependent whilst case study two showed that geometric discontinuities are machine-independent. However, not all errors are of these forms. A more realistic expectation is that errors are contributed from both components. Thus, given any machining error it is useful to quantify how much is caused by the machine-configuration and how much by the geometry to decide the best course of action. To ensure there are no machine-dependent errors, the Jacobian needs to be suitably bounded. To ensure there are no machine-independent errors, the $\dot{\mathbf{P}}_W$ data needs to be suitably bounded. The pose data $\mathbf{P}_W$ is inherited from the surface geometry of the CAD model. Therefore geometric properties of the surface affect the machining characteristics.

To avoid geometric discontinuities of the $\mathbf{P}_W$ poses, the underlying CAD surface must satisfy certain continuity constraints. This section derives these constraints and then discusses approaches to bound the $\dot{\mathbf{P}}_W$ data. To begin the following claim is proven: given a G^n surface, a curve can be obtained from the surface in which the normal vector is C^{n-1} smooth.

The unit normal can be defined in terms of first derivatives of curves on the surface. Given a non-degenerate parametrised surface, $\mathbf{S}(u, v)$, the normal, $\mathbf{N}(s)$, is derived from:

$$\mathbf{N}(s) = \frac{\mathbf{S}_u \wedge \mathbf{S}_v}{\|\mathbf{S}_u \wedge \mathbf{S}_v\|}, \text{ where } \mathbf{S}_u = \frac{\partial S}{\partial u} \text{ and } \mathbf{S}_v = \frac{\partial S}{\partial v}.$$

This corresponds to the normalised cross product of the tangent vectors for two iso-parametric curves at the point on the surface. Given any two curves that meet at the surface normal with non-parallel tangents a similar expression, replacing $\mathbf{S}_u$ and $\mathbf{S}_v$ with the new tangents, can be formed. Given that $\mathbf{S}(u, v)$ is G^n it is possible to find two G^n curves [12] crossing at the surface normal with non-parallel tangents. The cross product of these tangents has C^{n-1} continuity since the cross product is a C^∞ operation and thus preserves the C^{n-1} continuity inherited from the tangents. Thus any G^n surface admits a G^n curve that delivers C^{n-1} continuity of the surface normal.

Given the surface, $\mathbf{S}(u, v)$, a curve lying on it can be written as $\mathbf{C}(t) = \mathbf{S}(u(t), v(t))$. A tool path motion that traces out such a curve has the following properties. The CC data is obtained from the arc-length parametrised curve of $\mathbf{C}(t)$, referred to as $CC(s)$. A local orientation frame, $\mathbf{R}_F = \{F_x, F_y, F_z\}$, for this curve can be defined. The F_x direction is defined as $\mathbf{T}(s) = CC'(s)$, the F_z direction is aligned to the surface normal and the unit bi-normal is the cross product of these $\mathbf{B}(s) = \mathbf{T}(s) \wedge \mathbf{N}(s)$. This frame is commonly referred to as the Darboux frame [14].

The cutting tool can be orientated in any direction depending upon the desired machining strategy. The angles the cutter makes with the surface normal and tangent affect the chip pattern on the surface [4]. In five-axis machining efforts are usually made to preserve these angles to maintain a consistent surface finish. In this case, a fixed rotational offset of the Darboux frame can be defined as a unit vector, $\mathbf{O}_O$, that forms the respective angles with the F_x and F_z axes. A rotation matrix with the axes of the reference frame as rows, $\mathbf{R}_F = [\mathbf{T}, -\mathbf{B}, \mathbf{N}]$ can then be used to define the orientation vector in workpiece coordinates as:

$$\mathbf{O}_W = \mathbf{R}_F \mathbf{O}_O.$$

The last step involves finding the CL values to complete the pose data $\mathbf{P}_W$. The shape of the cutter can be interpreted as a surface of revolution and, when it touches a surface (that is it does not gouge), its normal intersects with the axis of revolution (see Fig. 2). The CL data, defined at the tip of the tool, can therefore be defined as a translation along the axis of revolution from the intersection point. This axis is defined from the orientation data and so it is possible to form the following equation:

$$CL(s) = CC(s) + d_1\mathbf{N}(s) + d_2\mathbf{O}_W,$$
$$= CC(s) + d_1\mathbf{N}(s) + d_2\mathbf{R}_F\mathbf{O}_O,$$

The values of d_1 and d_2 depend upon the cutter geometry and the orientation of the cutter with respect to the Darboux frame ($\mathbf{O}_O$) which are assumed to be constant throughout the motion.

Therefore the components of the pose data comprise the underlying curve along with first derivatives of the surface. Algebra of limit arguments can thus be employed to show that G^n continuity of $\mathbf{C}(t)$ delivers C^{n-1} continuity of the tool path motion $\mathbf{P}_W(s)$. Note that not every curve that can be obtained from a G^n surface is G^n continuous and hence sufficient care must be taken when tracing out the CC curve on the surface. This highlights another potential cause of error in selection of the tool paths. For example, machining the simplified blade in Sect. 5 along the orthogonal direction could have avoided traversing the G^2 discontinuity.

The occurrence of machine-independent errors, in the form of discontinuous axes movements, can thus be summarised with the following statement. Given a C^n smooth tool path from a G^n smooth surface the machine axes kinematics are C^{n-1} smooth. Note however that this assumes the Jacobian is C^{n-1} smooth and the angles between the surface normal and tangent direction are preserved.

In case study one the tool path was G^∞ smooth and there were no discontinuities in the $\mathbf{P}_W$ data. In case study two the curve was G^1 continuous and hence the machine coordinates were C^0 continuous. However, there was a G^2 discontinuity between flat and round sections and thus C^1 continuity of the machine coordinates could not be achieved. Using a G^2 curve and surface would ensure C^1 continuous machine coordinates and thus finite acceleration bounds.

Bounding the kinematic properties of the machine axes is possible by bounding pose data and the Jacobian. Bounds on the machine axis speeds and accelerations [6] can be determined from Eq. (3) as

$$||\dot{\mathbf{P}}_M|| \le ||\mathbf{J}|| \, ||\dot{\mathbf{P}}_W||, \quad ||\ddot{\mathbf{P}}_M|| \le ||\dot{\mathbf{J}}|| \, ||\dot{\mathbf{P}}_W|| + ||\mathbf{J}|| \, ||\ddot{\mathbf{P}}_W||.$$

It is not enough for these bounds to exist but rather they must be less than a predetermined amount. Thus simply eliminating discontinuities does not eradicate machine-independent errors. Bounding the maximum velocity and acceleration to specific values is required to ensure no machine-independent errors.

In the scenario where the velocity and acceleration of machine axes are bounded but exceed kinematic constraints one possible solution to eliminate machine errors is to reduce the cutting feed rate. This approach is undesirable because other issues can occur with a reduced cutting feed, such as rubbing of the workpiece. However an important concern of productivity is that a lower feed rate necessitates longer machining times. In the ideal situation a tool path should admit as high as possible feed rate before exceeding kinematic constraints. This motivates further research into determining what type of geometry could preclude machine-independent errors whilst accommodating high feed rates.

7 Conclusions

Machining flaws can be predicted by analysing the desired tool path motion. A tool path motion can be converted into machine coordinates representing the values of the machine axes. If the expected speed or acceleration of an individual machine axis is too high then it differs from the desired motion. Consequently a loss of control in the cutting conditions is experienced potentially leading to machining flaws. This forms the basis of a diagnostic tool for analysing tool path motions. If the velocity or acceleration of the machines axes exceed predetermined bounds a machining error is said to occur. These should be eliminated if potential machining flaws are to be avoided.

The mathematical processes required to form the position and orientation (poses) of the cutter relative to the workpiece was presented in Sect. 2. Given knowledge of the machine tool structure, in terms of a kinematic chain, the relationship between poses and machine axis values was then given. Further insight into this relationship was gleaned from the Jacobian which represents how the coordinate systems change with respect to each other.

Two case studies were then presented. In the first case study the effect of a Jacobian with a large spectral norm, relating to a machining singularity, was investigated. It was shown that by changing machine tool structure, or more specifically the kinematic chain, these machining errors can be eliminated. For this reason machining singularities were identified as a machine-dependent error. In the second case of machining a turbine blade, a geometric discontinuity caused an infinite acceleration of one of the machine axes. It was shown that this occurs independent of the kinematic chain. Such errors were referred to as machine-independent errors.

Section six then discussed the differences between machine-dependent and machine-independent errors. These are characterised by two separate factors. The Jacobian characterises machine-dependent errors whereas the pose data characterises machine-independent errors. Thus the geometric contribution of the machinability of a surface comes from the pose data. Discontinuities in the pose data can be eliminated if the underlying surface possesses sufficient continuity conditions. It was argued that given a G^n continuous surface the time derivatives of the machine axes, up to the $(n-1)^{th}$ derivatives, are also continuous (assuming a sufficiently smooth Jacobian).

A surface with a high machinability must admit curves that yield desirable machine kinematic profiles. The derivatives of the poses represent the geometric contribution to the machine axes velocity and acceleration. These in turn are defined from the surface and constrained by the geometry that defines them. For example, a bumpy surface intuitively seems to be more difficult to machine than a smooth surface. What needs to be done next is obtain criteria for surface geometry under which it possesses good machinability properties. This will be the subject of future investigations.

Acknowledgement. The research is supported by the EPSRC research council (EP/L010321/1 and EP/L006316/1). The authors also thank Delcam International PLC for supporting the research presented in this paper.

References

1. Kalpakjian, S., Schmid, S.: Manufacturing Engineering and Technology, 5th edn. Pearson Publishing Company, Upper Saddle River (2006)
2. Choi, B.K., Kim, B.H., Jerard, R.B.: Sculptured surface NC machining. In: Handbook of Computer Aided Geometric Design, pp. 543–574 (2002)
3. Powermill 2014 Delcam PLC, January 2016. www.powermill.com
4. Lavernhe, S., Quinsat, Y., Lartigue, C.: Model for the prediction of 3D surface topography in 5-axis milling. Int. J. Adv. Manuf. Technol. **51**, 915–924 (2010)
5. Suh, S., Kang, S., Ching, D., Stroud, I.: Theory and Design of CNC Systems. Springer, Heidelberg (2008). doi:10.1007/978-1-84800-336-1
6. Doughty, S.: Mechanics of Machines. Wiley, New York (1988)
7. Cripps, R., Cross, B., Hunt, M., Mullineux, G.: Singularities in five-axis machining: cause effect and avoidance. Int. J. Mach. Tools Manuf. **166**, 40–51 (2017)
8. Kincaid, D., Cheney, W.: Numerical Analysis, 2nd edn. Brooks/Cole Publishing Company, Pacific Grove (1996)
9. Zlatanov, D., Fenton, R.G., Benhabib, B.: Singularity analysis of mechanisms and robots via a velocity-equation model of the instantaneous kinematics. In: IEEE International Conference on Robotics and Automation (1994)
10. Hermle: Hermle C600 Series Brochure. Hermle, Gosheim (1999)
11. Alicona G5 InfiniteFocus Alicona Imaging GmbH, August 2016. http://www.alicona.com/products/infinitefocus/
12. Peters, J.: Geometric continuity. In: Farin, G., Hoschek, J., Kim, M. (eds.) Handbook on Computer Aided Geometric Design. Elsevier, Amsterdam (2002)
13. Powershape 2014 Delcam PLC, January 2016. www.powershape.com
14. Guggenheimer, H.W.: Differential Geometry. Dover Publications, New York (1997)

Simplicial Complex Entropy

Stefan Dantchev and Ioannis Ivrissimtzis[⊠]

Duarham University, Durham, UK
{s.s.dantchev,ioannis.ivrissimtzis}@durham.ac.uk

Abstract. We propose an entropy function for simplicial complices. Its value gives the expected cost of the optimal encoding of sequences of vertices of the complex, when any two vertices belonging to the same simplex are indistinguishable. We focus on the computational properties of the entropy function, showing that it can be computed efficiently. Several examples over complices consisting of hundreds of simplices show that the proposed entropy function can be used in the analysis of large sequences of simplicial complices that often appear in computational topology applications.

Keywords: Entropy · Simplicial complices · Ambiguous encoding · Graph entropy · Simplicial complex entropy

1 Introduction

In several fields of visual computing, such as computer vision, CAD and graphics, many applications require the processing of an input in the form of a set of unorganized points, that is, a finite subset of a metric space, typically $\mathbf{R}^2$ or $\mathbf{R}^3$. Often, the first step in the processing pipeline is the construction of a simplicial complex, or a series of simplicial complices capturing spatial relations of the input points. Such geometrically constructed simplicial complices commonly used in practice include the *Vietoris-Rips* and *Čech* complices, see for example [15], the *alpha shapes* [7] and the witness complices [6,9].

The two simplest constructions, giving the Vietoris-Rips and the Čech complices, emerged from studies in the field of algebraic topology. In the Vietoris-Rips construction, we connect two points with an edge if their distance is less than a fixed ε and the simplices of the complex are the cliques of the resulting graph. In the Čech construction, the simplices are the sets of vertices that lie inside a bounding sphere of radius ε.

Notice that the complices constructed in this way, apart from the input point set which gives their vertex set, also depend on the parameter ε. In applications where the goal is to extract topological information related to the input point set, it is quite common to consider sequences of complices corresponding to different values of ε and to study the evolution of their topological properties as ε varies [13,16]. Such investigations led to the development of the notion of *persistence*, in the form for example of persistent homology, as one of the main

© Springer International Publishing AG 2017
M. Floater et al. (Eds.): MMCS 2016, LNCS 10521, pp. 96–107, 2017.
https://doi.org/10.1007/978-3-319-67885-6_5

concepts in the field of computational topology [5,8]. Indicative of the need for computational efficiency, persistent homology calculations based on millions of distinct complices from the same input point set are now common and thus, the efficient computation of such series of complices is an active research area [4,15].

In this paper, our aim it to use information-theoretic tools to study sequences of geometrically constructed complices corresponding to different values of ε. In particular, we define an entropy function on simplicial complices; we show that it can be computed efficiently; and demonstrate that it can be used to find critical values of ε. Here, the value of ε is seen as a measure of spatial resolution and thus, we interpret the simplices of the geometrically constructed complices as sets of indistinguishable points.

The setting of our problem is very similar to one that gave rise to the concept of *graph entropy* [11] and *hypergraph entropy* [10]. There, a graph or a hypergraph describe indistinguishability relations between vertices and the sets of indistinguishable vertices are derived as the *independent sets* of the graph or hypergraph. In contrast, in our approach, the sets of indistinguishable vertices are readily given as the simplices of the complex. In the next section, immediately after introducing the proposed simplicial complex entropy, we discuss in more detail its relation to graph entropy.

2 Simplicial Complex Entropy

Let $V = \{v_1, \ldots, v_n\}$ be a point set consisting of n vertices. An abstract simplicial complex C over V is given by its maximal simplices $C_1, \ldots, C_m$. These are nonempty subsets of V whose union is the entire V and none of them is a subset of another.

We are also given a probability distribution P over V, i.e. non-negative numbers $p_1, \ldots, p_n$ and such that $\sum_{j=1}^{n} p_j = 1$. Assuming that all points that belong to the same simplex C_i for some i, $1 \le i \le m$ are indistinguishable, we define the *simplicial complex entropy* as

$$H(C, P) = \min_{\{q_i\}_{i=1}^{m}} \sum_{j=1}^{n} p_j \log \frac{1}{\sum_{i \in \mathrm{Simpl}(j)} q_i} \quad \text{s.t.} \tag{1}$$

$$\text{s.t.} \sum_{i=1}^{m} q_i = 1$$

$$q_i \ge 0 \quad 1 \le i \le m.$$

where $\mathrm{Simpl}(j)$ denotes the set of simplices containing vertex p_j.

The above simplicial complex entropy is similar to the graph entropy [11,12], defined over a graph G with a probability distribution P on its vertices, given by

$$H(G, P) = \min_{\mathbf{a} \in VP(G)} \sum p_j \log \frac{1}{a_j} \tag{2}$$

where $VP(G)$, the vertex packing polytope of a graph G, is the convex hull of the characteristic vectors of its independent sets.

In its information theoretic interpretation, the graph entropy gives the expected number of bits per symbol required in an optimal encoding of the information coming from a source emitting vertices of G under the probability distribution P, assuming that any two vertices are indistinguishable iff they are not connected with an edge [12]. In other words, the independent sets of G are the sets of mutually indistinguishable vertices.

The information theoretic interpretation of the proposed simplicial complex entropy is analogous to that of graph entropy. That is, it gives the expected cost, in bits per vertex, of the optimal encoding of a sequence of vertices under the assumption that the simplices indicate sets of mutually indistinguishable vertices. In other words, it gives the expected compression ratio of an optimal ambiguous encoder, which instead of actual vertices encodes simplices containing them. The entropy acquires geometric meaning when the simplices encode spatial information, as for example in the case of Čech complices. In that case, the entropy gives the information savings that we can achieve if we treat any points enclosed inside a sphere of radius ε as indistinguishable. We note however, that in this paper we do not describe any actual simplicial complex encoder, an efficient example of which can be found in [1].

In other information theoretic approaches into the study of simplicial complices or similar geometric structures, [3] defined an entropy function on convex corners of the non-negative orthant of the k-dimensional Euclidean space $\mathbf{R}^k_+$. Their construction is less general and as a result the minimising vector of the entropy function is uniquely determined, unlike the q_i's in Eq. 1. In [2] hierarchical systems modelled as simplicial complices are studied through the maximisation of a Kullback-Liebler divergence. While in a fashion similar to [2], Eq. 1 could be written as a mutual information minimisation problem, we note that in [2] the computational aspects of that optimisation problem are not considered.

While the proposed simplicial complex entropy can be seen as a simplification of the graph entropy, which however is at least as general. Indeed, on a graph G we can define a simplicial complex C on the same vertex set as G and its simplices being the independent sets of G. Then, the graph entropy of G is the simplicial complex entropy of C. On the other hand, given a simplicial complex C it is not immediately obvious how one can construct a graph G such that the simplicial complex entropy of C is the graph entropy of G.

In an abstract context, the proposed simplification might seem quite arbitrary: instead of deriving the sets of indistinguishable vertices from the connectivity of a graph, we consider them given in the form of simplices. However, in the context of geometrically constructed simplicial complices embedded in a metric space, the simplices are the natural choice of sets of indistinguishable points for a given spatial resolution ε and there is no need, or indeed an obvious way, to model the property of indistinguishability in terms of graph connectivity. One notable exception to this is the special case of Vietoris-Rips complices which we discuss next, aiming at further highlighting differences and similarities between simplicial entropy and graph entropy.

2.1 Example: Vietoris-Rips Simplicial Complex Entropy

In the case of Vietoris Rips complices, there is a straightforward interpretation of the simplicial complex entropy as graph entropy. Indeed, assume a probability distribution P on a set of vertices V embedded in a metric space, and assume that two vertices, are indistiguishable if their distance is less than ε. The graph G with its edges connecting pairs of distinguishable vertices is the complement of the underlying graph of the Vietoris-Rips complex constructed on V for the same ε.

It is easy to see that the independent sets of G are exactly the simplices of the Vietoris-Rips complex and thus, the graph entropy of G is the simplicial complex entropy of the Vietoris-Rips complex. Indeed, if there are no edges connecting points of a subset of V, it means that all distances between these points are less than ε, therefore they form a simplex of the Vietoris-Rips complex.

The simplicial entropy of the Vietoris-Rips complices has a straightforward graph entropy interpretation because Vietoris-Rips complices are completely defined by their underlying graph. Indeed, their simplices are the cliques of the underlying graph. However, this is not generally the case for geometrically constructed complices, with the Čech complex being a notable counterexample.

Indeed, consider as V the three vertices of an equilateral triangle of edge-length 1, embedded in $\mathbf{R}^2$. Any pair of vertices corresponds to an edge of the triangle and has a minimum enclosing sphere of radius $1/2$. The V itself has a minimum enclosing sphere of radius $\sqrt{3}/3$. Thus, for any $1/2 \leq \varepsilon \leq \sqrt{3}/3$ all three edges of the triangle are simplices of the Čech complex, i.e. pair-wise indistinguishable, but the triangle itself is not a simplex of the Čech complex.

3 Properties of Simplicial Complex Entropy

Solving the entropy minimisation turns out to be computationally tractable. Let us denote

$$S_j(q) \overset{\text{def}}{=} \sum_{i \in \text{Simpl}(j)} q_i$$

and rewrite Eq. 1 as a maximisation problem with an objective function

$$f(q) \overset{\text{def}}{=} \sum_{j=1}^{n} p_j \log S_j(q). \tag{3}$$

We can immediately prove the following

Proposition 1. *The objective function in Eq. 3 is concave. The sums $S_j(q)$ are unique (i.e. the same) for all vectors q where the maximum is attained, while the set of all maxima is a polyhedron.*

Proof. Let q' and q'' be two different feasible vectors. Clearly, the vector

$$q = \frac{1}{2}(q' + q'') \tag{4}$$

is also feasible and

$$S_j(q) = \frac{1}{2}(S_j(q') + S_j(q'')) \text{ for } 1 \leq j \leq n \tag{5}$$

We then have

$$\log S_j(q) \geq \frac{1}{2}(\log S_j(q') + \log S_j(q'')), \tag{6}$$

which proves the concavity of the objective function.

Imagine now that q' and q'' are two (different) optimal vectors (with $f(q') = f(q'')$) and moreover there is a j, $1 \leq j \leq n$ such that $S_j(q') \neq S_j(q'')$. For those particular q' and q'', Eq. 6 is a strict inequality and after summing up all inequalities, we get

$$f(q) > \frac{1}{2}(f(q') + f(q'')) \tag{7}$$

which contradicts the optimality of both q' and q''. Thus, the sums $\log S_j(q)$ are unique over all optimal vectors q.

Finally, if we denote these sums (at an optimum) by s_j, $1 \leq j \leq n$, we notice that the set of all optimal optimal vectors q is fully described by the following linear system:

$$\sum_{i \in \text{Simpl}(j)} q_i = s_j \; 1 \leq j \leq n$$

$$\sum_{i=1}^{m} q_i = 1$$

$$q_i \geq 0 \; 1 \leq i \leq m.$$

Let $\text{Pts}(i)$ denote the set of vertices of the simplex i. Another useful characterisation of an optimal vector q is given by

Proposition 2. *Any optimal vector q satisfies the following "polynomial complementarity" system:*

$$\sum_{j \in \text{Pts}(i)} \frac{p_j}{S_j(q)} \begin{cases} = 1 & \text{if } q_i > 0 \\ \leq 1 & \text{if } q_i = 0 \end{cases} \; 1 \leq i \leq m$$

$$\sum_{i=1}^{m} q_i = 1$$

$$q_i \geq 0 \; 1 \leq i \leq m$$

Proof. The gradient of the objective function, $\nabla f(q)$ is

$$\left(\sum_{j \in \text{Pts}(1)} \frac{p_j}{S_j(q)}, \cdots \sum_{j \in \text{Pts}(m)} \frac{p_j}{S_j(q)} \right)^T.$$

We start with Karush-Kuhn-Tucker conditions (for the maximisation problem) that an optimal vector q should satisfy:

$$\sum_{j \in \text{Pts}(i)} \frac{p_j}{S_j(q)} = \lambda - \mu_i \ \ 1 \le i \le m \tag{8}$$

$$\sum_{i=1}^{m} q_i = 1 \tag{9}$$

$$q_i, \mu_i \ge 0, \ q_i \mu_i = 0, \ 1 \le i \le m \tag{10}$$

for some λ and μ_i, $1 \le i \le m$.

We first expand the inner product

$$\langle q, \nabla f(q) \rangle = \sum_{i=1}^{m} q_i \sum_{j \in \text{Pts}(i)} \frac{p_j}{S_j(q)} \tag{11}$$

$$= \sum_{j=1}^{n} \frac{p_j}{S_j(q)} \sum_{i \in \text{Simpl}(j)} q_i = \sum_{j=1}^{n} p_j = 1. \tag{12}$$

On the other hand, from Eqs. 8, 9 and 10, we get

$$\sum_{i=1}^{m} q_i \sum_{j \in \text{Pts}(i)} \frac{p_j}{S_j(q)} = \sum_{i=1}^{m} q_i (\lambda - \mu_i) \tag{13}$$

$$= \lambda \sum_{i=1}^{m} q_i - \sum_{i=1}^{m} q_i \mu_i = \lambda, \tag{14}$$

and thus $\lambda = 1$.

3.1 Encoding/Decoding Accuracy Rate

The ambiguity in the description of a point set by a simplicial complex, results into an error when points are encoded as simplices and then simplices are decoded back to points. Here, the main motivation for studying the accuracy rate of specific encoding/decoding processes is the observation that while the optimisation problem in Eq. 1 has a unique solution, the optimising vector q is not necessarily unique. Thus, we will use the maximisation of the encoding/decoding accuracy rate, under maximisation of the entropy, as a second optimisation problem that will return a unique probability distribution q on the simplices.

We will describe two encoding/decoding strategies, one randomised, which is the one we will use to produce our examples in Sect. 4, and an adversarial which generally results to higher error rates.

Randomised Encoder. The encoder gets a point j, $1 \le j \le n$, produced by a memoryless random source under distribution p, and produces one of the cells that contains j, under the distribution $\frac{q_i}{S_j(q)}$ for all $i \in \text{Simpl}(j)$. The overall probability of seeing cell i as a result is

$$\sum_{j \in \text{Pts}(i)} p_j \frac{q_i}{S_j(q)} = q_i \sum_{j \in \text{Pts}(i)} \frac{p_j}{S_j(q)} = q_i(1 - \mu_i) = q_i \tag{15}$$

where μ_i is as in the proof of Proposition 2 above and taking into account that $\lambda = 1$, as expected.

The decoder sees a cell i and returns the vertex which according to the distribution p had the highest probability to have been encoded as i. Thus, the probability for accurate encoding/decoding, which we maximise through our choice of q_i's, is

$$\overline{acc} = \sum_{i=1}^{m} q_i \frac{\max_{j \in \text{Pts}(i)} p_j}{\sum_{j \in \text{Pts}(i)} p_j}. \tag{16}$$

The Adversarial Encoder. We can think of this encoding strategy as a game between the encoder and the decoder, in which whenever the decoder sees a simplex i, he responds with a guess of a point $j \in \text{Simpl}(i)$ according to probabilities r_{ij}, $r_{ij} \geq 0$ and such that

$$\sum_{j \in \text{Pts}(i)} r_{ij} = 1 \quad \text{for every } 1 \leq i \leq m. \tag{17}$$

These probabilities are known to the encoder, so if the source produced a point j, the encoder minimises the success rate of the decoder by picking a cell i that is $\arg\min_{i \in \text{Simpl}(j)} r_{ij}$. In turn, the decoder tries to maximise the accurate encoding/decoding rate as

$$\overline{acc} = \max \sum_{j=1}^{n} p_j r_j \quad \text{s.t.} \tag{18}$$

$$r_{ij} \geq r_j \quad 1 \leq j \leq n \text{ and } i \in \text{Simpl}(j) \tag{19}$$

$$\sum_{j \in \text{Pts}(i)} r_{ij} = 1 \quad 1 \leq i \leq m \tag{20}$$

$$r_{ij} \geq 0 \quad 1 \leq j \leq n \text{ and } i \in \text{Simpl}(j) \tag{21}$$

4   Examples

The computation of the simplicial complex entropy and the encoding/decoding accuracy rate were implemented in Matlab. Apart from some code for input/output operations and simplicial complex representation, *fmincon* and *linprog* were directly used to compute the entropy and the accuracy rate, respectively.

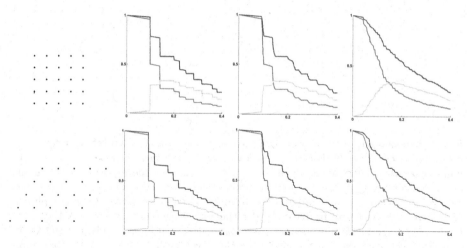

Fig. 1. The y-axis represents the normalised entropy (blue curve), the accuracy rate (red curve) and their difference (green curve). The x-axis represents the parameter ε (radius of the minimal enclosing sphere of a simplex) in the construction of the Čech complex. **Top:** The input point set is the 5×5 block of vertices of a square grid of edgelength 0.2 shown in the left. From left to right, uniform random noise $\pm 0.5\%, \pm 5\%$ and $\pm 50\%$ of the edgelength was added. The figures represent entropy and accuracy rate computations on all possible Čech complices for that range of ε, that is, 768, 746 and 685 distinct Čech complices, respectively. **Bottom:** As per the top, but for triangular grid points. The figures correspond to 725, 694 and 672 distinct Čech complices, respectively. (Color figure online)

In all examples, we report:

(i) the *normalised entropy*, that is, the simplicial complex entropy $H(C, P)$ divided by the entropy of the vertex set V under the same probability distribution P,

(ii) the accuracy rate, which correlates nicely with the entropy since the more bits we use the higher we expect the encoding/decoding accuracy,

(iii) the difference between these two values.

In a first example, Fig. 1 (Top) shows the values of these two functions on Čech complices constructed from vertex sets that are nodes of square grid of edgelength 0.2 with some added noise. Figure 1 (Bottom) shows a similar example with the vertices originally being nodes of a triangular grid. In all cases, the probability distribution P on the vertex set is uniform.

In the case of a square grid without any added noise, as the values of the parameter ε of the Čech complex construction parameter increase, they reach the first critical value at $\varepsilon = 0.1$, when edges, i.e. simplices of degree 2, are formed. The next critical value is $\varepsilon \simeq 0.141$, where the simplices of degree 4 are formed, and the next critical value is $\varepsilon = 0.2$ when simplices of degree 5 are formed. Similarly, the first critical values in the case of points from a triangular

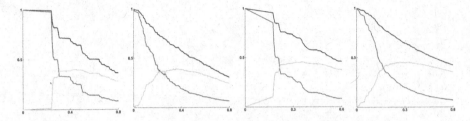

Fig. 2. The axes and the colour of the curves are as per Fig. 1. **Two left figures:** The input point set of size 50 is a computational solution to the Thomson problem with uniform noise of ±0.01 units added on each coordinate. In the right figure the input in an area uniform spherical random sample of the same size. The figures represent entropy and accuracy rate computations from 2523 and 2661 distinct Čech complices, respectively. **Two right figures:** As per the top, with point sets of size 100. Due to the very large number of distinct Čech complices, each figure represents 100 Čech complices, corresponding to a uniform sample of values of ε in [0, 0.6]. (Color figure online)

grid are $\varepsilon = 0.1$, when simplices of degree 2 are formed and $\varepsilon \simeq 0.115$ when simplices of degree 3 are formed.

These critical values are shown Fig. 1 as sudden drops in the entropy of the Čech complices constructed on the less noisy data sets. We also notice simultaneous drops of the accuracy rates since they, as expected, correlate well with entropy. As the level of noise increases the critical points become less visible on either of these two curves. However, their difference, shown in green, seems to be more robust against noise, and moreover, seems to peak at a favorable place. That is, it peaks in values of ε that would neither return a large number of non-connected components nor heavily overlapping simplices.

In the second example, the input set is a sample from the unit sphere in $\mathbf{R}^3$. Figure 2 (left) shows results from regular samples of size 50 (top) and 100 (bottom), computed in [14] as solutions to the Thomson problem, with added uniform noise of ±0.01 units. In [14], the minimum distances between a point and its nearest neighbour in an optimal solution are ~ 0.5 and ~ 0.35, respectively, and correspond to the steep entropy decreases at the half of these values, i.e. when the first edges of the Čech complices are formed. Figure 2 (right) shows results from random, area uniform samples of size 50 (top) and 100 (bottom). While the input is much less regular than at the left hand side of the figure, the peaks of the two green curves align well.

In the third example, the initial vertex sets are the nodes of a 4×4 square grid of edgelength 1/3 and of a $4 \times 4 \times 4$ cubic grid of the same edgelength. Figure 3 (left) shows the results for the square grid, first with added uniform noise of ±0.01% of the edgelength and then with ±0.1%. Figure 3 (right) shows the results for the cubic grid, again under noise addition of ±0.01% and ±0.01% of the edgelength. We notice that green line corresponding to the cubic grid

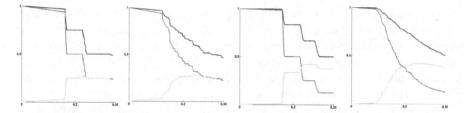

Fig. 3. The axes and the colour of the curves are as per Fig. 1. **Left:** The input point set is a 4×4 block of vertices of a square grid of edgelength $1/3$ with added uniform random noise equal to $\pm 0.01\%$ and $\pm 0.1\%$ of the edgelength. The figures represent entropy and accuracy rate computations on all possible Čech complices for a $[0,0.35]$ range of ε. **Right:** As per the left, but the input point set is a $4 \times 4 \times 4$ block of vertices of a cubic grid. (Color figure online)

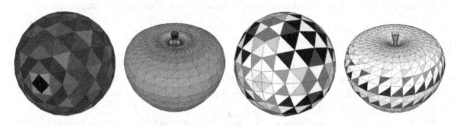

Fig. 4. Two left figures: The values q_i in Eq. 1 are color-mapped on the mesh triangles. Darker colors correspond to higher values. **Two right figures:** The probability distribution P on the vertices corresponds to the absolute values of the discrete Gaussian curvature of the vertices. The two meshes consist of 512 and 1704 triangle, respectively. (Color figure online)

peaks higher and later than the green line of the square grid, reflecting the higher dimensionality of the data.

In a fourth example, we solve the optimisation problem for the computation of the entropy on triangle meshes and show the obtained values of the vector q, as in Eq. 1, color-mapped on the mesh triangles. In Fig. 4 (left), the probability distribution on the mesh vertices is uniform, as it was in all previous examples. On the right hand side of the figure, the probability distribution follows the absolute value of the discrete Gaussian curvature of the vertices.

5   Conclusion

We presented an entropy function for simplicial complices which can be seen as a simplification and generalisation of the graph entropy since all the maximal sets of indistiguishable points are exactly the maximal simplices of the complex and do not have to be computed as the independent sets of the graphs, which, generally, are difficult to characterise. We show that this simplification makes the simplicial complex entropy a function that can be efficiently computed.

Even though the entropy is defined on abstract simplicial complexes, which are purely topological structures, in the examples we show that it can be relevant to geometric applications. For example, by computing the entropy of geometrically constructed simplicial complices, such as the Čech complices, or by using geometric properties of an embedded complex, such as a discrete curvature computed on the vertices to obtain a probability distribution on them.

In the future we would like to study in more detail the function given as the difference between normalised entropy and the decoding accuracy rates, which seems to be a robust to noise descriptor of an appropriate level of geometric detail defined by the variable ε of the Čech complex. We would also like to study the relationship between the accuracy rate of the a randomised encoder we used here and the that of the adversarial encoder discussed at the end of Sect. 3.1.

Acknowledgement. This research was partially supported by the EPRSC Grant EP/K016687/1 "Topology, Geometry and Laplacians of Simplicial Complexes".

References

1. Attali, D., Lieutier, A., Salinas, D.: Efficient data structure for representing and simplifying simplicial complexes in high dimensions. Int. J. Comput. Geom. Appl. **22**(04), 279–303 (2012)
2. Ay, N., Olbrich, E., Bertschinger, N., Jost, J.: A geometric approach to complexity. Chaos **21**(3), 22 (2011)
3. Csiszár, I., Körner, J., Lovász, L., Marton, K., Simonyi, G.: Entropy splitting for antiblocking corners and perfect graphs. Combinatorica **10**(1), 27–40 (1990)
4. Dantchev, S., Ivrissimtzis, I.: Efficient construction of the Čech complex. Comput. Graph. **36**(6), 708–713 (2012)
5. de Silva, V., Ghrist, R.: Coverage in sensor networks via persistent homology. Algebraic Geom. Topol. **7**, 339–358 (2007)
6. de Silva, V., Carlsson, G.: Topological estimation using witness complexes. In: Alexa, M., Rusinkiewicz, S. (eds.) Eurographics Symposium on Point-Based Graphics. ETH, Zürich (2004)
7. Edelsbrunner, H.: The union of balls and its dual shape. Discrete Comput. Geom. **13**(1), 415–440 (1995)
8. Edelsbrunner, H., Letscher, D., Zomorodian, A.: Topological persistence and simplification. In: FOCS 2000, p. 454. IEEE (2000)
9. Guibas, L.J., Oudot, S.Y.: Reconstruction using witness complexes. In: Proceedings of the Eighteenth Annual ACM-SIAM Symposium on Discrete Algorithms, SODA 2007, pp. 1076–1085, Philadelphia, PA, USA. SIAM (2007)
10. Korner, J., Marton, K.: New bounds for perfect hashing via information theory. Eur. J. Comb. **9**(6), 523–530 (1988)
11. Körner, J.: Coding of an information source having ambiguous alphabet and the entropy of graphs. In: 6th Prague Conference on Information Theory, pp. 411–425 (1973)
12. Simonyi, G.: Graph entropy: a survey. Comb. Optim. **20**, 399–441 (1995)
13. Vejdemo-Johansson, M.: Interleaved computation for persistent homology. CoRR, abs/1105.6305 (2011)

14. Wales, D.J., Ulker, S.: Structure and dynamics of spherical crystals characterized for the Thomson problem. Phys. Rev. B **74**(21), 212101 (2006)
15. Zomorodian, A.: Fast construction of the Vietoris-Rips complex. Comput. Graph. **34**, 263–271 (2010)
16. Zomorodian, A., Carlsson, G.: Computing persistent homology. Discrete Comput. Geom. **33**(2), 249–274 (2005)

Precise Construction of Micro-structures and Porous Geometry via Functional Composition

Gershon Elber[✉]

Department of Computer Science, Technion – IIT, 32000 Haifa, Israel
gershon@cs.technion.ac.il
http://www.cs.technion.ac.il/~gershon

Abstract. We introduce a modeling constructor for micro-structures and porous geometry via curve-trivariate, surface-trivariate and trivariate-trivariate function (symbolic) compositions. By using 1-, 2- and 3-manifold based tiles and paving them multiple times inside the domain of a 3-manifold deforming trivariate function, smooth, precise and watertight, yet general, porous/micro-structure geometry might be constructed, via composition. The tiles are demonstrated to be either polygonal meshes, (a set of) Bézier or B-spline curves, (a set of) Bézier or B-spline (trimmed) surfaces, (a set of) Bézier or B-spline (trimmed) trivariates or any combination thereof, whereas the 3-manifold deforming function is either a Bézier or a B-spline trivariate.

We briefly lay down the theoretical foundations, only to demonstrate the power of this modeling constructor in practice, and also present a few 3D printed tangible examples. We then discuss these results and conclude with some future directions and limitations.

Keywords: Freeform deformation · Trivariate splines · Symbolic computation · Freeform tiling

1 Introduction and Related Work

Deformations and metamorphosis captured the attention of the computer graphics and the geometric modeling communities for several decades, while in recent years this interest has reduced a bit. The idea of freeform deformations (FFD) was introduced around thirty years ago [25] as a global deformation mapping, $\mathcal{T} : \mathbb{R}^3 \to \mathbb{R}^3$, and was originally based on trivariate tensor-product Bézier vector functions. Trivariate splines were investigated by many and herein we will only survey the use of trivariates toward deformations.

A large body of work was presented on a variety of FFD techniques, following [25], including extensions that support the B-spline representation [14] and use of FFDs in animation [5]. While, in general, FFDs map a box-shaped domain into a deformed-box in Euclidean space, other topologies were considered and, for example, [4] introduces Extended FFDs to form a deformation that

© Springer International Publishing AG 2017
M. Floater et al. (Eds.): MMCS 2016, LNCS 10521, pp. 108–125, 2017.
https://doi.org/10.1007/978-3-319-67885-6_6

better resembles the shape of the input model. [4] suggested the use of prismatic and cylindrical FFD functions that can approximate some geometric models better than box-shaped tensor product FFDs. More general FFDs suggested the use of arbitrary topology FFDs based on subdivision volumes for free-form deformation [15].

Other, more recent, variations of FFDs considered the removal of certain topological restrictions from the deformed object. [11] considered torn surfaces that incorporated non-iso-parametric curves of C^{-1} discontinuity inside B-spline surfaces. Similarly, [23] suggested the exploitation of discontinuous FFDs to induce tears in the deformed models for animation and surgery incision simulations.

While the body of FFD work is significant, FFDs were always seen as manipulation tools of existing geometry. Almost exclusively, FFDs were applied to an existing model, resulting in a modified, deformed, model, that typically, and aside from the discussed torn surfaces abilities, preserved the topology. Also, in the last decade or so, surface detailing techniques where introduced in [8] and then in [9,22] that are volumetric but limited to a surface layer (between the surface and its small offset, with typically a linear interpolation in between) allowing the modeling of surface details like scales or thorns. Similar surface detailing abilities can also be found in commercial packages nowadays like Rhinoceros[1].

Other efforts toward the synthesis of porous geometry and modeling with porosity are known and, for example, include stochastic methods and the use of Boolean set operations [24], including voxels' based. In [26], the 3D Voronoi diagram of a set of points serves as the basis of the pore space, thickening the Voronoi edges and/or walls, and in [19], procedural (implicit) forms are employed toward the synthesis of micro-structures while also allowing for deformations and blendings. In [1], porous modeling of scaffolds is considered toward 3D printing, where a volumetric grid-like model is synthesized to follow the basic input scaffold. These methods typically synthesize piecewise constant (i.e. voxels) and linear porous geometry and are hence of limited continuity, and are further incapable of precisely controlling the geometry that is being synthesize. This, while herein the micro-structures' results can be fully piecewise-parametric.

In this work, we fuse the general FFD's idea with a surface detailing technique into a modeling constructor of porous and/or micro-structure geometry that smoothly and precisely builds the geometry. The constructor of the porous/micro-structure receives a volumetric model, $\mathcal{T}$, as a trivariate and a geometric tile, and paves the tile in the domain of $\mathcal{T}$ as desired, constructing a whole new topology of porous geometry in the general shape of $\mathcal{T}$. The paved tiles are then mapped to Euclidean space via a composition with the volumetric model $\mathcal{T}$.

The trivariate function $\mathcal{T}$ can be of any general shape. Techniques to build trivariate functions are known for a long time (i.e. [18] but also recently [16]). The vast majority of tensor product surface constructors can be made into a trivariate constructors, including volumetric extrusions, ruled volumes or vol-

---

[1] http://www.rhino3d.com.

umes of revolution, volumetric Boolean sums, and volumetric sweeps. With the clear ability to construct primitive shaped trivariates (i.e. cones and spheres), in [16] volumetric Boolean set operations over trivariates are now also demonstrated.

This work extends our previous recent work [10] that introduced the basic micro-structure construction idea, in several ways. While [10] only considered surface-trivariate compositions and 2-manifold micro-structures, here we expand and allow tiles consisting of univariates, (trimmed) bivariates, and even (trimmed) trivariates [16], including in combination thereof in the same tile (See Fig. 1). Hence, one can manage non-manifold tiles as well as tiles consisting of multi-dimensional shapes. With the additional ability of supporting trivariate-trivariate composition, we form a closure: the resulting elements of the micro-structure are again trivariates and hence, can be recursively used in the construction of nano-structure, etc. Finally, we formalize the conditions over the mapping trivariate, $\mathcal{T}$, and the tile so that the constructed model will be a viable k-manifold.

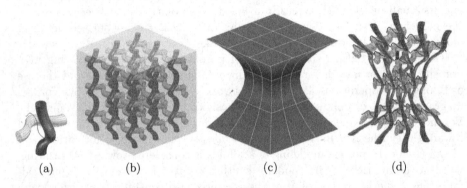

(a)   (b)   (c)   (d)

Fig. 1. A simple example of a tile, T, consisting of three different geometric types: a curve (in blue), a surface (in red), and a trivariate (in green) (a). T is paved ($3 \times 3 \times 3$) times in the domain of the mapping trivariate, $\mathcal{T}$, (b), that is shown in (c). (d) presents the precise smooth composition result of $\mathcal{T}(T)$. (Color figure online)

The rest of this work is organized as follows. Section 2 presents the different computational needs of this variant of a micro-structure constructor and the necessary foundations. In Sect. 3, some examples and results are presented, only to be discussed, in Sect. 4. Then, we conclude, in Sect. 5.

2   Algorithm

Let $\mathcal{T}$ be a trivariate Bézier vector function:

$$\mathcal{T}(x, y, z) = \sum_{i_1=0}^{n_1} \sum_{i_2=0}^{n_2} \sum_{i_3=0}^{n_3} P_{i_1, i_2, i_3} B_{i_1}^{n_1}(x) B_{i_2}^{n_2}(y) B_{i_3}^{n_3}(z), \qquad (1)$$

where P_{i_1,i_2,i_3} are the control points of the $3D$ mesh of T and $B_{i_1}^{n_1}(u)$ is the i_1'th Bézier basis function, of degree n_1.

Herein, we only discuss the necessary computation imposed by a trivariate-trivariate composition, while the cases of curve-trivariate and surface-trivariate are similar, yet obviously simpler.

Consider the trivariate-trivariate composition $\bar{T} = \mathcal{T}(T)$, where $\mathcal{T}$ is as in Eq. (1) and T is:

$$T(u,v,w) = \sum_{j_1=0}^{m_1} \sum_{j_2=0}^{m_2} \sum_{j_3=0}^{m_3} Q_{j_1,j_2,j_3} B_{j_1}^{m_1}(u) B_{j_2}^{m_2}(v) B_{j_3}^{m_3}(w). \tag{2}$$

One can map the control points, Q_{j_1,j_2,j_3}, of T through $\mathcal{T}$ as $\mathcal{T}(Q_{j_1,j_2,j_3})$, yielding

$$\bar{T} \approx \sum_{j_1=0}^{m_1} \sum_{j_2=0}^{m_2} \sum_{j_3=0}^{m_3} \mathcal{T}(Q_{j_1,j_2,j_3}) B_{j_1}^{m_1}(u) B_{j_2}^{m_2}(v) B_{j_3}^{m_3}(w), \tag{3}$$

as is typically done with the FFD of input polygonal data where only the vertices of the polygons are mapped through $\mathcal{T}$. However, Eq. (3) is only an approximation of $\bar{T} = \mathcal{T}(T)$. Further, continuity will not be preserved and the geometry will only loosely follow the micro-shape induced by T. Alternatively, a precise mapping of T through $\mathcal{T}$ (See Fig. 1) can be computed using function composition [6, 7]:

$$\bar{T} = \mathcal{T}(T)$$
$$= \mathcal{T}(t^x(u,v,w), t^y(u,v,w), t^z(u,v,w))$$
$$= \sum_{i_1=0}^{n_1} \sum_{i_2=0}^{n_2} \sum_{i_3=0}^{n_3} P_{i_1,i_2,i_3} B_{i_1}^{n_1}(t^x(u,v,w)) B_{i_2}^{n_2}(t^y(u,v,w)) B_{i_3}^{n_3}(t^z(u,v,w)), \tag{4}$$

where (t^x, t^y, t^z) are the coefficients of T. Equation (4) amounts to the computation of products of terms in the form of $B_{i_1}^{n_1}(t^x(u,v,w))$. If $B_{i_1}^{n_1}$ is a polynomial (Bézier) function and $Q_{j_1,j_2,j_3} = (q_{j_1,j_2,j_3}^x, q_{j_1,j_2,j_3}^y, q_{j_1,j_2,j_3}^z)$, then:

$$B_{i_1}^{n_1}(t^x(u,v,w)) = \binom{n_1}{i_1} t^x(u,v,w)^{i_1}(1-t^x(u,v,w))^{n_1-i_1}$$

$$= \binom{n_1}{i_1} \left(\sum_{j_1=0}^{m_1} \sum_{j_2=0}^{m_2} \sum_{j_3=0}^{m_3} q_{j_1,j_2,j_3}^x B_J^M(u,v,w) \right)^{i_1}$$

$$\left(1 - \sum_{j_1=0}^{m_1} \sum_{j_2=0}^{m_2} \sum_{j_3=0}^{m_3} q_{j_1,j_2,j_3}^x B_J^M(u,v,w) \right)^{n_1-i_1}, \tag{5}$$

where $B_J^M(u,v,w) = B_{j_1}^{m_1}(u) B_{j_2}^{m_2}(v) B_{j_3}^{m_3}(w)$.

Algorithms to directly evaluate the product (and summation) of splines, in both Bézier and B-spline forms, are known [7,12,13,17], and all the above formulation can be applied to either the Bézier or the B-spline representation, with one caution. If $\mathcal{T}$ is a B-spline trivariate, the tile T cannot, in general, cross knot lines in $\mathcal{T}$. Because tensor product splines can represent finite (dis)continuities only along knots, a general crossing of a knot line of $\mathcal{T}$ by T is likely to introduce a irrepresentable by tensor product splines diagonal (dis)continuity into $\mathcal{T}(T)$. Hence, T must be divided along the knot lines of $\mathcal{T}$. If T consists solely of univariates, these univariates could be divided at the knot lines of $\mathcal{T}$, only to be re-merged into $\mathcal{T}(T)$. However, if T is a surface or a trivariate it must be divided along the knot lines into smaller, not necessarily rectangular surface/trivariates patches. Further, those new patches must be again divided into rectangular/cuboid patches. While a feasible process, it is unfortunately a far more difficult process that also affects the regularity of the tiles' representation, in the micro-structure as a whole. Hence, herein we limit ourselves to bivariate and trivariate tiles that cross no knot lines in $\mathcal{T}$.

The continuity of $\bar{T} = \mathcal{T}(T)$ is directly governed by the lowest continuity between the continuities of $\mathcal{T}$ and T, a result that stems directly from the chain rule of differentiation of composition functions. A function is considered regular if its Jacobian vanishes in no place. Again, and using the chain rule of differentiation $\bar{T}$ can be shown to be regular if both $\mathcal{T}$ and T are regular. This means that a mapping of a regular k-manifold, $k = 1, 2, 3$ through a regular trivariate $\mathcal{T}$ will yield back a regular k-manifold, $k = 1, 2, 3$, albeit of a higher degree.

Having the ability to compute $\mathcal{T}(T)$ as well as the simpler cases of surface-trivariate (and polygon-trivariate) and curve-trivariate compositions, we consider periodic tiles in 3D that pave the domain of $\mathcal{T}$ ($c_x \times c_y \times c_z$) times (See also Fig. 1). A tile is considered C^n periodic if the boundaries of the tile for u_{min}, v_{min}, w_{min} match the boundaries of the same tile for u_{max}, v_{max}, w_{max}, with C^n continuity, respectively. That is,

$$\left.\frac{\partial^m t_a}{\partial p^m}\right|_{u_{min}} = \left.\frac{\partial^m t_a}{\partial p^m}\right|_{u_{max}} , \quad m = 0, ..., n, \tag{6}$$

in all axes $a = x, y, z$ and in all permutations of parameters $p \in \{u, v, w\}$.

We also need to consider $\mathcal{T}$'s boundary end conditions. Consider the pavement of the domain $\mathcal{T}$ by tile T ($c_x \times c_y \times c_z$) times. A tile T can be C^n periodic but tiles placed on the boundary of $\mathcal{T}$ must be closed along their boundary. That is, all tiles $(i, j, 1)$ and (i, j, c_z), $i \in (1, ..., c_x)$, $j \in (1, ..., c_y)$ must all be closed in the w_{min} and w_{max} directions, respectively, and the same holds for the u and v min/max boundaries. While one can consider handling these boundary openings after the mapping through $\mathcal{T}$, we propose a simpler remedy. Given a C^n periodic tile T, process it by computing its Boolean set operations with the six planes $u = u_{min}$, $u = u_{max}$, $v = v_{min}$, $v = v_{max}$, $w = w_{min}$, $w = w_{max}$, and their combination thereof, considering face, edge and vertex neighborhoods, 26 neighbors in all. As an example, tile $(1, 1, 1)$, that should be closed in u_{min}, v_{min} and w_{min}, will be applied with Boolean set operations and sealed against planes $u = u_{min}$, $v = v_{min}$ and $w = w_{min}$. Then, and based on the tile's indices in $\mathcal{T}$, the proper boundary or interior tile will be employed out of the 27 tiles we will have, in whole. While the output can include hundred of thousands if not millions of mapped tiles, by computing the boundary tiles a-priori, only 26 different, local to the tile, Boolean set operations are required.

Then, if T is a regular C^n (or better) trivariate and T is a regular C^n k-manifold periodic tile, a watertight C^n k-manifold model can be formed, except possibly at the boundaries, as the intersection curves along the Boolean set operations are typically only C^0. In the next section, the power of this modeling constructor is fully revealed and demonstrated.

3 Results and Examples

A modeling constructor based on trivariate functions can be quite powerful. It enables the fabrication of delicate geometry that is very difficult to construct in alternative ways. In Fig. 2, the domain of a trivariate duck is paved with piece-wise linear B-spline surfaces' tiles, only to be using precise surface-trivariate composition computation. Six bilinear B-spline surfaces define this hollowed tile, shown in Fig. 2(a). The result of the composition is shown in Fig. 2(b) whereas Fig. 2(c) presents a similar result when the surfaces of the tile are first converted to polygons while only the vertices of the polygons are mapped through T. Note the silhouettes near the belly area, in Fig. 2(c), that are clearly C^1 discontinuous where they should have been smooth, at common boundaries between two different tiles.

We seek viable models which means they should be watertight. If each tile is watertight and closed, the result will be watertight but consisting of numerous disjoint parts. If the tiles are periodic (and possibly smoothly periodic) between $(x_{min}, y_{min}, z_{min})$ and $(x_{max}, y_{max}, z_{max})$, the interior will be connected and hence sealed and watertight. However, we still need to close boundary openings along the boundary of T. This closure is simple to achieve as explained in Sect. 2 - every tile that is a boundary tile in some direction, in the domain of the trivariate, will be sealed with the plane of that boundary, possibly using a Boolean set operation. Figure 2(d) shows a watertight porous model that resulted from applying this boundary sealing operation to the model in Fig. 2(c). As stated in the previous section, 26 such Boolean operations will be needed.

Alternatively, one can provide a-priori sealed tiles for the proper boundary. 26 such sealed tiles could be provided for all possible neighboring boundaries. Figure 3 shows a tile consisting of three Borromean rings[2], tailored so they can also be linked to their neighbors. In Fig. 3(a), the interior as well as the primary $u_{min}, u_{max}, v_{min}, v_{max}, w_{min}, w_{max}$ boundary tiles are shown, left to right. Figure 3(b) and (c) shows two views of a $(3 \times 3 \times 3)$ tiling using these tiles, as a full watertight and smooth model, using the interior and six boundary tiles shown in Fig. 3(a).

In Fig. 4, we pave 3D twisted tubes in a domain of a trivariate in the shape of a knot. The knot surface was created as a regular sweep of a circular cross section along a 3D knot curve. Then, volumetric Boolean sum was used to convert the

---

[2] https://en.wikipedia.org/wiki/Borromean_rings.

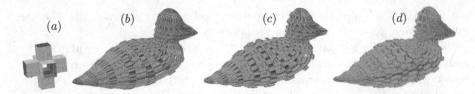

Fig. 2. Six bilinear B-spline surfaces form the tile in (a) that paves the domain of a trivariate in the shape of a duck. (b) presents the smooth and precise surface-trivar composition. In (c), the tile is converted to polygons and vertices are mapped through the trivariate, resulting in C^1 discontinuities (note the belly area). The interiors of the resulting surfaces are exposed, in magenta, in (b) and (c). (d) shows the result of the boundary sealing via Boolean set operations, creating a watertight model. See also Fig. 9(a).

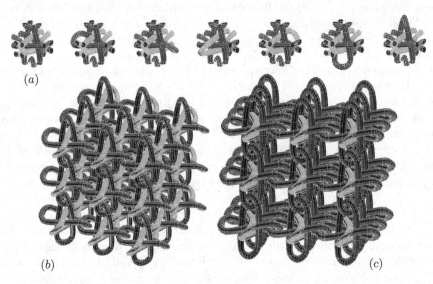

Fig. 3. One can create a sealed, watertight, model by providing sealed boundary tiles for the proper boundaries of $\mathcal{T}$. In (a), the interior tile and the u_{min}, u_{max}, v_{min}, v_{max}, w_{min}, w_{max} tiles are provided (left to right). (b) and (c) shows two different views of a $(3 \times 3 \times 3)$ tiling, using these linked Borromean rings' tiles.

sweep surface to the trivariate that is shown in Fig. 4(b). The tile in Fig. 4(a) consists of four bicubic helical B-spline surfaces, constructed using algebraic sum [21] between a quarter of a helical curve and a circle. In this example, we pave the tiles mostly in one direction - along the axis of the knot trivariate. The tile (smoothly) shifts between the four boundary openings, bottom to top, creating the twisting effect, in the C^1 continuous final result shown in Figs. 4(c) and (d). Figures 4(e) to (h) shows the same trivariate paved using increasingly higher resolutions, in all three axes.

The complexity of the final model depends on the resolution of the pavements but also on the complexity of the individual tile. Figure 5 shows an example where

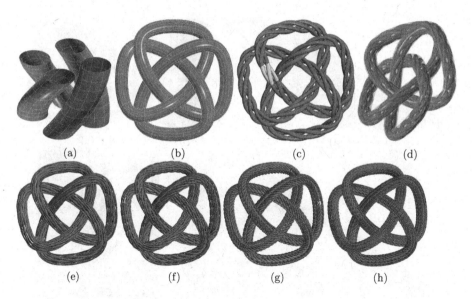

Fig. 4. A tile consisting of four B-spline helical-looking surfaces (a) is paved $(1 \times 1 \times 47)$ times in the domain of a B-spline trivariate of degrees $(3 \times 3 \times 3)$ and lengths $(4 \times 4 \times 50)$ in the shape of a knot (b), resulting in (c) (note one tile is highlighted in cyan). (d) shows a different view of the same final result, embedded in the transparent knot trivariate. (e) to (h) present different results using increasingly higher resolutions of pavements, with one tile highlighted in cyan. (Color figure online)

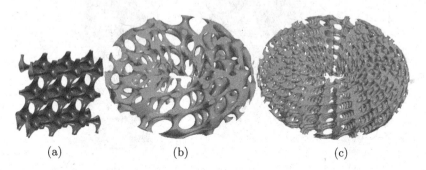

Fig. 5. A fairly complex polygonal periodic tile consisting of around 20k polygons in (a) is paving the domain of a torus trivariate using two different resolutions in (b) and (c). See also Fig. 9(b).

a fairly complex tile is exploited. The polygonal tile, in Fig. 5(a), is paving the domain of a torus trivariate in Fig. 5(b) and (c), using two different pavement's resolutions.

Because of the capability to conduct trivariate-trivariate composition, a closure is formed. A recursive application of the composition operator may be performed, and Fig. 6 demonstrates this ability. The wing model, in Fig. 6(a), is

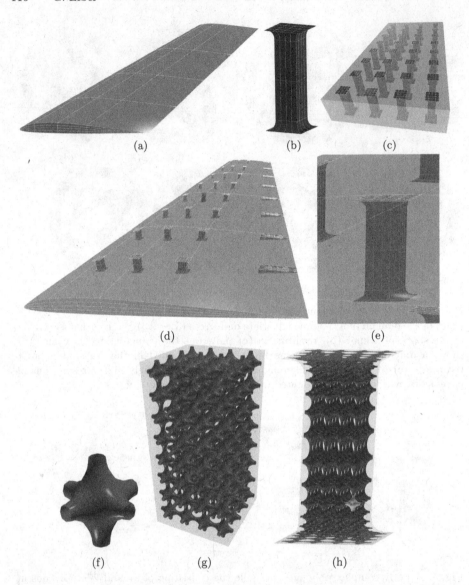

Fig. 6. A recursive application of the composition operator. The trivariate model of
the wing in (a) is trivariate-trivariate composed $(8 \times 4 \times 1)$ times with the vertical
support trivariate pillar shown in (b). The tiling in the domain of the wing is shown
in (c) and the result of the composition is shown in (d). Now each of these pillars (e),
which is a zoom-in on (d), is, in turn, surface-trivariate composed with a polygonal
surface tile (f). The tiling in the domain of the pillar is shown in (g) and the result
of the composition is shown in (h). Note one tile is highlighted in cyan, in (h). (Color
figure online)

constructed as a trivariate ruled volume. The supporting pillar structure, in Fig. 6(b), is another trivariate volume that is constructed by lofting along a set of square surfaces of different sizes. By composing the support pillar trivariate tile into the wing domain $(8 \times 4 \times 1)$ times (wing's domain is shown in Fig. 6(c)), 32 deformed pillars, shown in Fig. 6(d), result. Note the 32 composed pillars are all differently shaped trivariates. In Fig. 6(e), a zoom-in on one of these pillars is shown, only to recursively compose the polygonal surface tile, in Fig. 6(f), in the pillar trivariate $(8 \times 4 \times 4)$ times (pillar's domain is shown in Fig. 6(g)). The final, two-levels composition, result is shown in Fig. 6(h), with one tile highlighted in cyan.

This surface-trivariate (and trivariate-trivariate) approach can also serve to handle trimmed geometry. Herein, the tensor product geometry undergoes composition whereas the trimming information is simply propagated along, as the domain(s) of the surface(s) (or trivariate(s)) in the tile is (are) not affected. Figure 7 shows one example where three concentric trimmed through-cylinder B-spline surfaces are serving together as a tile that is composed $(4 \times 4 \times 2)$ times in the mapping trivariate.

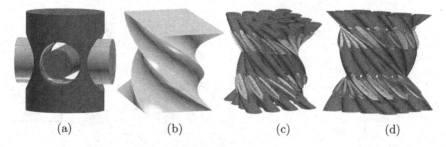

(a)   (b)   (c)   (d)

Fig. 7. Three concentric trimmed through-cylinder surfaces are serving as a tile (a) that is composed $(4 \times 4 \times 2)$ times in the trivariate shown in (b). The result is shown from two somewhat different views, in (c) and (d).

We now present another example that exploits surface-surface and surface-trivariate compositions, in two composition levels, toward the precise modeling of composite materials. The domain of the B-spline duck surface shown in Fig. 8(a) is tiled in Fig. 8(b) with parallel strip surfaces that are composed with the duck surface to yield Fig. 8(c). The strip surface in Fig. 8(c) is then offset a bit to yield some real thickness and a trivariate strip is formed as a ruled volume between the original strip and its offset, and is shown in Fig. 8(d). Finally, the trivariate strip, from Fig. 8(d), is populated with tiles in the shape of stitches to model the internal stitched fibers, as is shown in Fig. 8(e) and (f) (a zoom-in).

Finally, and as a testimony for the viability of the constructed models and their watertightness, Fig. 9 presents two of the presented examples 3D-printed using additive manufacturing.

While in this work we focused on surface-trivariate and trivariate-trivariate composition, we like to add that curve-trivariate composition can also play a

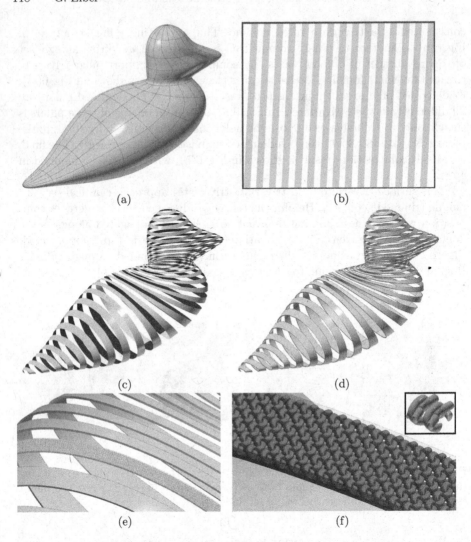

Fig. 8. A two level composition. The periodic strip in (b) is composed into the duck surface in (a) to yield (c). By offsetting the strip surface in (c) and ruling a volume between the strip and its offset, (d) is formed. The trivariate strip in (d) is made transparent (to enable the visibility of its interior) and then composed with a stitches like tile as a second composition, yielding (e) (and a zoom-in, in (f), with one tile enlarged), modeling a composite strip with detailed stitched fibers.

role in the placement of fibers, in the modeling of composites. Curves can be embedded in the domain of the mapping trivariate, $\mathcal{T}$, only to be mapped to Euclidean space via curve-trivariate composition. Because $\mathcal{T}$ is unlikely to be an isometric mapping, one might be required to compensate for the distances between adjacent curves in the domain of $\mathcal{T}$ so the mapped curves are more

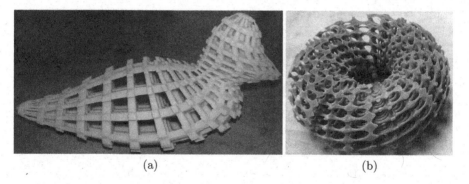

(a)   (b)

Fig. 9. A 3D printed porous duck (a) from Fig. 2 and a porous torus (b) from Fig. 5. Printing courtesy of Stratasys Israel.

equally spaced in Euclidean space. Figure 10 shows one such example, where helical curves are mapped through $\mathcal{T}$ to yield the Euclidean space curves. The result is a set of univariates, through which, one can, for example, sweep any cross section to precisely yield fibers of that cross section.

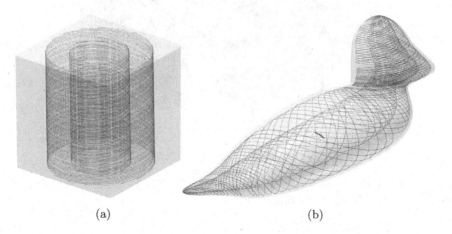

(a)   (b)

Fig. 10. Two sets of concentric helical curves are embedded in the domain of the trivariate (a) only to be curve-trivariate composed to yield the result shown in (b), in a trivariate duck model.

We have already shown that trivariate-trivariate composition is viable. One can also use trivariate-trivariate composition to construct volumetric micro-structures toward Isogeometric Analysis (IGA). Figure 11 shows two structures of trivariate B-spline functions, created using the presented trivariate-trivariate composition. Each trivariate in Fig. 11 is differently colored.

Finally, Fig. 12 shows a small-deformation linear elasticity analysis of an isotropic material (young modulus $E = 1$ and Poisson coefficient $nu = 0.3$),

(a)   (b)

Fig. 11. Two examples of micro-structures consisting of trivariate B-spline functions (each trivariate in a different color) that were created using trivariate-trivariate composition, possibly toward Isogeometric Analysis (IGA). In (a), 168 trivariate Bézier elements of orders $(4, 4, 7)$ are shown and in (b), 336 trivariate Bézier elements of orders $(5, 5, 9)$ are presented. (Color figure online)

Fig. 12. The result of elasticity isogeometric analysis for the micro-structures from Fig. 11(b), using the igatools library [20].

using the igatools library [20], for the micro-structure in Fig. 11(b). The boundary conditions of the problem include a lower internal ring that is completely blocked and a Dirichlet boundary condition that is applied for the top faces of the upper external ring, that is moved vertically a quantity 0.5, but cannot move horizontally. This IGA problem was solved in less than two minutes, on a modern PC workstation.

4   Discussion, Limitations, and Future Work

The complexity of the result deserves some considerations. The composition operations in Fig. 1 employed a tile that consists of a cubic curve, a cubic by quadratic surface, and a trivariate of degrees $(3, 3, 2)$. The mapping function was a trivariate Bézier of degrees $(1, 1, 3)$.

A direct mapping of control points will ends up with the same degrees as the input. On the other hand, the smooth result, of Fig. 1(d), of the precise composition mapping, resulted in deformed curves of degree 15, deform surfaces of degrees $(15, 10)$ and deformed trivariates of degrees $(15, 15, 10)$.

Recall Eq. (1). Given a trivariate $\mathcal{T}$ of degrees (n_1, n_2, n_3) and a surface S of degrees (m_1, m_2), the degrees of the composition surface $\mathcal{T}(S)$ are

$$(n_1 m_1 + n_2 m_1 + n_3 m_1, n_1 m_2 + n_2 m_2 + n_3 m_2). \tag{7}$$

For the surface case of Fig. 1, Eq. (7) indeed yields the degrees of

$$(1 \times 3 + 1 \times 3 + 3 \times 3, 1 \times 2 + 1 \times 2 + 3 \times 2) = (15, 10).$$

Clearly, the degrees can be higher and as a second example, in Fig. 4, the tile consists of four surfaces of degrees (3×3), and the trivariate knot is of degrees $(3 \times 3 \times 3)$. The composed surface elements are of degrees (27×27) (and lengths (109×163)). That said, and beyond the computation overhead that these high degrees poses, we have detected no instability difficulties for degrees below one hundred, including with rational forms. Rational input will yield rational output but in the same degrees as the polynomials case. Figure 13 exemplifies this observed stability. Given a unit cube polynomial trivariate, T, of a tri-order d and a rational cylinder surface of unit size, S, of bi-order d, we examined the error in the circularity of $T(S)$ (measured as the distance deviation from the axis of the cylinder) as a function of d. As can be seen in the figure, the error remains very small and its growth is approximately linear with the orders.

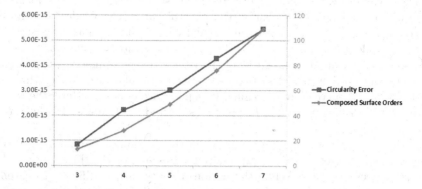

Fig. 13. Given a unit cube polynomial trivariate, T, of a tri-order d and a rational cylinder surface of unit size, S, of bi-order d, the error in the circularity of $T(S)$ is examined and shown in red as a function of d. Also shown is the bi-order of $T(S)$ in blue. (Color figure online)

Since the geometry that is synthesized in this work is typically deformed, the need for precise rational circular arcs is diminished. While we do support compositions of rational forms, in all other presented examples in this work, polynomial forms were used.

All the micro-structure models presented in this work were created in seconds to minutes. Measured on an 2.8 GHz $i7$ laptop running Windows 10, the example in Fig. 2 synthesized $256 = (4 \times 4 \times 16)$ tiles (each consisting of 6 bilinear surfaces) in a little over a minute. The example in Fig. 4(c) synthesized 47 tiles (each consisting of 4 helical surfaces) in around 20 s. Finally, the example in Fig. 11(b) synthesized 48 tiles (each consisting of 7 trivariates, 336 Bézier elements in all) in around 5 s.

Both inputs, the tile and/or the mapping trivariate, $\mathcal{T}$, can be singular, in which case the constructed micro-structure is likely to be singular as well. However, verifying the regularity of either the tile or $\mathcal{T}$ is fairly simple by computing and bounding the magnitude of their Jacobian. For $\mathcal{T}$, this amounts to the (spline) product of:

$$|J| = \left\langle \frac{\partial \mathcal{T}}{\partial x} \times \frac{\partial \mathcal{T}}{\partial y}, \frac{\partial \mathcal{T}}{\partial z} \right\rangle,$$

and verifying that $|J|$ never vanish, for example, by verifying that all the coefficients of $|J|$ are of the same sign.

In all examples presented and due to $\mathcal{T}$, different local scale factors are applied to different tiles in the output geometry. This clear limitation stems from the fact that the deformation trivariate function is rarely isometric. One can only establish bounds on the different scale factors, by computing the field of the first fundamental form [2] of the deforming trivariate, as spline functions, and bound their range.

Because every tile is likely to be deformed a bit differently from other tiles, if some local properties are to be preserved, some extra measures must be taken into considerations. Consider the example in Fig. 8. While the original tile could have been synthesized using a highly accurate circular cross section, the mapped/deformed result in Fig. 8(f) is likely to violate that circularity. However, one can deform an axial curve of the tile's circular geometry, using curve-trivariate composition (i.e. Fig. 10), only to sweep a circular cross section through the deform axial curve, using a regular sweep surface operator. Here, the result will be stitches that are circular as precise as desired (and allowed by the sweep operation).

In this work, we uniformly paved tiles in the domain of the mapping FFD. However, and as already stated, especially for B-spline FFDs, this can result in unequally stretched tiles. While the notion of arc-length is difficult to extend from curves to surfaces and trivariates, one can still devise a scheme to try to equalize the stretch in the mapping $\mathcal{T}$, via a domain reparametrization, before the paving process takes place. Indeed, the internal parameterization of $\mathcal{T}$ can have a grand affect on the distribution of the tiles in the structure and is a degree of freeform to further investigate and employ.

Having fairly complex tiles, paving and mapping them numerous times can require an intense amount of memory (and computing power). While unavoidable at times, one can consider a lazy synthesis of the geometry on the fly and as needed in real time, possibly with the help of parallel computing. Further, one can take advantage of the inherent hierarchy and given local geometric operations (like slicing), converge rapidly and process/synthesize only these micro-structure's tiles that are active and affect the local geometry (intersect with the slicing plane). Similarly, if tiles are (trimmed) freeform shapes, their tessellation into polygons might also be done on the fly and by demand, again with the possible help of GPUs or parallel computing. This potential difficulty of large memory and computing needs is likely to play a role also in analysis of porous geometry, while part of the difficulties in the analysis might be alleviated via homogenization of the structure [3].

In all examples, by definition, the deforming function was a tensor product trivariates and the tiles were cuboids aligned along the main axes, in a cube-like topology. In [16], use of trimmed trivariates was already proposed and herein, in order to support $\mathcal{T}$ mapping using trimmed trivariates, tiles that intersect with the trimming domain must be properly pruned or sealed. Further, considering an interior trimming boundary in some volumetric model, between two trimmed trivariates must also be addressed, following [16]. Conceptually, there might be no need to prune tiles that are completely inside the volumetric model, even when they cross interior trimming boundaries, while matching the boundaries of the tiles along these interior trimming boundaries will be required. Proper, whatever that means, treatment of tiles in such a trimmed volumetric environment is still an open question, including the proper management of continuity across interior trimming surfaces.

Non-tensor product FFDs were already proposed, for example, in [4], as Extended FFDs. Use of Extended FFDs or other mappings, instead of the tensor product trivariates employed herein, will allow one to support micro-structures that are not necessarily of cuboid topology. Further, the tiles themselves are not confined to cube-like topology, and any tile that periodically paves 3-space can be used. Examples include hexagonal prisms or tetrahedra tiles, possibly embedded in a large hexagonal prism or tetrahedra deformation function. Alternatives to tensor product trivariates should be explored as mapping function and those can include splines over general triangulations or box-splines. Moreover, any tiling of 3-space can be used and semi-regular tiling, where two (or more) differently shaped tiles are employed together, is another example.

The presented micro-structures' construction scheme can be further refined and improved in additional directions. Attributes like colors or texture can be mapped to the resulting geometry where the attributes' specifications can either be local, coming from the tile itself and repeated for all tiles, or be global as a specification over the mapping trivariate.

Herein, the same tile was used throughout the pavement of a deforming trivariate. Alternatively, one can select each tile out of a (predetermined or created on the fly) random (set of) tile, resulting in a randomly looking porous geometry. Further, if certain physical constraints apply, such as local stress fields, the synthesized tiles can obey such constraints and locally adapt their shape to optimally satisfy these constraints while preserving continuity conditions between tiles. The preservation of continuity can be performed incrementally and on the fly, by using, for random tile at indices (ijk), the boundary conditions of previously constructed random tiles $(i-1, j, k)$, $(i, j-1, k)$, and $(i, j, k-1)$, if any, along their shared boundaries.

5   Conclusions

In this work, we have presented a purely geometric modeling constructor for the synthesis of complex, porous or micro-structure objects, by drawing from FFD techniques, possibly recursively. We have presented constructors' using tiles that are either polygonal or spline based, as curves or (trimmed) surfaces and (trimmed) trivariates. If the input geometry is precise in the form of B-spline curves, (trimmed) surfaces, or (trimmed) trivariates, the mapped output will be precise as well, to within machine precision, also in the same form of B-spline shapes, albeit typically of higher degrees. The successful utilization of the presented micro-structure modeling constructor in application areas such as bio-engineering, mission critical engineering, or alternatively jewelry design, a few fields we already foresee as viable, or other areas, is yet to be seen.

All the implementation, include source code, is available as part of the IRIT geometric modeling environment (http://www.cs.technion.ac.il/~irit). Further, most micro-structure examples presented in this work are available, as obj (for polygons) and iges (for surfaces) files, in http://www.cs.technion.ac.il/~gershon/site/modeling.html.

Acknowledgments. This research was supported in part by the ISRAEL SCIENCE FOUNDATION (grant No. 278/13). I also like to thank Boris van Sosin for his help in implementing the trivariate-trivariate composition operator.

The IGA of the model in Fig. 12 has been performed with the help of Pablo Antolin (EPFL Lausanne), Annalisa Buffa (EPFL Lausanne and IMATI-CNR Pavia), Massimiliano Martinelli (IMATI-CNR Pavia); Giancarlo Sangalli (University of Pavia and IMATI-CNR Pavia)

References

1. Armillotta, A., Pelzer, R.: Modeling of porous structures for rapid prototyping of tissue engineering scaffolds. Int. J. Adv. Manuf. Technol. **39**, 501–511 (2008)
2. Carmo, M.P.D.: Differntial Geometry of Curves and Surfaces. Prentice-Hall, Englewood Cliffs (1976)
3. Chen, J., Shapiro, V.: Optimization of continuous heterogeneous models. In: Pasko, A., Adzhiev, V., Comninos, P. (eds.) Heterogeneous Objects Modelling and Applications. LNCS, vol. 4889, pp. 193–213. Springer, Heidelberg (2008). doi:10.1007/978-3-540-68443-5_8
4. Coquillart, S.: Extended free-form deformation: a sculpturing tool for 3D geometric modeling. In: Proceedings of the 17th Annual Conference on Computer Graphics and Interactive Techniques, vol. 24, pp. 187–196. ACM Press, August 1990

5. Coquillart, S., Jancne, P.: Animated free-form deformation: an interactive animation technique. In: Proceedings of the 18th Annual Conference on Computer Graphics and Interactive Techniques, vol. 25, pp. 23–26. ACM Press, July 1991
6. DeRose, T.D., Goldman, R.N., Hagen, H., Mann, S.: Functional composition algorithms via blossoming. ACM Trans. Graph. **12**(2), 113–135 (1993)
7. Elber, G.: Free form surface analysis using a hybrid of symbolic and numerical computation. Ph.D. thesis, University of Utah (1992)
8. Elber, G.: Geometric deformation-displacement maps. In: The Tenth Pacific Graphics, pp. 156–165, October 2002
9. Elber, G.: Geometric texture modeling. IEEE Comput. Graph. Appl. **25**(4), 66–76 (2005)
10. Elber, G.: Constructing porous geometry. In: FASE 2016, June 2016
11. Ellens, M.S., Cohen, E.: An approach to C^{-1} and C^0 feature lines. In: Mathematical Methods for Curves and Surfaces, pp. 121–132 (1995)
12. Farin, G.: Curves and Surfaces for Computer Aided Geometric Design. Academic Press Professional, Boston (1993)
13. Farouki, R.T., Rajan, V.T.: Algorithms for polynomials in Bernstein form. Comput. Aided Geom. Des. **5**(1), 1–26 (1988)
14. Griessmair, J., Purgathofer, W.: Deformation of solids with trivariate B-splines. Eurograpghic **89**, 137–148 (1989)
15. MacCracken, R., Joy, K.I.: Free-form deformations with lattices of arbitrary topology. In: Proceedings of the 23rd Annual Conference on Computer Graphics and Interactive Techniques, pp. 181–188. ACM Press (1996)
16. Massarwi, F., Elber, G.: A B-spline based framework for volumetric object modeling. Comput. Aided Des. **78**, 36–47 (2016)
17. Morken, K.: Some identities for products and degree raising of splines. Constr. Approx. **7**(1), 195–208 (1991)
18. Paik, K.L.: Trivariate B-splines. MSc. Department of Computer Science, University of Utah (1992)
19. Pasko, A., Fryazinov, O., Vilbrandt, T., Fayolle, P.-A., Adzhiev, V.: Procedural function-based modelling of volumetric microstructures. Graph. Models **73**(5), 165–181 (2011)
20. Pauletti, M.S., Martinelli, M., Cavallini, N., Antolin, P.: Igatools: an isogeometric analysis library. SIAM J. Sci. Comput. **37**(4), 465–496 (2015)
21. Piegl, L., Tiller, W.: The NURBS Book. Springer, Heidelberg (1997)
22. Porumbescu, S.D., Budge, B., Feng, L., Joy, K.I.: Shell maps. ACM Trans. Graph. **24**(3), 626–633 (2005)
23. Schein, S., Elber, G.: Discontinuous free form deformations. In: Proceedings of Pacific Graphics 2004, pp. 227–236 (2004)
24. Schroeder, C., Regli, W.C., Shokoufandeh, A., Sun, W.: Computer-aided design of porous artifacts. Comput. Aided Des. **37**, 339–353 (2005)
25. Sederberg, T.W., Parry, S.R.: Free-form deformation of solid geometric models. Comput. Graph. **20**, 151–160 (1986)
26. Xiao, F., Yin, X.: Geometry models of porous media based on Voronoi tessellations and their porosity-permeability relations. Comput. Math. Appl. **72**, 328–348 (2016)

Partially Nested Hierarchical Refinement of Bivariate Tensor-Product Splines with Highest Order Smoothness

Nora Engleitner, Bert Jüttler$^{(\boxtimes)}$, and Urška Zore

Institute of Applied Geometry, Johannes Kepler University, Linz, Austria
{nora.engleitner,bert.juettler}@jku.at

Abstract. The established construction of hierarchical B-splines starts from a given sequence of nested spline spaces. In this paper we generalize this approach to sequences formed by spaces that are only partially nested. This enables us to choose from a greater variety of refinement options while constructing the underlying grid. We identify assumptions that allow to define a hierarchical spline basis, to establish a truncation mechanism, and to derive a completeness result. Finally, we present an application to surface approximation that demonstrates the potential of the proposed generalization.

Keywords: Tensor-product B-splines · Hierarchical B-splines · Adaptive refinement

1 Introduction

Hierarchical tensor-product B-splines are one of the major approaches to perform local refinement in geometric modeling and isogeometric analysis, besides splines defined by control meshes with T-junctions (T-splines), locally refined (LR) splines and polynomial splines over hierarchical T-meshes (PHT-splines). See [3,16–18] and the references therein for more information on the latter three.

Hierarchical spline refinement can be traced back to the work of Forsey and Bartels [6] on surface design using locally defined control meshes. Based on a selection mechanism, a system of basis functions spanning the resulting hierarchical spline space was established by Kraft in his PhD thesis [14]. Another basis, which consists of *truncated* hierarchical B-splines, possesses improved properties (increased locality, partition of unity and strong stability) and has been established more recently [8]. Its properties regarding stability, completeness and approximation power have been analyzed in greater detail [9,19,22].

Hierarchical B-splines have found numerous applications due to their good mathematical properties. They were used for surface reconstruction in Computer-Aided Design [10,12]. Additionally, they were employed for performing numerical simulations using the powerful framework of isogeometric analysis [1,2,15,20]. The recent article [7] discusses the potential of the truncated basis for geometric design and isogeometric analysis.

© Springer International Publishing AG 2017
M. Floater et al. (Eds.): MMCS 2016, LNCS 10521, pp. 126–144, 2017.
https://doi.org/10.1007/978-3-319-67885-6_7

In addition to the work on applications, several authors proposed various extensions and generalizations of the hierarchical construction. These include extensions to Powell-Sabin splines [21], box splines and doubly hierarchical splines as instances of more general generating systems [24], B-splines on triangulations [11], hierarchical T-splines [5], and functions defined by subdivision algorithms [23,25].

The established construction of hierarchical B-splines starts from a given sequence of nested spline spaces. As a consequence, if the refinement process inserts knot lines at some level, then they will automatically be present at all higher levels, even if they are not needed in all parts of the domain. This may lead to an unnecessary increase of the number of degrees of freedom. It should be noted that this limitation is not present when using alternative constructions such as T-splines, LR splines or PHT-splines.

In order to overcome the limitation caused by the sequential nature of hierarchical B-spline refinement, while maintaining their good mathematical properties, we extend the construction to sequences formed by spaces that are only *partially* nested. The proposed generalization enables us to choose from a greater variety of refinement options while constructing the underlying grid. This additional flexibility is potentially useful when designing surfaces that possess creases or similar features, and a related technique has been developed in the context of subdivision surface modeling [13]. It might also open new perspectives for adaptivity in isogeometric analysis by providing the opportunity to use different refinement techniques (such as h- versus p-refinement) in different parts of the computational domain.

In order to keep the presentation simple, in this paper we limit ourselves to the discussion of partially nested refinement for bivariate spline spaces of uniform degrees. We identify a number of assumptions that enable the definition of a hierarchical spline basis, of a truncation operation to obtain the partition of unity property, and the derivation of a completeness result.

The remainder of the paper consists of seven sections. We describe the framework of our construction in the next section and establish a hierarchical spline basis in Sect. 3. We then derive a characterization of the space spanned by the basis and adapt the definition of the truncation operation to the non-nested setting in the next two sections. The completeness properties of the basis are analyzed in Sect. 6. We then present an application to least-squares approximation that demonstrates the power of the new construction before concluding the paper with suggestions for future work.

2 Preliminaries

We consider a finite sequence of bivariate tensor-product spline spaces

$$V^\ell = \mathrm{span} B^\ell, \quad \ell = 1, \dots, N,$$

which are spanned by spline bases B^ℓ. The upper index ℓ will be called the *level*. Each of the spline spaces is defined on the open unit square $(0,1)^2$.

The spline bases B^ℓ consist of tensor-product B-splines that are defined by two open knot vectors with boundary knots 0 and 1. We consider a uniform polynomial degree $\mathbf{p} = (p_x, p_y)$ and use only single knots except for the boundary knots that have multiplicity $p_x + 1$ and $p_y + 1$, respectively. The supports of the basis functions are axis-aligned boxes in $(0, 1)^2$.

We use the subspace relation to restrict the natural ordering of the levels to a partial ordering. We say that *level k precedes level ℓ*, denoted by $k \prec \ell$, if k is less than ℓ and V^k is a subspace of V^ℓ, i.e.

$$k \prec \ell \quad \Leftrightarrow \quad k < \ell \text{ and } V^k \subseteq V^\ell. \tag{1}$$

The spaces are not necessarily nested. If they are, however, then the finer space is assumed to have the higher level, i.e.

$$V^k \subset V^\ell \Rightarrow k \prec \ell. \tag{2}$$

Any finite sequence of spline spaces can be re-ordered such that this condition is satisfied.

We present an example that will be used throughout the paper to illustrate the discussion of notions and results.

Example. We consider C^1-smooth biquadratic tensor-product spline spaces ($p_x = p_y = 2$) on dyadically refined knots,

$$D^{r,s} = S^2\left(0, 0, 0, \frac{1}{2^r}, \ldots, \frac{2^r - 1}{2^r}, 1, 1, 1\right) \otimes S^2\left(0, 0, 0, \frac{1}{2^s}, \ldots, \frac{2^s - 1}{2^s}, 1, 1, 1\right),$$

where S^2 denotes the univariate spline space defined by a given knot sequence, with positive integers r, s. Among them we use the spaces

$$V^1 = D^{3,3}, \ V^2 = D^{4,3}, \ V^3 = D^{3,4},$$
$$V^4 = V^5 = D^{4,4}, \ V^6 = D^{5,4}, \ V^7 = D^{4,6}, \tag{3}$$

which define the partial ordering

$$
\begin{array}{ccc}
2 & & 6 \\
\nwarrow \ \nearrow & & \nearrow \\
1 & 4 \prec 5 & \\
\searrow \ \nearrow & & \searrow \\
3 & & 7
\end{array}
\tag{4}
$$

of the seven levels. $\Diamond$

The functions in all spline spaces V^ℓ are $C^{\mathbf{s}}$-smooth on $(0, 1)^2$, where the order of smoothness is given by

$$\mathbf{s} = (p_x - 1, p_y - 1). \tag{5}$$

More precisely, they possess continuous partial derivatives of order $p_x - 1$ and $p_y - 1$ with respect to x and y, respectively. We shall denote the set of all functions on an open subset $X \subseteq (0, 1)^2$ with this smoothness as $C^{\mathbf{s}}(X)$.

In addition to the spline spaces we consider an associated sequence of open sets

$$\pi^\ell \subseteq (0,1)^2, \quad \ell = 1, \ldots, N,$$

which will be called *patches*. We assume that these are mutually disjoint,

$$\pi^\ell \cap \pi^k \neq \emptyset \Rightarrow \ell = k.$$

We use the closures $\overline{\pi^\ell}$ of the patches to define the *domain*

$$\Omega = \text{int}\left(\bigcup_{\ell=1}^N \overline{\pi^\ell}\right) \subseteq (0,1)^2.$$

The part of the boundary of each patch that is shared with patches of a *lower* level,

$$\Gamma^\ell = \bigcup_{k=1}^{\ell-1} \overline{\pi^k} \cap \overline{\pi^\ell},$$

is called the *constraining boundary* of the patch π^ℓ. Note that the constraining boundary may be empty. In particular we have $\Gamma^1 = \emptyset$.

Example. We consider again the spaces (3), which are defined by the dyadically refined knot vectors. Figure 1a visualizes an associated sequence of patches, which defines a subdivision of the domain Ω. In this case, the domain is also the unit square. Additionally, Fig. 1b shows the knot lines of the spline spaces within each patch.

$\Diamond$

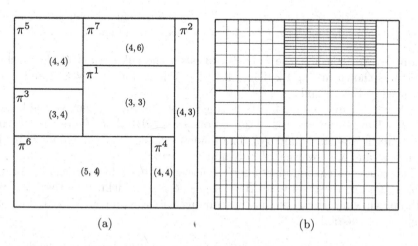

(a)   (b)

Fig. 1. The subdivision of the domain into patches (a). The numbers (r, s) in each patch specify the dyadically refined knot sequences that define the associated spline spaces. The corresponding partially nested hierarchical mesh (b).

We conclude this section by defining the *partially nested hierarchical spline space*

$$H = \{s \in \mathcal{C}^{\mathbf{s}}(\Omega) : s|_{\pi^\ell} \in V^\ell|_{\pi^\ell} \ \forall \ell = 1, \ldots, N\}. \tag{6}$$

It consists of all the $\mathcal{C}^{\mathbf{s}}$-smooth functions with the property that their restrictions $s|_{\pi^\ell}$ to the patches are contained in the associated spline spaces $V^\ell|_{\pi^\ell}$. In particular, the space of tensor-product polynomials of degree $\mathbf{p}$, restricted to the domain Ω, is a subspace of H.

3   Basis Functions

We define the basis by a *selection procedure*, which generalizes the definition of Kraft's basis for hierarchical B-splines. This procedure selects elements of each spline basis B^ℓ. Among all B-splines that do not vanish on the patch π^ℓ, we select the ones that take zero values on the constraining boundary Γ^ℓ of that patch, i.e.,

$$K^\ell = \{\beta^\ell \in B^\ell : \beta^\ell|_{\pi^\ell} \neq 0 \quad \text{and} \quad \beta^\ell|_{\Gamma^\ell} = 0\}.$$

Each set K^ℓ of selected functions defines the *shadow* of the associated patch π^ℓ,

$$\hat{\pi}^\ell = \operatorname{supp} K^\ell = \bigcup_{\beta^\ell \in K^\ell} \operatorname{supp} \beta^\ell.$$

We collect the selected B-splines of all levels into the set

$$K = \bigcup_{\ell=1}^{N} K^\ell. \tag{7}$$

We will denote this set of functions as *PNHB-splines*, since it consists of *hierarchical B-splines* defined by a *partially nested* sequence of spline spaces.

Example. We consider the PNHB-splines on the subdivision of the domain which was shown in Fig. 1a. The selected functions for the levels 2 and 6 are visualized in Fig. 2a and b.

The constraining boundary of π^2 consists of the line segment on the border with π^1. The set K^2 consists of 30 tensor-product B-splines (note the Greville points on the domain boundary). The shadow defined by them extends into the patches π^4 and π^7, covering π^4 fully and π^7 partially.

The constraining boundary of π^6 consists of three line segments. The set K^6 contains 144 tensor-product B-splines. The shadow defined by them is equal to the patch, since the only non-constraining patch boundary is located on the boundary of the domain Ω.   ◊

The following condition is essential for proving the linear independence of the PNHB-splines:

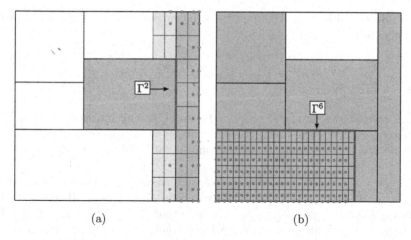

(a)   (b)

Fig. 2. Constraining boundaries (dark blue line segments), shadows (blue and light blue) and selected basis functions (represented by their Greville points, which are shown as red dots) of the patches (shown in blue) π^2 (a) and π^6 (b) for the domain subdivision shown in Fig. 1. Patches of lower levels are shown in green. For the latter patch, the shadow is equal to the patch itself. (Color figure online)

Assumption. If the shadow $\hat{\pi}^\ell$ of the patch of level ℓ intersects another patch π^k of a different level k, then the first level is lower than the second one,

$$\hat{\pi}^\ell \cap \pi^k \neq \emptyset \quad \Rightarrow \quad \ell \leq k. \tag{SOA}$$

This will be called the *Shadow Ordering Assumption* (SOA).

We will use this assumption in the remainder of the paper. Since we will make several further assumptions throughout the paper, we provide Table 1 containing their names and acronyms, in order to guide the reader.

Table 1. Assumptions and acronyms.

Name	Acronym	Defined on page
Shadow Ordering Assumption	SOA	5
Shadow Compatibility Assumption	SCA	7
Constraining Boundary Alignment	CBA	8
Full Boundary Alignment	FBA	14
Support Intersection Condition	SIC	15

SOA enables us to obtain our first result:

Theorem 1. *The PNHB-splines are linearly independent on Ω if SOA holds.*

Proof. The proof of linear independence follows an idea originally formulated in [14], see also [8]. However, we will repeat it here in order to keep this paper self-contained and in order to adapt it to the current setting. We need to prove the implication

$$\sum_{\ell=1}^{N} \sum_{\beta^{\ell} \in K^{\ell}} d_{\beta^{\ell}} \beta^{\ell} = 0 \ \Rightarrow \ d_{\beta^{\ell}} = 0 \ \forall \beta^{\ell} \in K^{\ell} \ \forall \ell = 1, \dots, N. \tag{8}$$

We first restrict the sum in (8) to π^1. Due to SOA only functions $\beta^1 \in K^1$ are non-zero on π^1. The local linear independence of the B-splines B^1 gives $d_{\beta^1} = 0$ for all $\beta^1 \in K^1$. This implies that the sum in (8) involves only functions with $\ell > 1$.

We now consider the restriction of the sum to π^2. Again, according to SOA only the functions $\beta^2 \in K^2$ take non-zero values there. As the B-splines in B^2 are locally linearly independent, we conclude that $d_{\beta^2} = 0$ for all $\beta^2 \in K^2$. By repeatedly using the above argument, we eventually exhaust all the terms in (8), which concludes the proof of linear independence.   □

While the selection mechanism and SOA guarantee linear independence, they do not ensure that the spline space spanned by PNHB-splines contains a class of functions that guarantees certain approximation properties, such as tensor-product polynomials of degree $\mathbf{p}$. This is shown in the following example:

Example. We consider the two biquadratic spline spaces

$$V^1 = D^{3,0}, \ V^2 = D^{1,1} \tag{9}$$

and the two patches

$$\pi^1 = (0, \tfrac{1}{2}) \times (0, 1), \quad \pi^2 = (\tfrac{1}{2}, 1) \times (0, 1). \tag{10}$$

The first set K^1 of selected functions consists of 18 tensor-product B-splines and defines the shadow $\hat{\pi}^1 = (0, \tfrac{3}{4}) \times (0, 1)$. The second set K^2 of selected functions contains only 4 functions. The functions in $K^1 \cup K^2$ are linearly independent but cannot represent any biquadratic function on Ω. This can be seen easily by analyzing the space which is spanned by the 4 functions in K^2 and noting that only these functions take non-zero values on $(\tfrac{3}{4}, 1) \times (0, 1)$.   ◇

4   The Spline Space

Consequently, we need to introduce further assumptions. We replace SOA by a stronger condition, which will be used in the remainder of this paper.

Assumption. If the shadow $\hat{\pi}^{\ell}$ of the patch of level ℓ intersects another patch π^k of a different level k, then the first level precedes the second one,

$$\hat{\pi}^{\ell} \cap \pi^k \neq \emptyset \ \Rightarrow \ \ell = k \ \text{ or } \ \ell \prec k. \tag{SCA}$$

This will be called the *Shadow Compatibility Assumption* (SCA).

In other words, the shadow $\hat{\pi}^\ell$ intersects only patches that correspond to spaces containing V^ℓ as a subspace.

Example. We consider again the situation shown in Figs. 1 and 2. The shadow $\hat{\pi}^2$ intersects π^4 and π^7. SCA is satisfied since $2 \prec 4$ and $2 \prec 7$, see (4). $\lozenge$

This condition obviously implies (SOA). However, it turns out that SCA does not yet suffice to prove that the space spanned by the PNHB-splines contains a class of functions, which would guarantee certain approximation properties (e.g. polynomials). We need to impose a condition on the location of the constraining boundaries.

Assumption. For each level ℓ, the constraining boundary Γ^ℓ of the patch π^ℓ is aligned with the knot lines of the spline space V^ℓ. This will be called the *Constraining Boundary Alignment* (CBA) condition.

More precisely, the constraining boundary Γ^ℓ is either empty or is formed by horizontal segments, vertical segments and isolated vertices, where not all these features need to be present. We assume that the segments are all located on knot lines of V^ℓ and that the vertices are intersections of knot lines.

We will use both assumptions CBA and SCA in the remainder of the paper. Under these assumptions we can characterize the spline space that is generated by the PNHB-splines:

Theorem 2. *The PNHB-splines span the partially nested hierarchical spline space H if both SCA and CBA are satisfied.*

We will need a technical lemma to prove this result. This lemma uses the notion of *homogeneous boundary conditions of order* **s**. A function f is said to satisfy these conditions at a point (x, y) if $(\vartheta f)(x, y) = 0$, where the operator

$$\vartheta = \left(\frac{\partial^i}{\partial x^i} \frac{\partial^j}{\partial y^j} \right)_{i=0,\ldots,s_x; j=0,\ldots,s_y}$$

transforms a function into a matrix of dimension **p** that contains all the partial derivatives up to order **s**. (Note here that 0 denotes the null matrix of dimension **p**, not a scalar.) In particular, this operator contains the evaluation of its argument as its first element.

Lemma 1. *The selected functions of level ℓ span the subspace*

$$\operatorname{span}K^\ell|_{\pi^\ell} = \{ f \in V^\ell : (\vartheta f)|_{\Gamma^\ell} = 0 \}|_{\pi^\ell}$$

of the associated spline space $V^\ell|_{\pi^\ell}$ on the patch π^ℓ, which consists of the restrictions $f|_{\pi^\ell}$ of all functions $f \in V^\ell$ that satisfy homogeneous boundary conditions of order **s** *on the constraining boundary Γ^ℓ, provided that CBA holds.*

Proof. First, we show that all selected functions of level ℓ satisfy the homogeneous boundary conditions of order $\mathbf{s}$ on the constraining boundary Γ^ℓ.

Consider a selected tensor-product B-spline $\beta^\ell \in K^\ell$. None of the points $(x, y) \in \Gamma^\ell$ of the constraining boundary belongs to the interior of the support suppβ^ℓ. This point thus either belongs to the support's boundary, or it is even farther away. The tensor-product B-spline β^ℓ satisfies homogeneous boundary conditions of order $\mathbf{s}$ at (x, y) in both cases, since it is $\mathcal{C}^{\mathbf{s}}$-smooth.

Second, we show that the restriction $f|_{\pi^\ell}$ of any function $f \in V^\ell$, which satisfies the homogeneous boundary conditions of order $\mathbf{s}$ on the constraining boundary Γ^ℓ, can be represented as a linear combination of the selected functions K^ℓ. Obviously, the restriction possesses a representation of the form

$$f(\mathbf{x}) = \sum_{\substack{\beta^\ell \in B^\ell \\ \text{supp}\beta^\ell \cap \pi^\ell \neq \emptyset}} c_{\beta^\ell} \beta^\ell(\mathbf{x}), \quad \mathbf{x} \in \pi^\ell. \tag{11}$$

We consider a function $\beta^\ell \in B^\ell \setminus K^\ell$ that does not vanish on π^ℓ. There exists an isolated vertex $\mathbf{v}$ or a (horizontal or vertical) segment L of the constraining boundary such that β^ℓ takes non-zero values there.

In the case of an isolated vertex, the matrix $(\vartheta f)(\mathbf{v})$ depends on $p_x \times p_y$ spline coefficients due to CBA, and one of them is c_{β^ℓ}. The matrix has the same dimensions, cf. (5), and the linear mapping that transforms the spline coefficients into the matrix elements has full rank, simply because the spline function can take any values of $(\vartheta f)(\mathbf{v})$. Thus we conclude that $c_{\beta^\ell} = 0$ if $(\vartheta f)(\mathbf{v}) = 0$.

In the case of a segment L we choose a (sub-) segment L' which is contained in only one knot span, and consider the tensor-product Bernstein–Bézier (BB) representation of f with respect to a sufficiently small axis-aligned box in π^ℓ with this segment on its boundary. More precisely, this box is chosen such that it is simultaneously located within π^ℓ and in one of the tensor-product knot spans of V^ℓ.

The elements of the matrix $(\vartheta f)|_{L'}$ depend on the $p_x + 1$ columns (each of height p_y) of adjacent BB coefficients for a horizontal segment, and on the $p_y + 1$ rows (each of width p_x) of adjacent BB coefficients for a vertical segment. The matrix is equal to the null matrix on L' if and only if all these BB coefficients are equal to zero.

Due to CBA, these BB coefficients depend on the same number of spline coefficients, and c_{β^ℓ} is one of them. The linear mapping that transforms the spline coefficients into the considered BB coefficients has full rank, since any tensor-product polynomial of degree $\mathbf{p}$ is contained in the spline space V^ℓ. Thus we conclude that $c_{\beta^\ell} = 0$ if $(\vartheta f)|_{L'} = 0$. □

We now proceed with the proof of the Theorem:

Proof (Theorem 2). Given a function $f \in H$, we consider its restriction to the patch of level 1 and find a representation

$$f(\mathbf{x}) = \sum_{\beta^1 \in K^1} c_{\beta^1} \beta^1(\mathbf{x}), \quad \mathbf{x} \in \pi^1. \tag{12}$$

It exists since $f|_{\pi^1} \in V|_{\pi^1}$ according to the definition of H and because the associated constraining boundary is empty. We use this local representation to derive the globally defined level 1 representation

$$f^1(\mathbf{x}) = \sum_{\beta^1 \in K^1} c_{\beta^1} \beta^1(\mathbf{x}), \quad \mathbf{x} \in \Omega.$$

We now proceed by iterating over the remaining levels $\ell = 2, \ldots, N$. In each level, we consider the restriction of

$$f - \sum_{k=1}^{\ell-1} f^k$$

to the patch π^ℓ and its local representation

$$f(\mathbf{x}) - \sum_{k=1}^{\ell-1} f^k(\mathbf{x}) = \sum_{\beta^\ell \in K^\ell} c_{\beta^\ell} \beta^\ell(\mathbf{x}), \quad \mathbf{x} \in \pi^\ell, \tag{13}$$

which leads to the globally defined level ℓ representation

$$f^\ell(\mathbf{x}) = \sum_{\beta^\ell \in K^\ell} c_{\beta^\ell} \beta^\ell(\mathbf{x}), \quad \mathbf{x} \in \Omega. \tag{14}$$

The existence of a local representation (13) with respect to the full basis B^ℓ is guaranteed by $f|_{\pi^\ell} \in V|_{\pi^\ell}$ according to the definition of H, and by using SCA. This confirms that the function on the left-hand side of (13) is contained in $V^\ell|_{\pi^\ell}$. Additionally, we use the fact that

$$f(\mathbf{x}) - \sum_{j=1}^{\ell-1} f^j(\mathbf{x}) = 0, \quad \mathbf{x} \in \pi^k, \ k < \ell, \tag{15}$$

which follows immediately from the definition of f^j. Combining this observation with the C^s-smoothness of f gives the homogeneous boundary conditions of order s on the constraining boundary Γ^ℓ. Finally, these conditions enable us to apply Lemma 1, which confirms that only the selected functions $K^\ell \subseteq B^\ell$ are needed in (13).

We conclude the proof by noting that (15) is satisfied since Eqs. (13) and (14) imply

$$f(\mathbf{x}) - \sum_{k=1}^{\ell-1} f^k(\mathbf{x}) = f^\ell(\mathbf{x}), \quad \mathbf{x} \in \pi^\ell,$$

while SOA (which is implied by SCA) means that increasing the level ℓ does not affect the values on patches of lower levels. Thus, we finally choose $\ell = N+1$ in (15) and arrive at

$$f(\mathbf{x}) = \sum_{k=1}^{N} f^k(\mathbf{x}), \quad \mathbf{x} \in \Omega.$$

$\square$

In particular, this proves that every tensor-product polynomial of degree $\mathbf{p}$ can be represented as a linear combination of PNHB-splines, since these polynomials belong to the partially nested hierarchical spline space.

5   Truncation

We define the *truncated PNHB-splines* by suitably generalizing the truncation mechanism, which has been established in [8]. These functions are linearly independent, form a partition of unity, and span the partially nested hierarchical spline space H.

We consider a fixed level $\ell > 1$ and a function

$$f \in \text{span} \bigcup_{k=1}^{\ell-1} K^k, \tag{16}$$

which is a linear combination of all tensor-product B-splines that have been selected at lower levels. SCA then implies that

$$f|_{\pi^k} \in V^k|_{\pi^k}, \quad k = 1, \dots, \ell,$$

for all levels that do not exceed ℓ. When restricted to the patch π^ℓ, this function possesses a unique local representation

$$f(\mathbf{x}) = \sum_{\substack{\beta^\ell \in B^\ell \\ \text{supp}\beta^\ell \cap \pi^\ell \neq \emptyset}} c_{\beta^\ell} \beta^\ell(\mathbf{x}), \quad \mathbf{x} \in \pi^\ell, \tag{17}$$

as a linear combination of tensor-product B-splines in B^ℓ. We now define the *truncation of f with respect to K^ℓ* as the globally defined function

$$(\text{trunc}^\ell f)(\mathbf{x}) = f(\mathbf{x}) - \sum_{\beta^\ell \in K^\ell} c_{\beta^\ell} \beta^\ell(\mathbf{x}), \quad \mathbf{x} \in \Omega, \tag{18}$$

where the coefficients c_β^ℓ are taken from the representation (17). Combining this definition with (16) implies that

$$\text{trunc}^\ell f \in \text{span} \bigcup_{k=1}^{\ell} K^k.$$

Consequently, we are now able to apply truncation of the next higher level $\ell+1$ to $\text{trunc}^\ell f$.

For future reference we note that the trunction with respect to level ℓ does not change the value of the function on patches of previous levels,

$$f|_{\pi^k} = (\text{trunc}^\ell f)|_{\pi^k} \quad \text{if} \quad k < \ell. \tag{19}$$

This is a direct consequence of SOA. We also note that

$$(\mathrm{trunc}^\ell f)|_{\pi^\ell} \in \mathrm{span}(B^\ell \setminus K^\ell)|_{\pi^\ell}. \tag{20}$$

This can be confirmed by combining the local representation (17) with the definition (18) of the truncation.

We define *truncated PNHB-splines of level* ℓ by applying the truncation repeatedly to the selected tensor-product splines in K^ℓ,

$$T^\ell = \mathrm{trunc}^N \cdots \mathrm{trunc}^{\ell+1} K^\ell = \{\, \mathrm{trunc}^N \cdots \mathrm{trunc}^{\ell+1} \beta^\ell : \beta^\ell \in K^\ell \,\}. \tag{21}$$

Collecting the contributions from all levels gives the set of *truncated PNHB-splines*

$$T = \bigcup_{\ell=1}^{N} T^\ell. \tag{22}$$

Lemma 2. *We assume SCA and consider a selected B-spline* $\beta^\ell \in K^\ell$ *of level* ℓ *and a lower level* $k \leq \ell$. *Then*

$$(\mathrm{trunc}^N \cdots \mathrm{trunc}^{\ell+1} \beta^\ell)|_{\pi^k} = \begin{cases} 0 & \text{if } k < \ell \\ \beta^\ell|_{\pi^k} & \text{if } k = \ell. \end{cases} \tag{23}$$

Moreover, for larger levels $k > \ell$ *we have*

$$(\mathrm{trunc}^N \cdots \mathrm{trunc}^{\ell+1} \beta^\ell)|_{\pi^k} \in \mathrm{span}(B^k \setminus K^k)|_{\pi^k}. \tag{24}$$

Proof. Due to SCA we have that

$$\beta^\ell|_{\pi^k} = \begin{cases} 0 & \text{if } k < \ell \\ \beta^\ell|_{\pi^k} & \text{if } k = \ell. \end{cases}$$

This implies (23) since the truncations with respect to the levels $\ell+1, \ldots, N$ do not change the values on π^k according to (19).

To prove (24) we first observe that (20) gives

$$(\mathrm{trunc}^k \cdots \mathrm{trunc}^{\ell+1} \beta^\ell)|_{\pi^k} \in \mathrm{span}(B^k \setminus K^k)|_{\pi^k},$$

and note that the remaining truncations with respect to the levels $k+1, \ldots, N$ do not change the values on π^k according to (19). □

Proposition 1. *The truncated PNHB-splines are linearly independent if SCA is satisfied.*

Proof. We use Eq. (23) and proceed as in the proof of Theorem 1. □

Proposition 2. *The truncated PNHB-splines span the partially nested hierarchical spline space* H *if both SCA and CBA hold.*

Proof. The definition of truncation implies that every function in T can be represented with respect to K. Indeed, a function $\mathrm{trunc}^N \cdots \mathrm{trunc}^{\ell+1}\beta^\ell$ for $\beta^\ell \in K^\ell$ is obtained by subtracting contributions of functions included in K^k for $k = \ell+1, \ldots, N$. Hence, it can be written as a linear combination of functions in K, and consequently, span $T \subseteq$ span K. Since both T and K are linearly independent, and since $|T| = |K|$, we conclude that span $T =$ span K. Finally, we use Theorem 2 to complete the proof. □

Similar to [9] we show that the functions in T *preserve the coefficients* of the corresponding selected functions in K^ℓ.

Theorem 3 (Preservation of coefficients). *Any function $f \in H$ possesses local representations*

$$f(\mathbf{x}) = \sum_{\substack{\beta^k \in B^k \\ \mathrm{supp}\beta^k \cap \pi^k \neq \emptyset}} c_{\beta^k} \beta^k(\mathbf{x}), \quad \mathbf{x} \in \pi^k, \tag{25}$$

on the patches. The representation with respect to the truncated PNHB-splines inherits the coefficients c_{β^k} from these local representations,

$$f(\mathbf{x}) = \sum_{\ell=1}^{N} \sum_{\beta^\ell \in K^\ell} c_{\beta^\ell} (\mathrm{trunc}^N \cdots \mathrm{trunc}^{\ell+1}\beta^\ell)(\mathbf{x}), \quad \mathbf{x} \in \Omega,$$

provided that SCA and CBA are valid.

Proof. Proposition 2 guarantees that there exists a representation of $f \in H$ with respect to T,

$$f(\mathbf{x}) = \sum_{\ell=1}^{N} \sum_{\beta^\ell \in K^\ell} d_{\beta^\ell} (\mathrm{trunc}^N \cdots \mathrm{trunc}^{\ell+1}\beta^\ell)(\mathbf{x}), \quad \mathbf{x} \in \Omega,$$

with certain coefficients d_{β^ℓ}. We consider the restriction of this representation to the patch π^k of level k. None of the terms obtained for $\ell > k$ contributes to this restriction, according to (23). This also implies that the PNHB-splines of level k are simply tensor-product splines on π^k. Using these two observations we obtain

$$f(\mathbf{x}) = \sum_{\ell=1}^{k-1} \sum_{\beta^\ell \in K^\ell} d_{\beta^\ell} (\mathrm{trunc}^N \cdots \mathrm{trunc}^{\ell+1}\beta^\ell)(\mathbf{x}) + \sum_{\beta^k \in K^k} d_{\beta^k} \beta^k(\mathbf{x}), \quad \mathbf{x} \in \pi^k. \tag{26}$$

We note that the first sum is contained in $(B^k \backslash K^k)|_{\pi^k}$, due to (24). Consequently we may use the linear inpendence of the tensor-product B-splines $(B^k)|_{\pi^k}$ on the patch of level k to conclude

$$d_{\beta^k} = c_{\beta^k}, \quad \forall \beta^k \in K^k,$$

by comparing the coefficients of (25) and (26). □

The property of preservation of coefficients implies that the functions in T form a partition of unity:

Corollary 1. *The sum of the truncated PNHB-splines is the constant function with value 1 if both SCA and CBA are valid.*

Proof. Since the constant function with value 1 is contained in H, we may consider its representations (25) on all patches π^k, with the coefficients $c_{\beta^k} = 1$. Theorem 3 confirms the partition of unity property of truncated PNHB-splines.   □

Similarly, the function $\text{trunc}^N \cdots \text{trunc}^{k+1} \beta^k$ has the same Greville abscissa as its corresponding function $\beta^k \in K^k$ from which it was derived by using truncation.

6 Completeness

The knot lines of each spline space V^ℓ define a subdivision of the unit square $(0,1)^2$ into the *mesh M^ℓ of level ℓ*. More precisely, the elements of M^ℓ are axis-aligned boxes, which are the Cartesian product of two closed intervals that represent knot spans of V^ℓ in x- and y-direction. These elements will be denoted as *cells of level ℓ*

Another assumption, which is stronger than CBA, is required to investigate the completeness of the PNHB-splines:

Assumption. The boundaries of the patches π^ℓ are aligned with the mesh of level ℓ. More precisely, each patch π^ℓ is obtained as the interior

$$\pi^\ell = \text{int} \bigcup_{c \in C^\ell} c \qquad \text{(FBA)}$$

of the union of a cell set $C^\ell \subseteq M^\ell$. This condition will be called the *Full Boundary Alignment* (FBA) condition.

The union of the cell sets C^ℓ over all levels forms the *partially nested hierarchical mesh*.

Example. We consider again the partially nested hierarchical spline space, which is defined by the patches and spaces shown in Fig. 1a. The partially nested hierarchical mesh $\bigcup_{\ell=1}^{N} C^\ell$ is shown in Fig. 1b.   ◊

In this section we are interested in the full spline space F of degree $\mathbf{p}$ and maximal smoothness $\mathbf{s} = \mathbf{p} - (1,1)$,

$$F = \{f \in C^{\mathbf{s}}(\Omega) : f|_c \in \Pi_{\mathbf{p}} \; \forall c \in C^\ell \; \forall \ell = 1,\ldots,N\}, \qquad (27)$$

where $\Pi_{\mathbf{p}}$ denotes the space of tensor-product polynomials of degree $\mathbf{p}$. This space contains the partially nested hierarchical spline space H, but it is generally larger. A simple condition implies that both spaces are equal:

Assumption. The support intersections of the basis functions of level ℓ with the associated patches π^ℓ are all connected,

$$\mathrm{supp}\beta^\ell \cap \pi^\ell \quad \text{is connected} \quad \forall \beta^\ell \in B^\ell, \ \ell = 1, \ldots, N. \tag{SIC}$$

This will be denoted as the *Support Intersection Condition* (SIC).

If this assumption is satisfied in addition to all previous ones, then PNHB-splines are *complete*.

Theorem 4. *The PNHB-splines span the full spline space F if FBA, SIC and SCA are satisfied.*

Proof. Given a function $f \in F$, we proceed exactly as in the proof of Theorem 2. There is one modification, however, since we need to use a different argument to confirm the existence of the local representations (12) and (13).

This is achieved with the help of a result from [19]: Each patch π^ℓ is a multi-cell domain due to FBA. Theorem 2.12 of that paper implies that we obtain local representations as linear combinations of tensor-product B-splines β^ℓ if one uses several copies for functions with more than one support intersection. More precisely, when considering the function in (13) we obtain

$$f(\mathbf{x}) - \sum_{k=1}^{\ell-1} f^k(\mathbf{x}) = \sum_{\beta^\ell \in B^\ell} \sum_{\substack{\sigma \\ \sigma \text{ is connected} \\ \text{component of} \\ \mathrm{supp}\beta^\ell \cap \pi^\ell}} c_{\beta^\ell, \sigma} \, \beta^\ell(\mathbf{x}) \, \chi_\sigma(\mathbf{x}), \quad \mathbf{x} \in \pi^\ell,$$

where $\chi_\sigma(\mathbf{x})$ is the characteristic function of the connected component σ.

SIC implies that only one instance of each B-spline β^ℓ is required, as the support intersections with π^ℓ possess at most one connected component σ. Lemma 1 can be applied again and confirms that only functions $\beta^\ell \in K^\ell$ need to be considered. Consequently, we can find a representation of the form (13) (and also (12) for the first level).

The remainder of the proof applies without any modifications. □

Since PNHB-splines span the partially nested hierarchical spline space H, we also proved that the full spline space is equal to the partially nested hierarchical spline space under the assumptions of the theorem. All these results apply to truncated PNHB-splines as well.

7   An Example: Least-Squares Fitting

We consider a surface approximation problem to compare PNHB-splines with classical tensor-product B-splines and hierarchical B-splines. We choose the function

$$f(x,y) = 0.6\left(\sum_{i=0}^{10} \sum_{j=0}^{10} d_{ij} b_i(x) b_j(y) + (x - 0.5)^2 \right),$$

which is constructed by multiplying the elements of the tensor-product basis constructed from the univariate Bernstein polynomials

$$b_k(t) = \binom{10}{k} t^k (1-t)^{10-k}, \text{ for } k = 0, \ldots, 10,$$

of degree 10 with the function-valued coefficients

$$D = [d_{ij}] = \begin{bmatrix} 1 + \sin(60x) & 1 \cdots 1 & 1 \\ 1 & 1 \ldots 1 & 1 \\ \vdots & \vdots \; \vdots & \vdots \\ 1 & 1 \ldots 1 & 1 \\ 1 & 1 \cdots 1 & 1 + \sin(60y) \end{bmatrix},$$

see Fig. 3. Its domain is the unit square $(0,1)^2$. This function is fairly flat in most parts of the domain, but has distinctive vertical and horizontal features in the southwest and the northeast corners of the domain, which motivates us to use partially nested spline refinement.

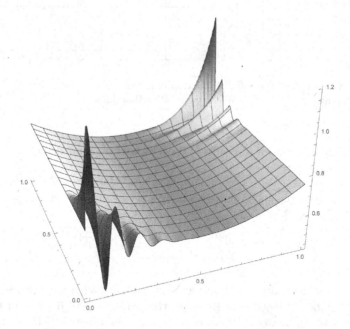

Fig. 3. The function considered in the fitting example.

We use a simple least-squares approximation to project this function into spline spaces spanned by

1. biquadratic tensor-product B-splines defined on the mesh shown in Fig. 4a,

2. hierarchical B-splines defined on the mesh shown in Fig. 4b, and

3. partially nested hierarchical B-splines defined on the mesh shown in Fig. 4c.

For tensor-product splines we consider a vertical and horizontal refinement in the southwest and northeast corner. These knot lines propagate to the northwest corner since tensor-product splines do not support local refinement. This is not the case for HB-splines, where we can perform local refinement. Nevertheless, one still needs to use nested splines spaces, which enforces the simultaneous refinement in both directions. Therefore, we add knot lines in x- and y-direction in both considered corners. Finally, we show the mesh used for PNHB-spline approximation. It seems to be perfectly suited for this task as the knot line segments are aligned with the features of the function.

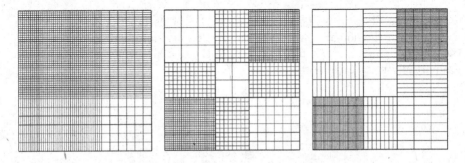

Fig. 4. The meshes used for defining the approximating spline functions. Left: tensor-product B-splines, middle: HB-splines, right: PNHB-splines.

Table 2. Numerical results of the least-squares approximation.

	No. of dof.	% of dof.	Max. error	Average error
Tensor-product B-splines	2304	100%	3.39e−3	3.81e−5
HB-splines	1633	71%	3.08e−3	4.37e−5
PNHB-splines	769	33%	8.12e−4	1.89e−5

The numerical results are reported in Table 2, which presents information about the number of degrees of freedom, the percentage of degrees of freedom (with respect to the tensor-product case), the maximum error between the original function and the fitting result and the average error.

The tensor-product splines provide the baseline for these tests. The number of control points is equal to 2304 and this is sufficient to obtain a reasonable result. By using hierarchical B-splines we saved some degrees of freedom and obtained a similar result. We used spline spaces $D^{i,i}$ for $i \leq 6$. Further refinement in the corners would substantially increase the number of degrees of freedom. Finally, the use of PNHB-splines leads to an additional improvement: a better approximation is obtained by using an even smaller number of degrees of freedom.

So far we constructed the meshes manually. Our current work is devoted to the use of error estimators for automating this process.

8   Conclusion

We proposed the new construction of partially nested hierarchical B-splines in order to overcome the limitations of the existing hierarchical spline constructions, which are based on sequences of fully nested spline spaces. Suitable assumptions enabled us to define a hierarchical spline basis, to establish a truncation mechanism, and to derive a completeness result. The application potential of the proposed generalization has been demonstrated by a first experimental result on least-squares approximation.

Future work will be devoted to extensions of this construction to the full multivariate case and to refinement strategies that can guide the process of local mesh refinement. Further, we will study alternative formulations of the generalized truncation mechanism, in order to analyze the non-negativity of the resulting spline basis. Also, we will investigate algorithms for assigning spaces to patches which ensure that the various assumptions are satisfied. Finally, we will continue to explore the application potential of our new construction. Some results on these topics will be presented in a forthcoming paper [4].

Acknowledgment. Supported by project NFN S117 "Geometry + Simulation" of the Austrian Science Fund and the EC projects "EXAMPLE", GA no. 324340 and "MOTOR", GA no. 678727.

References

1. Bornemann, P.B., Cirak, F.: A subdivision-based implementation of the hierarchical B-spline finite element method. Comput. Methods Appl. Mech. Eng. **253**, 584–598 (2013)
2. Buffa, A., Giannelli, C.: Adaptive isogeometric methods with hierarchical splines: error estimator and convergence. Math. Methods Appl. Sci. **26**, 1–25 (2016)
3. Dokken, T., Lyche, T., Pettersen, K.F.: Polynomial splines over locally refined box-partitions. Comput. Aided Geom. Des. **30**, 331–356 (2013)
4. Engleitner, N., Jüttler, B.: Patchwork B-spline refinement. In: Computer-Aided Design, vol. 90, pp. 168–179 (2017). ISSN 0010-4485. https://doi.org/10.1016/j.cad.2017.05.021
5. Evans, E.J., Scott, M.A., Li, X., Thomas, D.C.: Hierarchical T-splines: analysis-suitability, Bézier extraction, and application as an adaptive basis for isogeometric analysis. Comput. Methods Appl. Mech. Eng. **284**, 1–20 (2015)
6. Forsey, D.R., Bartels, R.H.: Hierarchical B-spline refinement. Comput. Graph. **22**, 205–212 (1988)
7. Giannelli, C., Jüttler, B., Kleiss, S.K., Mantzaflaris, A., Simeon, B., Špeh, J.: THB-splines: an effective mathematical technology for adaptive refinement in geometric design and isogeometric analysis. Comput. Methods Appl. Mech. Eng. **299**, 337–365 (2016)

8. Giannelli, C., Jüttler, B., Speleers, H.: THB-splines: the truncated basis for hierarchical splines. Comput. Aided Geom. Des. **29**, 485–498 (2012)

9. Giannelli, C., Jüttler, B., Speleers, H.: Strongly stable bases for adaptively refined multilevel spline spaces. Adv. Comput. Math. **40**(2), 459–490 (2014)

10. Greiner, G., Hormann, K.: Interpolating and approximating scattered 3D-data with hierarchical tensor product B-splines. In: Le Méhauté, A., Rabut, C., Schumaker, L.L. (eds.) Surface Fitting and Multiresolution Methods. Innovations in Applied Mathematics, pp. 163–172. Vanderbilt University Press, Nashville (1997)

11. Kang, H., Chen, F., Deng, J.: Hierarchical B-splines on regular triangular partitions. Graph. Models **76**(5), 289–300 (2014)

12. Kiss, G., Giannelli, C., Zore, U., Jüttler, B., Großmann, D., Barner, J.: Adaptive CAD model (re-)construction with THB-splines. Graph. Models **76**(5), 273–288 (2014)

13. Kosinka, J., Sabin, M., Dodgson, N.: Subdivision surfaces with creases and truncated multiple knot lines. Comput. Graph. Forum **23**, 118–128 (2014)

14. Kraft, R.: Adaptive und linear unabhängige Multilevel B-Splines und ihre Anwendungen. Ph.D. thesis, Universität Stuttgart (1998)

15. Kuru, G., Verhoosel, C.V., van der Zeeb, K.G., van Brummelen, E.H.: Goal-adaptive isogeometric analysis with hierarchical splines. Comput. Methods Appl. Mech. Eng. **270**, 270–292 (2014)

16. Li, X., Chen, F.L., Kang, H.M., Deng, J.S.: A survey on the local refinable splines. Sci. China Math. **59**(4), 617–644 (2016)

17. Li, X., Deng, J., Chen, F.: Polynomial splines over general T-meshes. Visual Comput. **26**, 277–286 (2010)

18. Li, X., Zheng, J., Sederberg, T.W., Hughes, T.J.R., Scott, M.A.: On linear independence of T-spline blending functions. Comput. Aided Geom. Des. **29**, 63–76 (2012)

19. Mokriš, D., Jüttler, B., Giannelli, C.: On the completeness of hierarchical tensor-product B-splines. J. Comput. Appl. Math. **271**, 53–70 (2014)

20. Schillinger, D., Dedé, L., Scott, M.A., Evans, J.A., Borden, M.J., Rank, E., Hughes, T.J.R.: An isogeometric design-through-analysis methodology based on adaptive hierarchical refinement of NURBS, immersed boundary methods, and T-spline CAD surfaces. Comput. Methods Appl. Mech. Eng. **249–252**, 116–150 (2012)

21. Speleers, H., Dierckx, P., Vandewalle, S.: Quasi-hierarchical Powell-Sabin B-splines. Comput. Aided Geom. Des. **26**, 174–191 (2009)

22. Speleers, H., Manni, C.: Effortless quasi-interpolation in hierarchical spaces. Numer. Math. **132**(1), 155–184 (2016)

23. Wei, X., Zhang, Y., Hughes, T.J.R., Scott, M.A.: Extended truncated hierarchical Catmull-Clark subdivision. Comput. Methods Appl. Mech. Eng. **299**, 316–336 (2016)

24. Zore, U., Jüttler, B.: Adaptively refined multilevel spline spaces from generating systems. Comput. Aided Geom. Des. **31**, 545–566 (2014)

25. Zore, U., Jüttler, B., Kosinka, J.: On the linear independence of truncated hierarchical generating systems. J. Comput. Appl. Math. **306**, 200–216 (2016)

Regression Analysis Using a Blending Type Spline Construction

Tatiana Kravetc[(✉)], Børre Bang, and Rune Dalmo

R&D Group Simulations, Department of Computer Science and Computational
Engineering, Faculty of Science and Technology,
UiT - The Arctic University of Norway, Narvik, Norway
`tatiana.kravetc@uit.no`

Abstract. Regression analysis allows us to track the dynamics of change
in measured data and to investigate their properties. A sufficiently good
model allows us to predict the behavior of dependent variables with
higher accuracy, and to propose a more precise data generation hypothesis.

By using polynomial approximation for big data sets with complex
dependencies we get *piecewise* smooth functions. One way to obtain a
smooth spline representation of an entire data set is to use local curves
and to blend them using smooth basis functions. This construction allows
the computation of derivatives at any point on the spline. Properties such
as tangent, velocity, acceleration, curvature and torsion can be computed,
which gives us the opportunity to exploit these data in the subsequent
analysis.

We can adjust the accuracy of the approximation on the different segments of the data set by choosing a suitable knot vector. This article
describes a new method for determining the number and location of the
knot-points, based on changes in the Frenet frame.

We present a method of implementation using generalized exporational B-splines (GERBS) for regression problems (in two and three
variables) and we evaluate the accuracy of the model using comparison
of the residuals.

1 Introduction

1.1 The Problem

We describe a general method suitable for use on stream based data. A wide
range of applications is possible, trend analysis for correlation of trends between
weather stations, and estimation of trends in online revenue streams are two
examples that will be considered for further work at a later time. In this paper,
the model implementation is based on weather data. Stages in the data form the
basis for a segmentation intended to be used in further classification, which gives

---

9th International Conference on Mathematical Methods for Curves and Surfaces,
Tønsberg, June 23rd–28th, 2016.

M. Floater et al. (Eds.): MMCS 2016, LNCS 10521, pp. 145–161, 2017.
https://doi.org/10.1007/978-3-319-67885-6_8

us the opportunity to identify state changes, or derived properties as for instance accumulations (mass/volume etc.). Identifying shifts in trends, i.e. stages in time-dependent open ended stream based data sets is considered to be of general interest in a wide area of real world applications.

1.2 Contribution

A natural way of treating possibly noise contaminated data which consists of more or less well defined stages, is to utilize local approximation in the relevant stages.

Our method changes the representation of the raw data to a form that accommodates both local approximation and adjustable criteria for identifying stages. Local piecewise Bézier curves are formed from the Least Squares Method of a limited number of data points that are close in time. Blending splines makes it possible to keep the original approximation and gives a gradual refinement that can be used to balance accuracy and computational effort. The Hermite properties of the interpolation method we use, where the local curves supply the additional information regarding derivatives, makes the resulting approximation work on localized intervals without keeping the complete data set (in memory).

1.3 Related Work

1.3.1 Statistical methods

In [14], James solves the registration problem between curves by "equating the moments of the curve while also shrinking toward a common shape". This could also be used to recognize "stages" or states, if information about the shapes of the stages are available. Statistical methods like the use of first and second moments of data for feature extraction, have been treated in [7]. For several application specific problems, an expansion in this direction of the approximation theory related to a blending type spline construction, would be appropriate.

1.3.2 Geometrical Methods

Real-time approximation and smoothing of geometry data, like position and orientation under constraints, arises frequently in path planning and control theory for self-guided vehicles. In [6], Brezak and Petrovic consider locally smoothing of paths by clothoids, using table lookup to implement a numerical integration that otherwise could be prohibitive in computational cost.

The problem of finding suitable knot vectors has been studied in the literature [5,8,12]. It seems to be an active area of research, however, the methods are depending on the data. In [3], Bittner and Brachtendorf utilize the Oslo algorithm for knot insertion in a wavelet context and provides an adaptive approximation method.

In [18] a locally supported spline quasi-interpolant is developed, with the assumption that the B-spline knots are chosen to lie midway between consecutive sampling points, or chosen to coincide with the sampling points. Then by blending of this and a local Hermite interpolation scheme, a method for fast interpolation is achieved, which has some resemblance to our method. Since our global setting is open ended, and we choose to approximate our data with local curves before blending, we also gain a data reduction for the current stage.

The implemented approximation method in [15], called MARS, provides searching for a suitable knot vector and approximates the data using the B-spline. It has parameters such us maximum number of basis functions, and an accuracy of this approximation depends on the size of the data set. The method presented in this paper provides a flexible approximation curve, independent of the complete data set. We do not need to keep all stages. This feature shows the distinction between local curves and control points. In MARS, if we remove the first knots, we completely lose the connection with the previous data points.

2   Method Overview

We seek to obtain a real-time continuous smooth approximation of noisy data, automatization of searching for non-stochastic deviation, and approximation with sufficient accuracy near sudden changes in the data.

The main goal of the algorithm presented in this paper is searching for features of the data. In other words, the algorithm provides preparation for clusterization of the data. Based on that, we consider to use data with the following definition: the 2- or 3-dimensional point set, which is the time dependence of some value, and can be divided into "stages". For example, change of temperature caused by the weather can be divided by times of day or seasons, depending on the length of the data set. In this paper we have considered open data from the Norwegian weather service yr.no at a specific location [20]. For the 2D data set we have used the maximum temperature between 01.12.2016 and 01.03.2017, measured once per hour, as shown in Fig. 1(a). For the 3D data set we have used the data from the weather radar, as the pixel's coordinates of the maximum amount of precipitation movement, as shown in Fig. 1(b).

In order to choose our approximation method we first notice some important properties: the data set has been collected throughout a long time and is has unpredictable dynamics. Additionally, the number of points increases with time, i.e., we based our approximation on the amount of points available at the current time step.

From existing methods for building regression models [3, 10, 13] we will take the Multivariate Adaptive Regression Splines method (also known as MARS) [15] for comparison with our implemented method. The MARS method makes cubic B-spline approximation and provides optimization of knots positions and the number of knots. The maximum number of basis functions can be set as a parameter.

The approach is to recognize the changes between "stages" on the data action, when they occur, where the duration of one "stage" is unknown. Thus, we need to provide the approximation of each "stage". We can use polynomial approximation for this purpose because the dynamics within one "stage" does not change

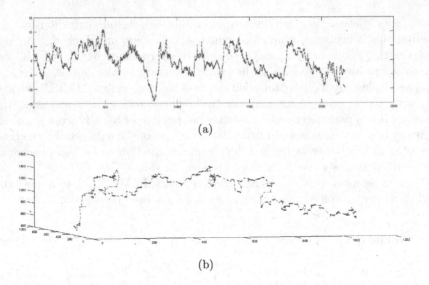

(a)

(b)

Fig. 1. The data is taken from yr.no [20]. The dashed lines between the data points are generated to obtain a more clear view. Figure 1(a) shows time dependence of the temperature in Narvik during winter months, the timestep is one hour; Fig. 1(b) shows time dependence of the pixel's coordinate of the maximum amount of precipitation in Nordland taken from the weather radar.

sharply. Then we blend these local polynomial curves together continuously and smoothly.

The algorithm considered in the present paper combines several approximation methods. We initiate the approximation with a rough estimate and improve it to obtain clustered sets of points. The motivation is to keep the accuracy independent of the length of on the point set, to provide stable real-time approximation, since the size of the data is unknown.

Consider the following short outline of the sequence of algorithms constituting the main algorithm proposed in the paper. Transition from one step to the next occurs only if the step returned a value. Throughout this text, kv is an array of the elements of the knot vector, ℓ is an array of local curves, A is the blending spline approximation curve, i is the index of a local curve, and t is time.

(i) Statistical searching of a "stage" or, in other words, element of a knot vector, **return** $kv(i)$; (see Algorithm 1)
(ii) Compute the $(i-1)^{\text{th}}$ local curve, **return** ℓ_{i-1}; (using formula (1))
(iii) Blend ℓ_{i-1} together with ℓ_{i-2}, **return** parametric curve $A(t)$; (see Algorithm 2 and formula (3))

Extension of the knot vector:
(a) Find candidates for knot insertion from the knot intervals; (see Algorithm 3)

(b) Find the positions for new knots, **return** recomputed arrays kv, ℓ, and curve $A(t)$; (see Algorithm 4)

The sequence above is realized for each time step.

3   GERBS

In this section we shall briefly consider some of the theory of blending type spline constructions, which is relevant for this work. A comprehensive study of GERBS can be found in [9,16].

In contrast to classical B-splines [4,17], where the coefficients typically are control points in some Eucledian space $\mathbb{R}^n$, blending spline coefficients can be local geometry. Here we have opted to use Bézier curves [2] defined by

$$\ell_i(t) = \sum_{k=0}^{d} c_k b_{d,k}(t), \ \text{ if } t_{i-1} \le t < t_{i+1}, \tag{1}$$

where $b_{d,k}$ are the Bernstein polynomials [1] of degree d, and $c_k \in \mathbb{R}^2$ are coefficients (or control points), t_{i-1} and t_{i+1} are blending spline interpolation knots, as described below.

We find the coefficients c_k by using the Least Squares Method [19]. According to this method, the vector of coefficients $c = (c_0, c_1, ..., c_d)^T$ is the solution of the common equation:

$$c = (W^T W)^{-1} W^T y,$$

where the vector y consist of values of the time dependent variable, and W is the matrix of basis functions.

The general formula for an expo-rational B-spline (ERBS) [9] curve over the knots $(t_i)_{i=0}^{n+1}$ is:

$$f(t) = \sum_{i=1}^{n} \ell_i(t) B_i(t), \ \text{ if } t_1 \le t \le t_n, \tag{2}$$

where the coefficients ℓ_i in our case are the local Bézier curves (1), and $B_i(t)$ are the expo-rational basis functions [16].

The domain of the local curves has to correspond with the domain of the respective ERBS basis function. This means that ℓ_i in (2) is defined on the domain (t_{i-1}, t_{i+1}), as shown in (1). Blending of local curves in the global domain $[0, 1]$ is performed via the formula

$$f(t) = \left(1 - B_{i-1} \circ \omega_{i-1}(t) \ B_i \circ \omega_i(t)\right) \begin{pmatrix} \ell_{i-1}(t) \\ \ell_i(t) \end{pmatrix}, \ 0 \le t \le 1, \tag{3}$$

where $\omega_k(t) = \frac{t-t_k}{t_{k+1}-t_k}$, t_k and t_{k+1} are knots, and our choice of B is the logistic expo-rational B-function [21]:

$$B(t) = \frac{1}{e^{\frac{1}{t} - \frac{1}{1-t}} + 1}.$$

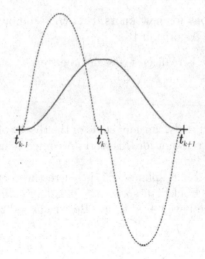

Fig. 2. A graph of the expo-rational basis function (solid red) and its first derivative (dotted blue). (Color figure online)

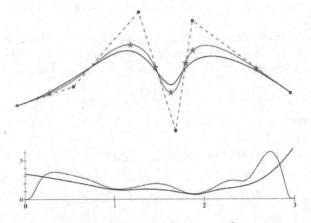

Fig. 3. The top picture shows a plot of a 3rd degree polynomial B-spline (blue). The dashed green lines are the control polygon. The red curve shows an expo-rational B-spline using the same knot vector and control points as the polynomial B-spline. The red and green stars denotes extreme values of the speed. The bottom picture shows the comparison between the speeds of these curves. Here the blue function shows the speed of the polynomial B-spline whereas the red function denotes the speed of the expo-rational B-splines. (Color figure online)

The choice of using expo-rational basis functions is founded on some important properties of these functions [16]. One main difference when compared to polynomial B-splines is the continuity properties. The continuity at the knots will increase with the order of the B-function. The expo-rational basis functions are C^∞-smooth at the respective exterior knot and all derivatives are zero at their interior knot. The blending splines in (2) possess a transfinite Hermite property, namely, $\ell_i(t)$ and all their existing derivatives are interpolated at the interior knots t_i. Figure 2 shows that the first derivative of the basis function is zero at the start and at the end of its support.

The speed of expo-rational B-splines occurs in a more "natural" way than the speed of polynomial B-splines, as shown in Fig. 3. Its minimums are at the start and at the end of curve, local minimums at the turns and local maximums at the straights. We will use these speed properties further as criteria for inserting new knots and finding knot positions.

4   Statistical Method for Real-Time Approximation

Our data are noisy, so we can divide them into noise and a non-stochastic part. We assume that the noise tends towards a normal distribution. We will construct probability distribution functions [11] for some initial number of points, and successively add points to the point set. The deviation from the normal distribution yields a new stage of activity. Such a method facilitates the so-called *real-time*, that is, we run the algorithm while we receive new data.

Algorithm 1 provides searching for knots by comparing the normal distribution with the probability distribution function for an open ended stream of data.

A virtual example of Algorithm 1 is shown in Fig. 4. Here, the green curves are normal distributions, where the red curves are probability distribution functions (PDF) for all points between each pair of blue lines. The blue lines illustrate the *stages* array, which corresponds to the knot vector.

Algorithm 1. Statistical searching of "stage"

1: **procedure** SEARCH STAGE
2: $start = 1$;
3: $step = 4$; //*take a* step *number of points for computation of the next PDF*
4:
5:     **for** $t = start + 1 : step : T$ **do**
6:
7:         compute $\sigma(start : t)$, $\mu(start : t)$; //*σ is the standard deviation and μ is the mean value*
8:
9:         $nd = \frac{1}{\sigma\sqrt{2\Pi}}e^{-(x-\mu)^2/2\sigma^2}$; //*normal distribution*
10:
11:         $f = \text{pdf}(y(start : t))$; //*compute probability distribution function*
12:
13:         **if** f has > 1 local maximums **then**
14:
15:             $start = t$;
16:
17:             append *start* to *stages*;
18:       **return** *stages*

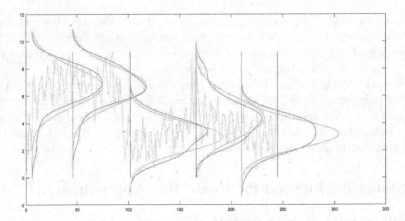

Fig. 4. Illustration of statistical "stages", i.e., initial knots, searching by Algorithm 1 for 2D data (see Fig. 1(a)). The blue lines show recognized knots, red lines show probability distribution functions of groups of data points separated by the blue lines, and the green curves are the normal distribution functions for those data points. (Color figure online)

Let us imagine that the set of points is the random translation of one point. Then for each time step it makes a displacement with an average value of 1. The initial value for displacement is 0 for both axes. Since the PDF can be unstable for small numbers of points, we have used a constant interval as a minimum distance between the knots.

For the $\mathbb{R}^3$ case we define translation of the point as a continuous displacement of the $\mathbb{R}^3 \to \mathbb{R}^2$ projection.

The distance between two points is

$$|\mathbf{d}_j| = |\mathbf{p}_{j+1} - \mathbf{p}_j|. \tag{4}$$

The initial displacement D_0 is zero, and accumulates the distance (4) between the first and the second point. Then we choose the direction D_{t+1} of translation as the sign of the difference between the y coordinates as shown in formula (5).

$$D_{t+1} = D_t + \text{sign}(y_{t+1} - y_t)\sqrt{(x_{t+1} - x_t)^2 + (y_{t+1} - y_t)^2}. \tag{5}$$

By involving these steps we get the picture of a random walk, see Fig. 5, to which we can apply Algorithm 1.

Figure 5 shows translation of the point as a projection from 3D to 2D, using formula (5), combined with an illustration of the searching for "stages".

Algorithm 2 provides construction of local curves, as Bézier curves of degree 3, and blending them together. A description of such a spline construction was addressed in Sect. 2 (see formula (2)). In Algorithm 2, the input value *stages* is adjusted by applying Algorithm 1. N denotes the number of stages, which changes with time, and $N + 2$ is the number of knots. We begin to compute the ith local curve when we have $i + 2$ knots. The blending starts when we have a minimum of two local curves. Finally, by repeating the procedure, we obtain the

Fig. 5. Illustration of searching for "stages" for 3D data (see Fig. 1(b)). The distance between neighbor points in this figure is equal to the distance between neighbor points in 3D, and the direction corresponds to the projection on the plane determined by the t and y axes. The blue lines show the knots found by applying Algorithm 1. (Color figure online)

Algorithm 2. Blending of local curves

1: **procedure** GERBS($stages$, N)
2: $kv = [stages(0), stages(0), stages(1), ...$
3:     $..., stages(N-1), stages(N), stages(N)]$; //kv *is the knot vector*
4:
5:    **for** $i = 1 : N$ **do**
6:
7:        $dom = kv_{i+1} - kv_{i-1}$; //*define the domain of the ith local curve*
8:
9:        $curve_i = \text{Bezier}(y(dom, 3))$; //*compute the Bézier local curve of degree 3 on the domain* dom *using formula (1)*
10:
11:    **if** $N > 2$ **then**
12:
13:        **for** $t = 0 : kv(N)$ **do**
14:
15:            $A(t) = \begin{pmatrix} 1 - B_{i-1} \circ \omega_{i-1}(t) & B_i \circ \omega_i(t) \end{pmatrix} \begin{pmatrix} curve_{i-1}(t) \\ curve_i(t) \end{pmatrix}$; //*blending using formula (3)*
16:        **return** A

---

blending spline $A(t)$. We make the observation that since our initial knot vector will increase with time, we can draw local curves and blend them in real-time.

5   Adding Knots

In the procedure above we defined the initial knot vector for a current time step. If this vector contains at least two knot intervals, we can improve the approximation of the data, i.e., increase the accuracy, by inserting new knots at certain positions.

Let $A(t)$ be a spline function. X is the discrete set of points, $\mathbf{x}_t \in X$, $\mathbf{x}_t = (t, y)$ for the 2D case and $\mathbf{x}_t = (t, x, y)$ for the 3D case, where t is the time variable. Thus, we have a point for each t and a continuous approximation of this set of points. The knot vector is denoted by $\boldsymbol{\tau} = (t_i)_{i=0}^N$, where N is number of knots.

We introduce *a moving frame*, denoted by its tangent and normal vectors, η and ξ for the 2D case, and tangent, normal and binormal vectors η, τ, ξ, for the 3D case, respectively. Such a frame represents a moving local coordinate system.

Definition 1. *The "scope" is the interior of the geometric boundary outlined by the pair of straight lines for the 2D case, or planes for the 3D case, which are defined via the knot interval. Each knot interval on $\boldsymbol{\tau}$ yields the top and the bottom borders of the "scope", which go through the points from the set $X^i \subset X$, $X^i = [\mathbf{x}_{t_i}, \mathbf{x}_{t_{i+1}}]$, which have the maximum distance from the local origin along the axis ξ of the moving frame, and are parallel to the axis η, for the 2D case, or rectifying plane denoted by the axes η and τ, for the 3D case.*

The process of inserting new knots consists of two steps: finding candidates for knot insertion from intervals in the knot vector, and defining the position for knot insertion.

(a) The first step is based on the detection of points which are outside of the "scope". The "scope" should contain all of the points. Algorithm 3 and Figs. 6 and 8 describe a process for finding the indices of intervals where new knots should be inserted.

   We shall now consider separately line 15 from Algorithm 3 for the 3D case and describe how to find the points which are outside of the "scope".

   The parametric expression of the plane which belongs to the "scope" is:

$$\mathbf{q}(u, v) = \mathbf{p}_0 + u\mathbf{t} + v\mathbf{b},$$

   where $\mathbf{t}$ is the tangent vector, $\mathbf{b}$ is the binormal vector, and $\mathbf{p}_0$ is the local origin.

   Let the vector between $\mathbf{p}_0$ and the point to check be $\mathbf{d}_{point}$, the vector between $\mathbf{p}_0$ and A be $\mathbf{d}_{curve}$, and $\mathbf{n}$ be the normal vector of the curve at $\mathbf{p}_0$ (see Fig. 8(c)). If the inner product between $\mathbf{n}$ and $\mathbf{d}_{point}$ has the same sign as the inner product between $\mathbf{n}$ and $\mathbf{d}_{curve}$, then the point and the curve lie on one side of the plane:

if
$$\langle \mathbf{d}_{point}, \mathbf{n} \rangle \langle \mathbf{d}_{curve}, \mathbf{n} \rangle < 0, \tag{6}$$
then add a new knot

(b) The second step is based on the properties of ERBS curve, as considered in Sect. 2 (see Fig. 3). We seek to divide the knot interval by inserting a new knot at the position where we can measure the largest change in the point set. We define this to be the position with the highest speed, thus, we need to find local maximums of the first derivative of the curve. Algorithm 4 and Fig. 7 describe the process of inserting new knots. x and y are point coordinates, kv is the knot vector, and ms is an array of local maximums of the first derivative of the curve A. in is an array of indices of knot intervals, and there are two types of indexation, i is the index of the knot interval and t is an index along the time axis.

Algorithm 3. Finding candidates for knot insertion from knot vector

---

1: **procedure** FINDINTERVALS(kv, A) //kv *is the knot vector,* A *is the curve*
2: $\rho = []$; //*an array for extremal points, which denote the border of the "scope"*
3:
4: *Define the moving frame to* $A(t)$; //Fig. 6(a), 8(a)
5:
6:     **for** i = 0 : length(kv)-1 **do**
7:       //*add to* ρ *extremal points along* ξ *on the interval* $[kv_i, kv_{i+1})$
8:       ρ.add($X(\max(\xi_t))$);
9:       ρ.add($X(\min(\xi_t))$);
10:
11: *Construct lines through points from* ρ, *parallel to the* η *axis, and we get the "scope" for the 2D case;* //Fig. 6(b)
12: *OR*
13: *Construct planes through points from* ρ, *parallel to the rectifying plane denoted by the* η *and* τ *axes, and we get the "scope" for the 3D case;* //Fig. 8(b)
14:
15:     **if** ∃ points from X on the interval $[kv_i, kv_{i+1})$ which lie below of the bottom border or above of the upper border **then**
16:       *insert a new knot using Algorithm 4*;
17:       *recompute the knot vector and the curve and* **goto 2**; // (Fig. 6(c), 8(d))

Algorithm 4. Inserting new knots

1: **procedure** NEWKV(kv, A, N)
2: $in = []$, $ms = []$;
3:
4: //*find knot intervals where we need to add new knots, using Algorithm 3*
    in = FINDINTERVALS(kv, A); //in *consists of indices of knot intervals*
5:
6: ms = LOCALMAXIMUMS(A'); //*populate the array* ms *with indices of local maximums of the* $A'(t)$
7:
8: Add to kv the indices (t) from ms, which lie on knot intervals with indices (i) from in; //*in the case when values of* ms *or* in *does not lead to add new knot, then we* **do not** *add any indices*
9:
10: **return** kv

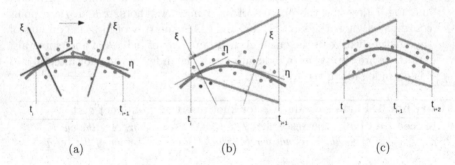

(a)   (b)   (c)

Fig. 6. Describes Algorithm 3. (a) illustrates the detection of a moving frame on the curve (Algorithm 3, line 4). The blue lines in (b) are the tangent lines of the curve through the red points, corresponding to maximum and minimum distances along the ξ axis on the considered knot interval, which are used to obtain the "scope" (Algorithm 3, line 11). Since there are points outside of the "scope" (Algorithm 3, line 15), we insert a new knot and recompute the curve as shown in (c) (Algorithm 3, lines 16–17). (Color figure online)

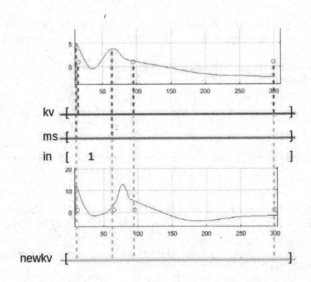

Fig. 7. Relation between the arrays, described in Algorithm 4, which uses Algorithm 3. The blue curves show the derivative of the curve $A(t)$ before (top) and after (bottom) knot insertion. The red circles illustrate the positions of the knots. kv is the knot vector before knot insertion, ms is the array of positions on the time-axis of the maximums of the curve's first derivative, in is the array of indices of the knot intervals where we need to insert new knots found by using Algorithm 3, and $newkv$ is the new knot vector, found via comparing the kv, ms and in arrays, as described in Algorithm 4. (Color figure online)

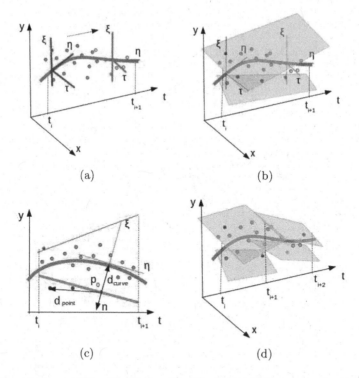

Fig. 8. Description of Algorithm 3 for the 3D case. (a) illustrates the detection of a moving frame on the curve (Algorithm 3, line 4). The blue planes in (b) are the tangent planes to the curve through the red points, corresponding to the maximum and minimum values along the ξ axis on the considered knot interval, constituting the "scope" (Algorithm 3, line 13). (c) illustrates how we recognize the points which are outside of the "scope" in the projection on a plane denoted by the t and y axes: $\mathbf{n}$ is the normal to the curve at $\mathbf{p}_0$. We are checking the position of $\mathbf{d}_{point}$ relative to the plane by applying formula (6). Since there are points which are outside of the "scope" (Algorithm 3, line 15), we insert a new knot and recompute the curve as shown in (d) (Algorithm 3, lines 16–17). (Color figure online)

6   Results and Concluding Remarks

Figures 9 and 11 show some steps of the proposed algorithm applied to representative examples in $\mathbb{R}^2$ and $\mathbb{R}^3$, respectively.

The approximation is a continuous smooth function which is generated by the automated algorithm. We can use intrinsic parameters of the resulting curve for further analysis. The settings of the algorithm are implicit: sensitivity to the detection of stages by changing the *step*, tolerance to the points which are outside of the "scope" (see Algorithm 3, line 15), and adjusting the degree of the local curves (in our case $d = 3$). We do not need to set the initial number of knots, or initial length of knot-intervals, or number of iterations.

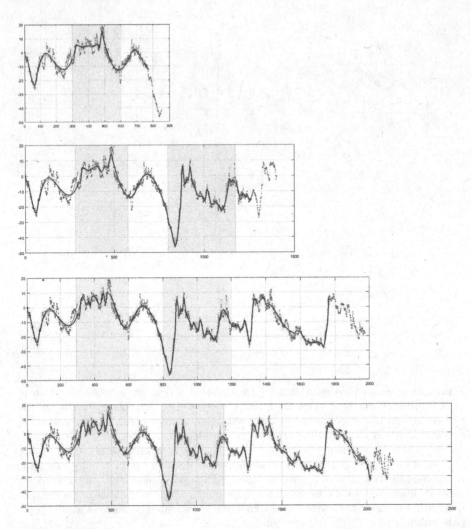

Fig. 9. The process of approximation for 2D data (see Fig. 1(a)). The red curves are local curves, the blue curve is the approximation curve. (Color figure online)

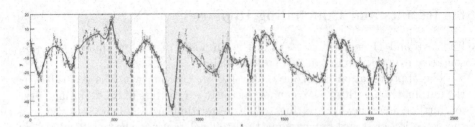

Fig. 10. Result of the MARS [15] algorithm. The blue curve is the resulting curve, black circles and dashed lines show knots. (Color figure online)

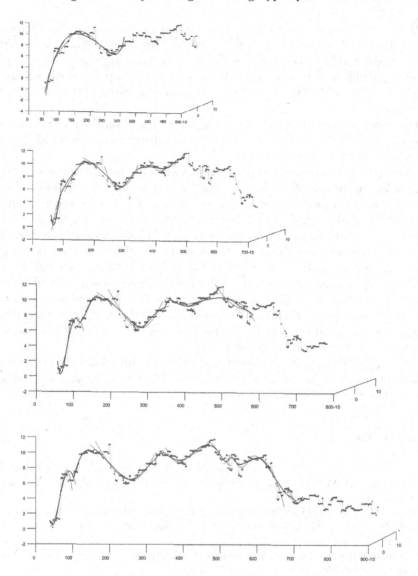

Fig. 11. The process of approximation for 3D data (see Fig. 1(b)). The red curves are local curves, the blue curve is the approximation curve. (Color figure online)

For comparison, we consider the MARS algorithm [15]. Parameters for the MARS model have been set as

```
params = aresparams2('maxFuncs', 70);
```

which limits of the maximum number of basis functions used. Note that the implementation is not a real-time version, so, assuming that we can realize a comparable approach, we simply run MARS for each time step.

By making a visual comparison of Figs. 9 and 10 one can see the differences and similarities between the results of the two algorithms. We note that the length of the resulting knot vectors for both methods are equal, but the values are very different. For example, the accuracy of the approximation changes within the range 300–600. Also one can compare the range 800–1200 for both algorithms.

One can not conclude which one is the better, based on the curves, since we do not have an original curve. However, we can discuss which method is more fit for our task and for our data.

The method presented in this paper provides flexible approximation of curves, independent of the complete data set. The curve is an affine combination of two and only two local functions on each knot interval. We do not need to keep all local curves or all "stages". This feature shows the distinction between local curves and control points. For comparison, with MARS, if we remove the first knots, we completely lose the connection with the "earlier" data points. But by using local curves, we keep the previous "stage" only as long as we need it.

We observe a constant (but not established) accuracy in our method for any time step. Conversely, with MARS, if $maxFuncs$ is large, then we obtain one accuracy at first, which will decrease when increasing the number of data points. This can be addressed by changing the maximum number of basis functions, but requires an extension of the algorithm.

Thus, we conclude that the presented algorithm is suitable for data possessing similar properties as our model data, whereas MARS is an established method with flexible settings for specific tasks.

References

1. Bernstein, S.: Démonstration du théoréme de Weierstrass fondée sur le calcul des probabilités, Communications de la Société Mathématique de Kharkow, 2-ée série, vol. 13(1), pp. 1–2 (1912)
2. Bézier, P.: Numerical Control: Mathematics and Applications. Wiley series in Computing. Wiley, London, New York (1972). English language edition
3. Bittner, K., Brachtendorf, H.G.: Fast algorithms for adaptive free-knot spline approximation using non-uniform biorthogonal spline wavelets. SIAM J. Sci. Comput. **37**(2), 283–304 (2015)
4. de Boor, C.: A Practical Guide to Splines. Springer, New York (1978)
5. Bratlie, J., Dalmo, R., Zanaty, P.: Fitting of discrete data with GERBS. In: Lirkov, I., Margenov, S., Waśniewski, J. (eds.) LSSC 2013. LNCS, vol. 8353, pp. 577–584. Springer, Heidelberg (2014). doi:10.1007/978-3-662-43880-0_66
6. Brezak, M., Petrovic, I.: Real-time approximation of clothoids with bounded error for path planning applications. IEEE Trans. Robot. **30**, 507–515 (2014)
7. Clarenz, U., Rumpf, M., Telea, A.: Robust feature detection and local classification for surfaces based on moment analysis. IEEE Trans. Vis. Comput. Graph. **10**, 516–524 (2004)
8. Dalmo, R.: Expo-Rational B-Splines in geometric modeling, methods for computer aided geometric design. Ph.D. thesis, University of Oslo (2016)

9. Dechevsky, L.T., Lakså, A., Bang, B.: Expo-rational B-splines. Int. J. Pure Appl. Math. **27**(3), 319–367 (2006)
10. Friedman, J.H.: Multivariate Adaptive Regression Splines. Stanford linear accelerator center, Stanford, California, vol. 19, pp. 1–67 (1990)
11. Guttman, I.: Introductory Engineering Statistics. Wiley, Hoboken (1965)
12. Hartley, P.J., Judd, C.J.: Parametrization of Bézier-type B-spline curves and surfaces. Comput. Aided Des. **10**, 130–134 (1978)
13. Härdle, W.: Applied nonparametric regression **34**(2), 341–342 (1989)
14. James, G.M.: Curve alignment by moments. Ann. Appl. Stat. **1**, 480–501 (2007)
15. Jekabsons, G.: ARESLab: Adaptive Regression Splines toolbox for Matlab/Octave. Riga Technical University, Riga, Latvia, Institute of Applied Computer Systems (2009)
16. Lakså, A.: Blending Technics for Curve and Surface Constructions. Narvik University College, Narvik (2012)
17. Shumaker, L.L.: Spline Functions, 3rd edn. Cambrige University Press, Cambridge (2007)
18. Van der Walt, M.D.: Real-time, local spline interpolation schemes on bounded intervals. Appl. Math. Sci. **10**, 205–234 (2015)
19. Vorontsov, K.V.: Lectures about algorithms for dependencies reconstruction (2007)
20. Weather service yr.no. Norwegian Meteorological Institute and Norwegian Broadcasting Corporation, 2007–2017, 01.12.2016–03.01.2017. www.yr.no/place/Norway/Nordland/Narvik/Narvik/almanakk.html, www.yr.no/place/Norway/Nordland/Narvik/Narvik/radar.html
21. Zanaty, P., Dechevsky, L.T.: On the numerical performance of FEM based on piecewise rational smooth resolutions of unity on triangulations. In: AIP Conference Proceedings, vol. 1570, p. 191 (2013)

On the Coupling of Decimation Operator with Subdivision Schemes for Multi-scale Analysis

Zhiqing Kui[1]([⊠]), Jean Baccou[2], and Jacques Liandrat[1]

[1] Centrale Marseille, I2M, UMR 7353, CNRS, Aix-Marseille University,
13451 Marseille, France
{zhiqing.kui,jacques.liandrat}@centrale-marseille.fr
[2] Institut de Radioprotection et de Sûreté Nucléaire (IRSN),
PSN-RES/SEMIA/LIMAR, CE Cadarache, 13115 Saint Paul Les Durance, France
jean.baccou@irsn.fr

Abstract. Subdivision schemes [5,11] are powerful tools for the fast generation of refined sequences ultimately representing curves or surfaces. Coupled with decimation operators, they generate multi-scale transforms largely used in signal/image processing [1,3] that generalize the multi-resolution analysis/wavelet framework [8]. The flexibility of subdivision schemes (a subdivision scheme can be non-stationary, non-homogeneous, position-dependent, interpolating, approximating, non-linear...) (e.g. [3]) is balanced, as a counterpart, by the fact that the construction of suitable consistent decimation operators is not direct and easy.

In this paper, we first propose a generic approach for the construction of decimation operators consistent with a given linear subdivision. A study of the so-called prediction error within the multi-scale framework is then performed and a condition on the subdivision mask to ensure a fast decay of this error is established. Finally, the cases of homogeneous Lagrange interpolatory subdivision, spline subdivision, subdivision related to Daubechies scaling functions (and wavelets) and some recently developed non stationary non interpolating schemes are revisited.

Keywords: Multi-scale analysis · Decimation · Subdivision

1 Introduction

Since the eighties [5,10,11] and even earlier [9], subdivision schemes have been developed, analyzed and used with very popular applications such as curve generation, image processing or animation movies. One of their advantages stands in the flexibility of the construction of their masks. In many ways, subdivision schemes are connected to the refinement process associated to the wavelet framework [6,8,15] and have to be coupled with a so called decimation operator [12] to fully generalize the classical multi-resolution approach. In practice, this coupling has been successfully used for image compression [1,3] since it offers an efficient

M. Floater et al. (Eds.): MMCS 2016, LNCS 10521, pp. 162–185, 2017.
https://doi.org/10.1007/978-3-319-67885-6_9

compromise between sparsity of the decomposition and quality of the reconstruction of an image. However, if the decimation operator is trivially defined as
a subsampling when working with interpolatory subdivision schemes, it is not
the case in other situations. As an example, one can mention the non-stationary
penalized Lagrange subdivision scheme [14] for which there is, up to now, no
available decimation operator in the literature due to the mixing between interpolating and non interpolating subdivisions. Many approximating subdivision
schemes can not be used for compression for the same reason. More generally,
even if a decimation operator is known, it is important in practice to be able to
derive a large choice of decimation masks in order to exhibit the most relevant
one according to specific objectives. It is the case for example in image compression where a criterion based on the stability of the multi-resolution transforms
could be used to select decimation operators.

Therefore, the goal of our work is to propose a generic approach for the
construction of decimation operators related to a given subdivision scheme. In
this paper, we focus on linear and homogeneous subdivision schemes as a starting
step. We first provide an overview on the multi-scale and subdivision frameworks
(Sect. 2). Then we revisit in Sect. 3 the fundamental aspects of consistency and
provide a generic construction of decimation operators of minimal length mask
that we call elementary decimation operators. We then describe how to generate
all the possible decimation operators from them. Section 4 is then devoted to
a theoretical analysis of the so-called prediction error. We show that the decay
of this quantity is fully controlled by the subdivision operator. Finally, several
examples and applications are considered in Sect. 5.

2   Combining Multi-scale Transforms and Subdivision Schemes

2.1   Multi-scale Analysis and Multi-scale Transform of Harten

A multi-scale analysis is characterized by the introduction of a family of separable
spaces $(V^j)_{j\in\mathbb{Z}}$ (j is a scale parameter) and two families of operators $(D^j_{j+1})_{j\in\mathbb{Z}}$
and $(P^{j+1}_j)_{j\in\mathbb{Z}}$ connecting two successive spaces V^j and V^{j+1}. For each value
of j, the decimation operator, D^j_{j+1}, maps $f^{j+1} = \left(f^{j+1}_k\right)_{k\in\mathbb{Z}} \in V^{j+1}$ to an
element $f^j = \left(f^j_k\right)_{k\in\mathbb{Z}} \in V^j$; the prediction operator, P^{j+1}_j, maps $f^j \in V^j$ to
an element of V^{j+1}. If f^j is obtained after decimation of f^{j+1}, $P^{j+1}_j f^j$ does
not usually coincide with f^{j+1}. However the following consistency condition is
required:

$$D^j_{j+1} P^{j+1}_j = I_{V^j} \tag{1}$$

where I_{V^j} stands for the identity operator in V^j. In order to recover f^{j+1} after a decimation and a prediction, a sequence of prediction errors $e^{j+1} = \left(e_k^{j+1}\right)_{k \in \mathbb{Z}}$ is introduced and defined as:

$$e_k^{j+1} = f_k^{j+1} - \left(P_j^{j+1} D_{j+1}^j f^{j+1}\right)_k = \left(\left(I_{V^{j+1}} - P_j^{j+1} D_{j+1}^j\right) f^{j+1}\right)_k . \quad (2)$$

The multi-scale transform of Harten [12] is then constructed as follows. Focussing on a level j, we note $\tilde{h} = D_{j+1}^j$ and $h = P_j^{j+1}$. Thanks to (1) and (2) that can be reformulated as $\tilde{h}h = I_{V^j}$ and $e^{j+1} = \left(I_{V^{j+1}} - h\tilde{h}\right) f^{j+1}$, it comes out that $\tilde{h}e^{j+1} = 0$. When $\tilde{h}$ is linear, we get $e^{j+1} \in Ker(\tilde{h}) = W^j$. We call $\tilde{g}$ the operator that associates e^{j+1} to its decomposition on a basis of W^j and g the canonical injection from W^j to V^{j+1}. We note $d^j = \tilde{g}e^{j+1}$ and $e^{j+1} = gd^j$. Then $\left(h, \tilde{h}, g, \tilde{g}\right)$ satisfy:

$$\begin{cases} \tilde{g}g = I_{W^j}, \\ h\tilde{h} + g\tilde{g} = I_{V^{j+1}}, \\ \tilde{g}h = 0, \\ \tilde{h}g = 0. \end{cases}$$

One-scale decomposition and reconstruction transforms are then classically sketched as follows [15]:

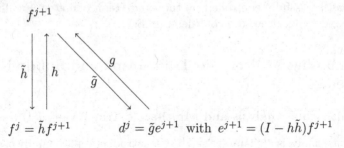

Iterating this process and denoting $J_0 < j$, two multi-scale decomposition and reconstruction transforms can be finally constructed as:

$$\text{decomposition:} \quad f^{j+1} \mapsto \{f^{J_0}, d^{J_0}, \dots, d^j\}, \quad (3)$$

$$\text{reconstruction:} \quad \{f^{J_0}, d^{J_0}, \dots, d^j\} \mapsto f^{j+1} . \quad (4)$$

These two transforms are of prime importance in data analysis and compression. In this context, their stability with respect to small perturbations is a key property that is recalled in the following definition.

Definition 1

The decomposition transform is said to be stable with regards to the norm $|| \cdot ||$ if there exists a constant C such that for all j and for all (f^j, f_ϵ^j), if $f^j \mapsto \{f^{J_0}, d^{J_0}, ..., d^{j-1}\}$ and $f_\epsilon^j \mapsto \{f_\epsilon^{J_0}, d_\epsilon^{J_0}, ..., d_\epsilon^{j-1}\}$, then,

$$sup\{||f_\epsilon^{J_0} - f^{J_0}||, \{||d_\epsilon^m - d^m||, m < j\}\} \leq C||f_\epsilon^j - f^j|| . \tag{5}$$

The reconstruction transform is said to be stable with regards to the norm $|| \cdot ||$ if there exists a constant C such that for all $j > J_0$ and for all $\{f^{J_0}, d^{J_0}, ..., d^{j-1}\}$ and $\{f_\epsilon^{J_0}, d_\epsilon^{J_0}, ..., d_\epsilon^{j-1}\}$, if $\{f^{J_0}, d^{J_0}, ..., d^{j-1}\} \mapsto f^j$ and $\{f_\epsilon^{J_0}, d_\epsilon^{J_0}, ..., d_\epsilon^{j-1}\} \mapsto f_\epsilon^j$, then,

$$||f_\epsilon^j - f^j|| \leq C \ sup\{||f_\epsilon^{J_0} - f^{J_0}||, \{||d_\epsilon^m - d^m||, m < j\}\} . \tag{6}$$

From the previous definition, it turns out that the stability of the multi-scale transforms fully depends on the choice of the four operators $(h, \tilde{h}, g, \tilde{g})$. Indeed the following results hold as a direct consequence of the definition of $(h, \tilde{h}, g, \tilde{g})$.

Proposition 1

Assuming that the one-scale decomposition and reconstruction transforms are constructed from the linear operators $(h, \tilde{h}, g, \tilde{g})$, if $f^j \mapsto \{f^{j-1}, d^{j-1}\}$ and $f_\epsilon^j \mapsto \{f_\epsilon^{j-1}, d_\epsilon^{j-1}\}$, then,

$$||f^{j-1} - f_\epsilon^{j-1}|| \leq ||\tilde{h}|| \ ||f^j - f_\epsilon^j||, \tag{7}$$
$$||d^{j-1} - d_\epsilon^{j-1}|| \leq ||\tilde{g}|| \ ||I_{V^j} - h\tilde{h}|| \ ||f^j - f_\epsilon^j||, \tag{8}$$

moreover, if $\{f^{j-1}, d^{j-1}\} \mapsto f^j$ and $\{f_\epsilon^{j-1}, d_\epsilon^{j-1}\} \mapsto f_\epsilon^j$, then

$$||f^j - f_\epsilon^j|| \leq (||h|| + ||g||) \ sup\{||f^{j-1} - f_\epsilon^{j-1}||, ||d^{j-1} - d_\epsilon^{j-1}||\}, \tag{9}$$

where for simplicity, $|| \cdot ||$ denotes the vector or operator norm.

The family of prediction operators $(P_j^{j+1})_{j \in \mathbb{Z}}$ plays a key role in the efficiency of the multi-scale process (3)–(4). In this paper, we focus on linear local operators. There exists several approaches to construct them. The first one exploits the classical multi-scale analysis and wavelet framework [8]. In this case, the coefficients involved in the linear combination are directly deduced from the scaling relation connecting scaling functions at different levels. Starting from functional (continuous) spaces then moving to separable (discrete) ones is then required. On the contrary, a second approach consists in defining explicitly the connection between V^j and V^{j+1} without specifying scaling functions and wavelets. This can be performed using subdivision schemes that are briefly described in the next section.

2.2   Subdivision and Decimation Schemes

The definition of a binary subdivision scheme [11] is first recalled.

Definition 2

A (univariate) subdivision scheme S is defined as a linear operator $S : l^\infty(\mathbb{Z}) \to l^\infty(\mathbb{Z})$ constructed from a real-valued sequence $(h_k)_{k \in \mathbb{Z}}$ having a finite number of non zero values such that

$$(f_k)_{k \in \mathbb{Z}} \in l^\infty(\mathbb{Z}) \mapsto ((Sf)_k)_{k \in \mathbb{Z}} \in l^\infty(\mathbb{Z}) \quad with \quad (Sf)_k = \sum_{l \in \mathbb{Z}} h_{k-2l} f_l \ .$$

The set of non zero values of $(h_k)_{k \in \mathbb{Z}}$ is called the mask of S and is denoted M_h.

From the definition of a subdivision scheme, it is easy to verify that a subdivision scheme reproduces constant[1] if and only if

$$\sum_{l \in \mathbb{Z}} h_{2l} = \sum_{l \in \mathbb{Z}} h_{2l+1} = 1 \ . \tag{10}$$

This property is also called shift invariance for constant and is assumed to be verified for all the subdivision schemes considered in this paper.

Subdivision can be iterated from an initial sequence $(f_k^{J_0})_{k \in \mathbb{Z}}$ to generate $(f_k^j)_{k \in \mathbb{Z}}$ for $j \geq J_0$ as

$$f^{j+1} = Sf^j, j \geq J_0 \ . \tag{11}$$

The mask plays a key role in the subdivision process and there exists many ways to construct it ([2,4,7,11,14] for example). In this work, we focus on homogeneous (the mask does not depend on k) and stationary (the mask does not depend on j) schemes.

Expression (11) can be interpreted as a prediction relation. There is therefore a one-to-one correspondence between subdivision and local prediction operators that will be exploited in this work.

The connection with the multi-scale framework is then fully achieved by introducing in the following definition the notion of binary decimation scheme.

Definition 3

A (univariate) decimation scheme D is defined as a linear operator $D : l^\infty(\mathbb{Z}) \to l^\infty(\mathbb{Z})$ constructed from a real-valued sequence $(\tilde{h}_k)_{k \in \mathbb{Z}}$ having a finite number of non zero values such that

$$(f_k)_{k \in \mathbb{Z}} \in l^\infty(\mathbb{Z}) \mapsto ((Df)_k)_{k \in \mathbb{Z}} \in l^\infty(\mathbb{Z}) \quad with \quad (Df)_k = \sum_{l \in \mathbb{Z}} \tilde{h}_{l-2k} f_l \ .$$

The set of non zero values of $(\tilde{h}_k)_{k \in \mathbb{Z}}$ is called the mask of D and denoted $M_{\tilde{h}}$.

Moreover, similarly to subdivision schemes, a decimation scheme reproduces constant if and only if

$$\sum_{k \in \mathbb{Z}} \tilde{h}_k = 1 \ . \tag{12}$$

---

[1] a scheme U is said to reproduce constant if $\forall k, f_k = C \implies \forall k, (Uf)_k = C$.

The subdivision framework leads to a large choice for the family $(P_j^{j+1})_{j\in\mathbb{Z}}$ thanks to the flexibility in the construction of the mask. This is not the case when considering wavelet multi-resolution analysis, since the prediction and the decimation are fixed once scaling functions and wavelets are specified. However, for a given subdivision scheme, the construction of a decimation mask leading to a family of consistent decimation operators satisfying (1) is more involved. This topic is addressed in the next section where a new method to generate decimation masks associated to a fixed subdivision is proposed.

3   Construction of Decimation Operators

The first part of this section (Sect. 3.1) is devoted to the derivation of a condition on the subdivision and decimation masks to ensure the consistency property for the associated operators. Then, we propose in Sect. 3.2 a generic approach to construct decimations consistent with a fixed subdivision.

3.1   Consistency Condition

Proposition 2
Let h be a prediction operator constructed from the mask of a subdivision scheme i.e. $\forall j, \forall \left(f_l^j\right)_{l\in\mathbb{Z}}$,

$$\left(hf^j\right)_k = \sum_{l\in\mathbb{Z}} h_{k-2l} f_l^j \ . \tag{13}$$

Let $\tilde{h}$ be a decimation operator constructed from the mask of a decimation scheme i.e. $\forall j, \forall \left(f_k^{j+1}\right)_{k\in\mathbb{Z}}$,

$$\left(\tilde{h}f^{j+1}\right)_l = \sum_{k\in\mathbb{Z}} \tilde{h}_{k-2l} f_k^{j+1}. \tag{14}$$

Then h and $\tilde{h}$ satisfy the consistency relation (1) if and only if

$$\forall j \in \mathbb{Z}, \ \sum_{i\in\mathbb{Z}} h_i \tilde{h}_{i+2j} = \delta_{j,0} \ . \tag{15}$$

where $(\delta_{j,0})_{j\in\mathbb{Z}}$ is the Dirac sequence.

Proof
According to (13) and (14), the consistency condition (1) implies that $\forall \left(f_m^j\right)_{m\in\mathbb{Z}}$,

$$\forall m \in \mathbb{Z}, \quad f_m^j = \sum_{k\in\mathbb{Z}} \tilde{h}_{k-2m} \sum_{l\in\mathbb{Z}} h_{k-2l} f_l^j = \sum_{l\in\mathbb{Z}} (\sum_{k\in\mathbb{Z}} \tilde{h}_{k-2m} h_{k-2l}) f_l^j$$

which is equivalent to

$$\forall m \in \mathbb{Z}, \quad \sum_{k \in \mathbb{Z}} \tilde{h}_{k-2m} h_{k-2l} = \delta_{m,l}$$

that leads to (15). □

Remark 1
According to Eqs. (10), (12) and (15), if a subdivision scheme and a decimation scheme are consistent and if the subdivision scheme reproduces constants, then the decimation scheme also reproduces constants.

We end up this section by providing results related to convex combination and translation of consistent decimation operators.

Definition 4
If $\tilde{h}$ is a decimation operator constructed from the sequence $(\tilde{h}_k)_{k \in \mathbb{Z}}$, for all $t \in \mathbb{Z}$ we call $T_t(\tilde{h})$ the decimation operator related to the sequence $(\tilde{h}_{k-t})_{k \in \mathbb{Z}}$.

Then, the following proposition holds:

Proposition 3
Let h be a prediction operator constructed from the sequence $(h_k)_{k \in \mathbb{Z}}$,

1. *if $\tilde{h}^0, \tilde{h}^1$ are two decimation operators of sequences $(\tilde{h}_k^0)_{k \in \mathbb{Z}}$ and $(\tilde{h}_k^1)_{k \in \mathbb{Z}}$ consistent with h, then $\forall \lambda \in \mathbb{R}$, $\lambda \tilde{h}^0 + (1 - \lambda)\tilde{h}^1$ is consistent with h;*
2. *if $\tilde{h}^0, \tilde{h}^1, \tilde{h}^2$ are three decimation operators of sequences $(\tilde{h}_k^0)_{k \in \mathbb{Z}}$, $(\tilde{h}_k^1)_{k \in \mathbb{Z}}$ and $(\tilde{h}_k^2)_{k \in \mathbb{Z}}$ consistent with h, then $\forall \lambda \in \mathbb{R}$ and $\forall t \in \mathbb{Z}$, $\tilde{h}^0 + \lambda T_{2t}(\tilde{h}^1) - \lambda T_{2t}(\tilde{h}^2)$ is consistent with h.*

Proof

1. We have

$$\sum_k h_{k+2j}(\lambda \tilde{h}_k^0 + (1 - \lambda)\tilde{h}_k^1) = \lambda \delta_{j,0} + (1 - \lambda)\delta_{j,0} = \delta_{j,0} \ .$$

2. Similarly,

$$\sum_k h_{k+2j}(\tilde{h}_k^0 + \lambda \tilde{h}_{k-2t}^1 - \lambda \tilde{h}_{k-2t}^2) = \delta_{j,0} + \lambda \delta_{j+t,0} - \lambda \delta_{j+t,0} = \delta_{j,0} \ .$$

□

Corollary 1
Subdivision operator h being fixed, let $\{\tilde{h}^i\}_{i \in \mathcal{I}}$ be a set of consistent decimation operators, a general consistent decimation operator can be constructed as

$$\sum_{t \in \mathcal{T}} \sum_{i \in \mathcal{I}} c_{i,t} T_{2t}(\tilde{h}^i) \tag{16}$$

with

$$\forall t \in \mathcal{T}, \sum_{i \in \mathcal{I}} c_{i,t} = \delta_{t,0}, 0 \in \mathcal{T} \subset \mathbb{Z} \ .$$

3.2   Generic Approach

In this section, we focus on the construction of all the decimation operators consistent with a given subdivision operator. We first construct elementary operators with masks of minimal number of non zero values. Then, we show how all consistent decimation operators can be recovered using linear combinations of translated versions of elementary operators.

The two following propositions provide the construction of the elementary operators. Proposition 4 gives a generic approach to get consistent decimation operators for a subdivision mask of even or odd length. Then, Proposition 5 deals with further consideration for subdivision masks of odd length.

Proposition 4
Let h be a prediction operator constructed from the mask

$$M_h = \{h_{n-2\alpha}, h_{n-2\alpha+1}, \ldots, h_n, h_{n+1}\}$$

of length $2(\alpha + 1)$ with $h_{n-2\alpha}h_{n+1} \neq 0$ or of length $2\alpha + 1$ with $h_{n-2\alpha} = 0$ and $h_{n-2\alpha+1}h_{n+1} \neq 0$.
We note H_{M_h} the following matrix,

$$H_{M_h} = \begin{bmatrix} h_n & h_{n-2} & \cdots & h_{n-2\alpha} & 0 & \cdots & 0 \\ h_{n+1} & h_{n-1} & \cdots & h_{n-2\alpha+1} & 0 & \cdots & 0 \\ 0 & h_n & h_{n-2} & \cdots & h_{n-2\alpha} & \cdots & 0 \\ 0 & h_{n+1} & h_{n-1} & \cdots & h_{n-2\alpha+1} & \cdots & 0 \\ \vdots & & & \vdots & & & \\ 0 & 0 & \cdots & h_n & h_{n-2} & \cdots & h_{n-2\alpha} \\ 0 & 0 & \cdots & h_{n+1} & h_{n-1} & \cdots & h_{n-2\alpha+1} \end{bmatrix}.$$

If $det(H_{M_h}) \neq 0$, there exists 2α consistent elementary decimation operators which masks are of length not larger than 2α. These masks are given by each row of $H_{M_h}^{-1}$.

Proof
First, let us assume that M_h is of even length ($h_{n-2\alpha}h_{n+1} \neq 0$) and denote formally for any integer $m \in \mathbb{Z}$,

$$M_{\tilde{h}} = \{\tilde{h}_{n-m}, \tilde{h}_{n-m+1}, \ldots, \tilde{h}_{n-m+2\alpha-2}, \tilde{h}_{n-m+2\alpha-1}\},$$

the mask of a consistent decimation operator of length not larger than 2α. Here the parameter n controls the centering of the mask M_h. The parameter m is related to the shift between the masks M_h and $M_{\tilde{h}}$.

If $\tilde{h}$ is consistent with h then the consistency condition (15) is verified. It can be written as

$$[\tilde{h}_{n-m}, \tilde{h}_{n-m+1}, \ldots, \tilde{h}_{n-m+2\alpha-2}, \tilde{h}_{n-m+2\alpha-1}] \begin{bmatrix} h_{n-m-2j} \\ h_{n-m+1-2j} \\ \vdots \\ h_{n-m+2\alpha-2-2j} \\ h_{n-m+2\alpha-1-2j} \end{bmatrix} = \delta_{j,0} \ . \ (17)$$

To ensure that (17) makes sense with a given M_h, we should have

$$\{h_{n-m-2j}, h_{n-m+1-2j}, \ldots, h_{n-m+2\alpha-1-2j}\} \bigcap \{h_{n-2\alpha}, h_{n-2\alpha+1}, \ldots, h_{n+1}\} \neq \varnothing,$$

which means

$$\begin{cases} n - m + 2\alpha - 1 - 2j \geq n - 2\alpha \\ n - m - 2j \leq n + 1 \end{cases}$$

and leads to

$$-\frac{m+1}{2} \leq j \leq -\frac{m+1}{2} + 2\alpha .$$

When m is odd, (17) corresponds to $2\alpha + 1$ linear equations for $j \in \{-\frac{m+1}{2}, \ldots, -\frac{m+1}{2} + 2\alpha\}$ including

$$\tilde{h}_{n-m} h_{n+1} = \delta_{m,-1} \quad \text{for} \quad j = \frac{m+1}{2}$$

and

$$\tilde{h}_{n-m+2\alpha-1} h_{n-2\alpha} = \delta_{m,4\alpha-1} \quad \text{for} \quad j = \frac{m+1}{2} + 2\alpha.$$

Since $h_{n+1} h_{n-2\alpha} \neq 0$, it necessarily leads to $\tilde{h}_{n-m} \tilde{h}_{n-m+2\alpha-1} = 0$. If $\tilde{h}_{n-m} = 0$, then $M_{\tilde{h}}$ is equivalent to $\{\tilde{h}_{n-m'}, \tilde{h}_{n-m'+1}, \ldots, \tilde{h}_{n-m'+2\alpha-2}, \tilde{h}_{n-m'+2\alpha-1}\}$ where m' is even by considering

$$\{\tilde{h}_{n-m+1}, \ldots, \tilde{h}_{n-m+2\alpha-2}, \tilde{h}_{n-m+2\alpha-1}, 0\},$$

and replacing $m-1$ by m'. The same kind of argument holds when $\tilde{h}_{n-m+2\alpha-1} = 0$ by considering

$$\{0, \tilde{h}_{n-m}, \ldots, \tilde{h}_{n-m+2\alpha-2}, \tilde{h}_{n-m+2\alpha-2}\} .$$

Therefore, m can always be considered as even without losing generality. Since m is even, (17) leads to 2α linear equations for $j \in \{-\frac{m}{2}, -\frac{m}{2}+1, \ldots, -\frac{m}{2} + 2\alpha - 1\}$ that can be written as

$$[\tilde{h}_{n-m}, \tilde{h}_{n-m+1}, \ldots, \tilde{h}_{n-m+2\alpha-2}, \tilde{h}_{n-m+2\alpha-1}] H_{M_h} = [\delta_{m,0}, \delta_{m-2,0}, \ldots, \delta_{m-4\alpha+2,0}]$$

with

$$H_{M_h} = \begin{bmatrix} h_n & h_{n-2} & \cdots & h_{n-2\alpha} & 0 & \cdots & 0 \\ h_{n+1} & h_{n-1} & \cdots & h_{n-2\alpha+1} & 0 & \cdots & 0 \\ 0 & h_n & h_{n-2} & \cdots & h_{n-2\alpha} & \cdots & 0 \\ 0 & h_{n+1} & h_{n-1} & \cdots & h_{n-2\alpha+1} & \cdots & 0 \\ \vdots & & \vdots & & & & \\ 0 & 0 & \cdots & h_n & h_{n-2} & \cdots & h_{n-2\alpha} \\ 0 & 0 & \cdots & h_{n+1} & h_{n-1} & \cdots & h_{n-2\alpha+1} \end{bmatrix},$$

where the column index corresponds to parameter j.

For $m \in \{0, 2, \ldots, 4\alpha - 4, 4\alpha - 2\}$, Eq. (17) can be written as

$$\tilde{H}_{M_h} H_{M_h} = I_{2\alpha}$$

with

$$\tilde{H}_{M_h} = \begin{bmatrix} M_{\tilde{h}^0} \\ M_{\tilde{h}^2} \\ \vdots \\ M_{\tilde{h}^{4\alpha-4}} \\ M_{\tilde{h}^{4\alpha-2}} \end{bmatrix} = \begin{bmatrix} \tilde{h}_n^0 & \tilde{h}_{n+1}^0 & \cdots & \tilde{h}_{n+2\alpha-1}^0 \\ \tilde{h}_{n-2}^2 & \tilde{h}_{n-1}^2 & \cdots & \tilde{h}_{n+2\alpha-3}^2 \\ \vdots & \vdots & & \vdots \\ \tilde{h}_{n-4\alpha+4}^{4\alpha-4} & \tilde{h}_{n-4\alpha+5}^{4\alpha-4} & \cdots & \tilde{h}_{n-2\alpha+3}^{4\alpha-4} \\ \tilde{h}_{n-4\alpha+2}^{4\alpha-2} & \tilde{h}_{n-4\alpha+3}^{4\alpha-2} & \cdots & \tilde{h}_{n-2\alpha+1}^{4\alpha-2} \end{bmatrix} .$$

Each row of $\tilde{H}_{M_h}$ corresponds to a value of m and to a consistent decimation operator. Note that, specifically for the elementary decimation operators defined above, the superscript k for $\tilde{h}^k$ controls the shift between M_h and $M_{\tilde{h}^k}$. Since $det(H_{M_h}) \neq 0$, $\tilde{H}_{M_h} = H_{M_h}^{-1}$, that concludes the proof when M_h is of even length.

In the case of subdivision mask of odd length, the same proof can be conducted assuming $h_{n-2\alpha} = 0$ and the same matrix $\tilde{H}_{M_h}$ can be deduced if $det(H_{M_h}) \neq 0$. $\qquad \square$

When the subdivision masks are of even length, the previous proposition provides 2α consistent elementary decimation operators. When the subdivision masks are of odd length, it turns out that the last row of $\tilde{H}_{M_h}$ can be obtained by a linear combination of translated versions of the decimation masks associated with the other rows. This leads to only $2\alpha - 1$ elementary decimation operators. It is stated by the next proposition.

Proposition 5
Let h be a prediction operator constructed from the mask

$$M'_h = \{h_{n-2\alpha+1}, h_{n-2\alpha+2}, \ldots, h_n, h_{n+1}\}$$

of length $2\alpha + 1$, $\alpha \geq 2$ with $h_{n-2\alpha+1} h_{n+1} \neq 0$.
We note $H'_{M'_h}$ the following matrix

$$H'_{M'_h} = \begin{bmatrix} h_n & h_{n-2} & \cdots & 0 & 0 & \cdots & 0 \\ h_{n+1} & h_{n-1} & \cdots & h_{n-2\alpha+1} & 0 & \cdots & 0 \\ 0 & h_n & h_{n-2} & \cdots & 0 & \cdots & 0 \\ 0 & h_{n+1} & h_{n-1} & \cdots & h_{n-2\alpha+1} & \cdots & 0 \\ \vdots & & \vdots & & & & \\ 0 & 0 & \cdots & h_n & h_{n-2} & \cdots & h_{n-2\alpha+2} \end{bmatrix} .$$

If $det(H'_{M'_h}) \neq 0$, there exists $2\alpha - 1$ consistent elementary decimation operators which masks are of length not larger than $2\alpha - 1$. These masks are given by each row of $H'^{-1}_{M'_h}$.

Proof
Following Proposition 4, we construct a similar matrix with $h_{n-2\alpha} = 0$

$$
H_{M_h'} = \begin{bmatrix}
h_n & h_{n-2} & \cdots & 0 & 0 & \cdots & 0 \\
h_{n+1} & h_{n-1} & \cdots & h_{n-2\alpha+1} & 0 & \cdots & 0 \\
0 & h_n & h_{n-2} & \cdots & & 0 & \cdots & 0 \\
0 & h_{n+1} & h_{n-1} & \cdots & h_{n-2\alpha+1} & \cdots & 0 \\
\vdots & & & \vdots & & & \\
0 & 0 & \cdots & h_n & h_{n-2} & \cdots & 0 \\
0 & 0 & \cdots & h_{n+1} & h_{n-1} & \cdots & h_{n-2\alpha+1}
\end{bmatrix} .
$$

Since $h_{n-2\alpha+1} \neq 0$ and $det(H_{M_h'}') \neq 0$, we have $det(H_{M_h'}) \neq 0$ and we can introduce

$$
\tilde{H}_{M_h'} = \begin{bmatrix}
M_{\tilde{h}^0} \\
M_{\tilde{h}^2} \\
\vdots \\
M_{\tilde{h}^{4\alpha-4}} \\
M_{\tilde{h}^{4\alpha-2}}
\end{bmatrix}
= \begin{bmatrix}
\tilde{h}_n^0 & \tilde{h}_{n+1}^0 & \cdots & \tilde{h}_{n+2\alpha-2}^0 & 0 \\
\tilde{h}_{n-2}^2 & \tilde{h}_{n-1}^2 & \cdots & \tilde{h}_{n+2\alpha-4}^2 & 0 \\
\vdots & & \vdots & & \vdots \\
\tilde{h}_{n-4\alpha+4}^{4\alpha-4} & \tilde{h}_{n-4\alpha+5}^{4\alpha-4} & \cdots & \tilde{h}_{n-2\alpha+2}^{4\alpha-4} & 0 \\
\tilde{h}_{n-4\alpha+2}^{4\alpha-2} & \tilde{h}_{n-4\alpha+3}^{4\alpha-2} & \cdots & \tilde{h}_{n-2\alpha}^{4\alpha-2} & \tilde{h}_{n-2\alpha+1}^{4\alpha-2}
\end{bmatrix}
$$

with $\tilde{H}_{M_h'} = H_{M_h'}^{-1}$.

Note that the last row of $\tilde{H}_{M_h'}$ denoted $M_{\tilde{h}^{4\alpha-2}}$ is the only mask with a non-zero last term. Therefore $\tilde{h}_{n-2\alpha+1}^{4\alpha-2} \neq 0$ according to the consistency condition.

In the sequel, we show that the last row of $\tilde{H}_{M_h'}$ can be obtained by linear combinations of the translated versions of the above ones.

First, note that the set

$$
\{\tilde{h}_{n+2\alpha-2}^0, \tilde{h}_{n+2\alpha-4}^2, \ldots, \tilde{h}_{n-2\alpha+4}^{4\alpha-6}\}
$$

has at least one non-zero term, otherwise, according to the consistency condition, all terms in

$$
\{\tilde{h}_{n+2\alpha-3}^0, \tilde{h}_{n+2\alpha-5}^2, \ldots, \tilde{h}_{n-2\alpha+3}^{4\alpha-6}\}
$$

would be also zero which implies $det(\tilde{H}_{M_h'}) = 0$.

So there exists $\tilde{h}_{n-2\alpha+2+2t}^{4\alpha-4-2t} \neq 0$ for $t \in \{1, 2, \ldots, 2\alpha - 2\}$. Introducing $\lambda = \tilde{h}_{n-2\alpha+2}^{4\alpha-4}/\tilde{h}_{n-2\alpha+2+2t}^{4\alpha-4-2t}$, we note

$$
\tilde{h}^* = \tilde{h}^{4\alpha-4} + \lambda T_{-2t}(\tilde{h}^{4\alpha-2-2t}) - \lambda T_{-2t}(\tilde{h}^{4\alpha-4-2t})
$$

which can have non-zero value from index $n - 4\alpha + 2$ to $n - 2\alpha + 2$. Calculating the last term gives

$$
\tilde{h}_{n-2\alpha+2}^* = \tilde{h}_{n-2\alpha+2}^{4\alpha-4} + \lambda \tilde{h}_{n-2\alpha+2+2t}^{4\alpha-2-2t} - \lambda \tilde{h}_{n-2\alpha+2+2t}^{4\alpha-4-2t} = 0 .
$$

It means that $\tilde{h}^*$ can have non-zero value from index $n - 4\alpha + 2$ to $n - 2\alpha + 1$.

Since $det(H_{M'_h}) \neq 0$, $\tilde{h}^{4\alpha-2}$ is the unique consistent operator with a mask of length not larger than 2α and admitting non-zero values from index $n - 4\alpha + 2$ to $n - 2\alpha + 1$. It therefore implies that $\tilde{h}^* = \tilde{h}^{4\alpha-2}$.

Thus, eliminating the last row and column of $H_{M'_h}$, we construct the matrix

$$H'_{M'_h} = \begin{bmatrix} h_n & h_{n-2} & \cdots & 0 & 0 & \cdots & 0 \\ h_{n+1} & h_{n-1} & \cdots & h_{n-2\alpha+1} & 0 & \cdots & 0 \\ 0 & h_n & h_{n-2} & \cdots & & 0 & \cdots & 0 \\ 0 & h_{n+1} & h_{n-1} & \cdots & h_{n-2\alpha+1} & \cdots & 0 \\ \vdots & & & \vdots & & \\ 0 & 0 & \cdots & h_n & h_{n-2} & \cdots & h_{n-2\alpha+2} \end{bmatrix}.$$

Since $det(H'_{M'_h}) \neq 0$, we then get elementary consistent decimation operator masks by considering the rows of $\tilde{H}'_{M'_h} = H'^{-1}_{M'_h}$. □

By construction, $\tilde{H}'_{M'_h}$ in Proposition 5 is a $(2\alpha - 1) \times (2\alpha - 1)$ sub matrix of $\tilde{H}_{M_h}$ introduced in Proposition 4. Therefore, for any subdivision scheme, $\tilde{H}_{M_h}$ provides all consistent elementary decimation masks. However, in practice, in the case of odd length (more than 3), the elementary operators will be obtained by restricting to $\tilde{H}'_{M'_h}$ that is denoted for simplicity $\tilde{H}'_{M_h}$ since a subdivision mask of odd length can be considered as a mask of even length by adding zero in front or behind.

The previous elementary decimation operators can then be used to construct any consistent decimation operator. Indeed we have the following proposition.

Proposition 6
Given a subdivision scheme h satisfying the hypotheses of Propositions 4 or 5, combining elementary decimation operators with formula (16) generates all the consistent decimation operators.

Proof
Let's first consider $M_h = \{h_{n-2\alpha}, h_{n-2\alpha+1}, \ldots, h_n, h_{n+1}\}$ the mask of a given operator h of length $2(\alpha + 1)$ with $h_{n-2\alpha}h_{n+1} \neq 0$. Then, Proposition 4 provides 2α consistent elementary decimation operators of length 2α which can be denoted as

$$M_{\tilde{h}^{2i}} = \{\tilde{h}^{2i}_{n-2i}, \tilde{h}^{2i}_{n-2i+1}, \ldots, \tilde{h}^{2i}_{n-2i+2\alpha-2}, \tilde{h}^{2i}_{n-2i+2\alpha-1}\}$$

with $i = 0, 1, 2, \ldots, 2\alpha - 1$.

Let $M_{\tilde{h}} = \{\tilde{h}_{m-2\beta}, \tilde{h}_{m-2\beta+1}, \ldots, \tilde{h}_m, \tilde{h}_{m+1}\}$ be the mask of an arbitrary decimation operator $\tilde{h}$ consistent with h. The length of $M_{\tilde{h}}$, $2\beta + 2$, is supposed to be larger than 2α that is to say $\beta > \alpha - 1$, otherwise $\tilde{h}$ is an elementary operator itself and the proof is completed. Moreover, m in $M_{\tilde{h}}$ is always chosen to ensure $n - m$ even by assuming that $\tilde{h}_{m-2\beta}$ and $\tilde{h}_{m+1}$ can be zero. However, $\{\tilde{h}_{m-2\beta}, \tilde{h}_{m-2\beta+1}\} \neq \{0,0\}$ and $\{\tilde{h}_m, \tilde{h}_{m+1}\} \neq \{0,0\}$ are always guaranteed.

The consistency of h and $\tilde{h}$ implies directly $n - 2\alpha - 1 < m < n + 2\beta + 1$.

The aim is to prove that $\tilde{h}$ can be represented as a linear combination of translated version of $(\tilde{h}^{2i})_{0 \leq i \leq 2\alpha-1}$.

This will be proved in two steps. The first step consists in writing $\tilde{h}$ as the sum of a term involving some $\tilde{h}^{2i}$ or its translated versions and of another one denoted $\tilde{h}^*$ that is a consistent decimation operator with a shorter mask than $M_{\tilde{h}}$. The second step is an iteration of this process until $\tilde{h}^*$ is an elementary decimation operator. We restrict the proof to the first step since the second one is straightforward.

The starting point is the consistency condition (15). Considering $j = \frac{m-n}{2} - \beta$ and $j = \frac{m-n}{2} + \alpha$, it leads to

$$h_n \tilde{h}_{m-2\beta} + h_{n+1} \tilde{h}_{m-2\beta+1} = \delta_{\frac{m-n}{2}-\beta,0} , \tag{18}$$

$$h_{n-2\alpha} \tilde{h}_m + h_{n-2\alpha+1} \tilde{h}_{m+1} = \delta_{\frac{m-n}{2}+\alpha,0} . \tag{19}$$

According to α and β, at least one of the two above RHS term is equal to zero. Let us suppose that the RHS of (18) is zero, i.e.

$$h_n \tilde{h}_{m-2\beta} + h_{n+1} \tilde{h}_{m-2\beta+1} = 0. \tag{20}$$

Since $h_{n+1} \neq 0$, we cannot have $\tilde{h}^{2i}_{n-2i} = 0$ for all $i \in \{1, 2, \ldots, 2\alpha - 1\}$ according to the consistency condition.

Let us introduce the two following operators with a mask of length 2α with $\tilde{h}'_{m-2\beta} \neq 0$,

$$M_{\tilde{h}'} = \{\tilde{h}'_{m-2\beta}, \tilde{h}'_{m-2\beta+1}, \ldots, \tilde{h}'_{2\alpha-2\beta+m-2}, \tilde{h}'_{2\alpha-2\beta+m-1}\} = T_{m-2\beta-n+2i}(\tilde{h}^{2i}),$$

$$M_{\tilde{h}''} = \{\tilde{h}''_{m-2\beta+2}, \tilde{h}''_{m-2\beta+3}, \ldots, \tilde{h}''_{2\alpha-2\beta+m}, \tilde{h}''_{2\alpha-2\beta+m+1}\} = T_{m-2\beta-n+2i}(\tilde{h}^{2i-2}),$$

which are elementary operators with the same translation. The consistency condition implies

$$h_n \tilde{h}'_{m-2\beta} + h_{n+1} \tilde{h}'_{m-2\beta+1} = 0. \tag{21}$$

Considering (20) and (21), $\tilde{h}_{m-2\beta} = 0$ leads to $\tilde{h}_{m-2\beta+1} = 0$ which is not allowed. Moreover, $\tilde{h}_{m-2\beta+1} = 0$ implies $h_n = 0$ and then $\tilde{h}'_{m-2\beta+1} = 0$. Therefore there exists $\lambda \in \mathbb{R}/\{0\}$ such that

$$\lambda[\tilde{h}'_{m-2\beta}, \tilde{h}'_{m-2\beta+1}] = [\tilde{h}_{m-2\beta}, \tilde{h}_{m-2\beta+1}] . \tag{22}$$

According to Proposition 3, $\tilde{h}^* = \tilde{h} - \lambda \tilde{h}' + \lambda \tilde{h}''$ is consistent with h. Since $M_{\tilde{h}}$ has length $2\beta + 2$, M_{h^*} has length 2β from index $m - 2\beta + 2$ to $m + 1$.

If $\beta = \alpha$, $M_{\tilde{h}^*}$ and $M_{\tilde{h}''}$ have the same length and indices. According to Proposition 4, $\tilde{h}^* = \tilde{h}''$ and

$$\tilde{h} = \lambda \tilde{h}' + (1 - \lambda) \tilde{h}'' ,$$

which leads to the expected result with a zero second term.

If $\beta > \alpha$, $M_{\tilde{h}*}$ has a shorter length than $M_{\tilde{h}}$ and

$$\tilde{h} = \lambda\tilde{h}' - \lambda\tilde{h}'' + \tilde{h}^* \ ,$$

that allows us to iterate by replacing $\tilde{h}$ with $\tilde{h}^*$ and then to conclude.

The above process actually eliminate the first two terms $\tilde{h}_{m-2\beta}, \tilde{h}_{m-2\beta+1}$ of $M_{\tilde{h}}$ using elementary decimation operators. If we suppose that the RHS of (19) is zero, a symmetrical similar process can be performed and the last two terms $\tilde{h}_m, \tilde{h}_{m+1}$ of $M_{\tilde{h}}$ will be eliminated.

To complete the proof in the case of subdivision mask of odd length, we suppose $h_{n+1} = 0$ in (20) and (21). It is straightforward that $\tilde{h}_{m-2\beta} = 0$, $\tilde{h}_{m-2\beta+1} \neq 0$ and then $\tilde{h}'_{m-2\beta} = 0$. Moreover, introducing $M_{h'}$ and $M_{h''}$ with $\tilde{h}'_{m-2\beta} = 0$ and $\tilde{h}'_{m-2\beta+1} \neq 0$, there exists $\lambda \neq 0$ verifying (22). □

Coming back to the multi-scale framework, the next section is devoted to a theoretical study of the prediction error involving consistent subdivision and decimation operators.

4   Analysis of the Prediction Error

The following result holds:

Proposition 7
Let $\{(V^j, h, \tilde{h})\}_{j\in\mathbb{Z}}$ define a multi-scale analysis with h a prediction operator constructed from the real sequence $(h_k)_{k\in\mathbb{Z}}$ and $\tilde{h}$ a decimation operator constructed from the real sequence $(\tilde{h}_k)_{k\in\mathbb{Z}}$. We assume that the associated multi-scale transform is applied from a fine scale J_{max} to a coarse one J_0.
If there exists $L \in \mathbb{N}$ such that $\forall n \in \{0, 1, \ldots, L\}$,

$$\forall k \in \mathbb{Z}, \quad \sum_{l\in\mathbb{Z}} h_{k-2l} \sum_{m\in\mathbb{Z}} \tilde{h}_{m-2l}(m-k)^n = \delta_{n,0} \tag{23}$$

then for sufficiently large $j \in [J_0, J_{max} - 1]$,

$$||e^j|| \leq C2^{-(L+1)j}, \tag{24}$$

where C does not depend on j.

Proof
Condition (23) is equivalent to

$$\forall 1 \leq n \leq L, \quad k^n = \sum_l h_{k-2l} \sum_m \tilde{h}_{m-2l} m^n,$$

and implies that

$$\forall j \in \mathbb{Z}, \forall 1 \leq n \leq L, \quad (k2^{-j})^n - \sum_l h_{k-2l} \sum_m \tilde{h}_{m-2l}(m2^{-j})^n = 0 \ . \tag{25}$$

Moreover, for any j, one can introduce $f_j \in C^{L_0}(\mathbb{R})$ with $L_0 \gg L$ such that $f_k^j = f_j(k2^{-j})$. We postpone to the end of the proof the construction of a particular f_j to get the expected result of the proposition.

Using Taylor expansion, it then comes out,

$$f_k^j = f_j(k2^{-j}) = \sum_{n=0}^{L+1} \frac{1}{n!} f_j^{(n)}(0)(k2^{-j})^n + o((2^{-j})^{L+1})$$

and the prediction error (2) can be rewritten

$$e_k^j = \sum_{n=1}^{L+1} \frac{1}{n!} f_j^{(n)}(0)\Big((k2^{-j})^n - \sum_l h_{k-2l} \sum_m \tilde{h}_{m-2l}(m2^{-j})^n\Big) + o((2^{-j})^{L+1}).$$

According to (25),

$$e_k^j = \frac{1}{(L+1)!} f_j^{(L+1)}(0)\Big((k2^{-j})^{L+1} - \sum_l h_{k-2l} \sum_m \tilde{h}_{m-2l}(m2^{-j})^{L+1}\Big) + o(2^{-j(L+1)}). \quad (26)$$

To finish the proof we introduce a particular f_j such that $\forall i \leq L_0$, $||f_j^{(i)}||_\infty$ is controlled independently of j that leads to a constant C independent of j such that $||e^j|| \leq C2^{-j(L+1)}$.

For any $j \in [J_0, J_{max} - 1]$, f_j is constructed from $f_{J_{max}} \in C^{L_0}(\mathbb{R})$.

More precisely, starting from $(f_k^{J_{max}})_{k \in \mathbb{Z}}$ with $f_k^{J_{max}} = f_{J_{max}}(k2^{-J_{max}})$, $(f_k^{J_{max}-1})_{k \in \mathbb{Z}}$ is written

$$f_k^{J_{max}-1} = (D_{J_{max}}^{J_{max}-1} f^{J_{max}})_k = \sum_{l \in \mathbb{Z}} \tilde{h}_{l-2k} f_l^{J_{max}} = \sum_{l \in \mathbb{Z}} \tilde{h}_l f_{l+2k}^{J_{max}} = \sum_{l \in \mathbb{Z}} \tilde{h}_l f_{J_{max}}((l+2k)2^{-J_{max}}),$$

$f_{J_{max}-1}$ can therefore be defined as $\forall x \in \mathbb{R}$

$$f_{J_{max}-1}(x) = \sum_l \tilde{h}_l f_{J_{max}}(l2^{-J_{max}} + x),$$

and it is straightforward that $\forall i \leq L_0$,

$$||f_{J_{max}-1}^{(i)}||_\infty \leq \Big(\sum_l |\tilde{h}_l|\Big)||f_{J_{max}}^{(i)}||_\infty . \quad (27)$$

Iterating this process, $\forall j \in [J_0, J_{max} - 1]$,

$$||f_j^{(i)}||_\infty \leq \Big(\sum_l |\tilde{h}_l|\Big)^{J_{max}-j}||f_{J_{max}}^{(i)}||_\infty . \quad (28)$$

Since $\sum_l |\tilde{h}_l| \geq 1$,

$$||f_j^{(i)}||_\infty \leq \Big(\sum_l |\tilde{h}_l|\Big)^{J_{max}-J_0}||f_{J_{max}}^{(i)}||_\infty;$$

and (26) leads to

$$||e^j|| \leq C2^{-(L+1)j}$$

with C independent of j. $\square$

Remark 2
The bound in Inequality (27) depends on the L^∞ norm of the decimation operator that is also a key quantity controlling the stability of the multi-scale decomposition (Expression (7) of Proposition 1). Therefore, an heuristic criterion to select a decimation scheme after applying our generic approach can be based on the L^∞ norm of its mask. An optimal choice corresponds to a minimal norm equal to 1 that leads to a bound in Inequality (28) independent of J_{max} and J_0. We show in Sect. 5 some numerical examples related to the stability of the decimation operators associated to classical subdivision schemes.

Condition (23) involves both subdivision and decimation masks. Under the consistency Condition (15), it can be reformulated in a simpler way.

Proposition 8
Let h and $\tilde{h}$ be two consistent operators. Condition (23) is satisfied if and only if $\forall n \in \{0, 1, 2, \ldots, L\}$

$$\sum_{l \in \mathbb{Z}} h_{2l}(2l)^n = \sum_{l \in \mathbb{Z}} h_{2l+1}(2l+1)^n \ . \tag{29}$$

Proof
Let us first introduce the following notations,

$E_{e,e}^n = \sum_l h_{2l} \sum_k \tilde{h}_{2k}(2k - 2l)^n,$
$E_{o,o}^n = \sum_l h_{2l+1} \sum_k \tilde{h}_{2k+1}(2k - 2l)^n,$
$E_{e,o}^n = \sum_l h_{2l} \sum_k \tilde{h}_{2k+1}(2k + 1 - 2l)^n,$
$E_{o,e}^n = \sum_l h_{2l+1} \sum_k \tilde{h}_{2k}(2k - 2l - 1)^n.$

Condition (23) becomes $\forall 0 \le n \le L$,

$$E_{e,e}^n + E_{e,o}^n = \delta_{n,0}, \ E_{o,e}^n + E_{o,o}^n = \delta_{n,0} \ . \tag{30}$$

First we will prove that, the consistency constraint implies

$$\forall n \in \mathbb{N}, \ E_{e,e}^n + E_{o,o}^n = \delta_{n,0} \ . \tag{31}$$

It is easy to verify that for $n = 0$, $E_{e,e}^0 + E_{o,o}^0 = 1$. Moreover, for any $n \in \mathbb{N}^*$, the consistency condition leads to

$$\sum_j (\sum_i h_{i-2j}\tilde{h}_i)(2j)^n = \sum_j \delta_{j,0}(2j)^n,$$

$$\sum_i (\sum_j h_{i-2j}(2j)^n)\tilde{h}_i = 0 \ .$$

Splitting the previous sum with respect to even and odd indices, we get

$$\sum_i (\sum_j h_{2i-2j}(2j)^n)\tilde{h}_{2i} + \sum_i (\sum_j h_{2i+1-2j}(2j)^n)\tilde{h}_{2i+1} = 0,$$

$$\sum_l h_{2l} \sum_k \tilde{h}_{2k}(2k - 2l)^n + \sum_l h_{2l+1} \sum_k \tilde{h}_{2k+1}(2k - 2l)^n = 0,$$

which is precisely,

$$E_{e,e}^n + E_{o,o}^n = 0 \ .$$

Considering (30) and (31), Condition (23) is then equivalent to $\forall n \in \{0, 1, 2, \ldots, L\}$

$$\sum_l h_{2l} \sum_k \tilde{h}_{2k}(2k - 2l)^n = \sum_l h_{2l+1} \sum_k \tilde{h}_{2k}(2k - 2l - 1)^n,$$

$$\sum_l h_{2l} \sum_k \tilde{h}_{2k+1}(2k + 1 - 2l)^n = \sum_l h_{2l+1} \sum_k \tilde{h}_{2k+1}(2k - 2l)^n,$$

which can be written as

$$\sum_l h_{2l} \sum_k \tilde{h}_{2k} \sum_{i=0}^n \binom{n}{i}(-1)^i(2k)^{n-i}(2l)^i = \sum_l h_{2l+1} \sum_k \tilde{h}_{2k} \sum_{i=0}^n \binom{n}{i}(-1)^i(2k)^{n-i}(2l+1)^i,$$

$$\sum_l h_{2l} \sum_k \tilde{h}_{2k+1} \sum_{i=0}^n \binom{n}{i}(-1)^i(2k+1)^{n-i}(2l)^i = \sum_l h_{2l+1} \sum_k \tilde{h}_{2k+1} \sum_{i=0}^n \binom{n}{i}(-1)^i(2k+1)^{n-i}(2l+1)^i,$$

leading to

$$\sum_{i=0}^n \binom{n}{i}(-1)^i \left(\sum_l h_{2l}(2l)^i - \sum_l h_{2l+1}(2l+1)^i\right) \sum_k \tilde{h}_{2k}(2k)^{n-i} = 0,$$

$$\sum_{i=0}^n \binom{n}{i}(-1)^i \left(\sum_l h_{2l}(2l)^i - \sum_l h_{2l+1}(2l+1)^i\right) \sum_k \tilde{h}_{2k+1}(2k+1)^{n-i} = 0,$$

which is equivalent to (29).   □

In fact, (29) in Proposition 8 and (23) in Proposition 7 are related to polynomial quasi-reproduction and to polynomial reproduction. Full discussion on these topics is beyond the scope of this paper.

5   Examples and Applications

Several applications of the previous results are provided in this section. They are focussed on different goals. The first one is to revisit, with our generic approach, classical decimation operators associated with well-known subdivision schemes (Sects. 5.1, 5.2 and 5.3). A special attention is devoted to the stability of the generated decimations. We illustrate how our approach can be used to improve stability constants. The second one is to apply our method to the practical case of a newly developed subdivision scheme for which there is no available decimation operator in the literature (Sect. 5.4).

5.1   Lagrange Subdivision

Lagrange interpolation provides interpolatory subdivision as follows. Given $(l, r) \in \mathbb{N}^{*2}$ and introducing for each value of k, $\sum_{m=-l+1}^{r} L_m(x) f_{k+m}^j$, the polynomial of degree $(l+r-1)$ that interpolates the points $((m, f_{k+m}^j), -l+1 \leq m \leq r)$, the subdivision is given by

$$\begin{cases} f_{2k}^{j+1} = f_k^j, \\ f_{2k+1}^{j+1} = \sum_{m=-l+1}^{r} L_m(\tfrac{1}{2}) f_{k+m}^j. \end{cases}$$

Here, $\{L_m(x)\}_{m=-l+1,\ldots,r}$ stands for the Lagrange functions associated to the stencil $\{-l+1,\ldots,r\}$. The coefficients of the subdivision mask are then given by $h_0 = 1$ and $h_{2i+1} = L_{-i}(1/2), -r \leq i \leq l-1$.

We then have

Proposition 9

Given $(l, r) \in \mathbb{N}^{*2}$, the Lagrange subdivision mask satisfies

$$\sum_{i=-r}^{l-1} h_{2i}(2i)^n = \sum_{i=-r}^{l-1} h_{2i+1}(2i+1)^n = \delta_{n,0}, \quad n = 0, 1, \ldots, l+r-1, \quad (32)$$

which can be regarded as an enhanced condition (29) with $L = l+r-1$. Moreover, it is the only family $\{h_k\}_{k=-2r,\ldots,2l-1}$ satisfying (32).

Proof

Since $h_{2i} = \delta_{i,0}$ the left hand side term of (32) is $\delta_{n,0}$. Moreover, using the interpolatory property of the Lagrange subdivision for polynomials of degree less than n gives directly that $\sum_{i=-r}^{l-1} h_{2i+1}(2i+1)^n = \delta_{n,0} \; \forall n \in \{0, 1, 2, \ldots, l+r-1\}$. To prove uniqueness, we rewrite (32) as

$$A \begin{bmatrix} h_{-2r} \\ h_{-2r+2} \\ \vdots \\ h_{2l-4} \\ h_{2l-2} \end{bmatrix} = \begin{bmatrix} 1 \\ 0 \\ \vdots \\ 0 \\ 0 \end{bmatrix} \quad and \quad B \begin{bmatrix} h_{-2r+1} \\ h_{-2r+3} \\ \vdots \\ h_{2l-3} \\ h_{2l-1} \end{bmatrix} = \begin{bmatrix} 1 \\ 0 \\ \vdots \\ 0 \\ 0 \end{bmatrix},$$

with

$$A = \begin{bmatrix} 1 & 1 & \cdots & 1 & 1 \\ -2r & -2r+2 & \cdots & 2l-4 & 2l-2 \\ \vdots & \vdots & \vdots & \vdots & \vdots \\ (-2r)^{l+r-2} & (-2r+2)^{l+r-2} & \cdots & (2l-4)^{l+r-2} & (2l-2)^{l+r-2} \\ (-2r)^{l+r-1} & (-2r+2)^{l+r-1} & \cdots & (2l-4)^{l+r-1} & (2l-2)^{l+r-1} \end{bmatrix},$$

$$B = \begin{bmatrix} 1 & 1 & \cdots & 1 & 1 \\ -2r+1 & -2r+3 & \cdots & 2l-3 & 2l-1 \\ \vdots & \vdots & \vdots & \vdots & \vdots \\ (-2r+1)^{l+r-2} & (-2r+3)^{l+r-2} & \cdots & (2l-3)^{l+r-2} & (2l-1)^{l+r-2} \\ (-2r+1)^{l+r-1} & (-2r+3)^{l+r-1} & \cdots & (2l-3)^{l+r-1} & (2l-1)^{l+r-1} \end{bmatrix}.$$

Since A and B are invertible Vandermonde matrices the uniqueness is proved.   □

Example
The 4-point centred Lagrange interpolating subdivision scheme is given by

$$\begin{cases} f_{2k}^{j+1} = f_k^j \\ f_{2k+1}^{j+1} = -\frac{1}{16}f_{k-1}^j + \frac{9}{16}f_k^j + \frac{9}{16}f_{k+1}^j - \frac{1}{16}f_{k+2}^j \end{cases}$$

corresponding to

$$M_h = \{h_3, h_2, h_1, h_0, h_{-1}, h_{-2}, h_{-3}, h_{-4}\} = \{-\frac{1}{16}, 0, \frac{9}{16}, 1, \frac{9}{16}, 0, -\frac{1}{16}, 0\} .$$

Applying Proposition 5 gives

$$\tilde{H}'_{M_h} = \begin{bmatrix} \tilde{h}_2^0 & \tilde{h}_3^0 & \tilde{h}_4^0 & \tilde{h}_5^0 & \tilde{h}_6^0 \\ \tilde{h}_0^2 & \tilde{h}_1^2 & \tilde{h}_2^2 & \tilde{h}_3^2 & \tilde{h}_4^2 \\ \tilde{h}_{-2}^4 & \tilde{h}_{-1}^4 & \tilde{h}_0^4 & \tilde{h}_1^4 & \tilde{h}_2^4 \\ \tilde{h}_{-4}^6 & \tilde{h}_{-3}^6 & \tilde{h}_{-2}^6 & \tilde{h}_{-1}^6 & \tilde{h}_0^6 \\ \tilde{h}_{-6}^8 & \tilde{h}_{-5}^8 & \tilde{h}_{-4}^8 & \tilde{h}_{-3}^8 & \tilde{h}_{-2}^8 \end{bmatrix} = \begin{bmatrix} 9 & -16 & 9 & 0 & -1 \\ 1 & 0 & 0 & 0 & 0 \\ 0 & 0 & 1 & 0 & 0 \\ 0 & 0 & 0 & 0 & 1 \\ -1 & 0 & 9 & -16 & 9 \end{bmatrix} .$$

Remark 3
The five elementary decimation operators defined above correspond to sub-sampling (lines $2, 3$ and 4 of $\tilde{H}'_{M_h}$) and polynomial extrapolations of degree 3 (from the positions $(0, .5, 1, 2)$ to $x = -1$ for line 1 and from the positions $(-2, -1, -.5, 0)$ to $x = 1$ for line 5). Note that as for all interpolatory subdivision, sub-sampling provides an optimally stable decimation with $\sum_l |\tilde{h}_l| = 1$.

5.2   Compactly Supported Wavelet Subdivision

Wavelets and, more precisely, scaling functions for multi-resolutions [8], are known to provide subdivision operators.

Orthogonality and zero moment conditions translate on the scaling coefficient $M_{h'} = \{h'_0, h'_1, \dots, h'_{2N-1}\}$ as [8]

$$\begin{cases} \sum_i h'_i h'_{i+2j} = 2\delta_{j,0} \\ \sum_i (-1)^i h'_i i^p = 0 \end{cases} \tag{33}$$

for $j \in \mathbb{Z}, p = 0, 1, \dots, N-1$.

According to orthogonal compactly supported wavelet theory, the rescaled operators $h = \sqrt{2}h'$ and $\tilde{h} = \frac{1}{\sqrt{2}}h'$ are consistent subdivision/decimation operators. More precisely, compact support wavelets of length $2N$ constructed in [8] lead to the unique subdivision/decimation operators with the same mask (up to $\sqrt{2}$ rescaling) with exponential decay of the error (Proposition 7) corresponding to $L = N - 1$.

Example

For $N = 2$ we get from [8]

$$[h'_0 \quad h'_1 \quad h'_2 \quad h'_3] = [\frac{1+\sqrt{3}}{4\sqrt{2}} \quad \frac{3+\sqrt{3}}{4\sqrt{2}} \quad \frac{3-\sqrt{3}}{4\sqrt{2}} \quad \frac{1-\sqrt{3}}{4\sqrt{2}}] \, .$$

Applying Proposition 4 for $h = \sqrt{2}h'$ we get

$$\tilde{H}_{M_h} = \begin{bmatrix} \tilde{h}^0_2 & \tilde{h}^0_3 \\ \tilde{h}^2_0 & \tilde{h}^2_1 \end{bmatrix} = H^{-1}_{M_h} = \begin{bmatrix} \frac{3-\sqrt{3}}{4} & \frac{1+\sqrt{3}}{4} \\ \frac{1-\sqrt{3}}{4} & \frac{3+\sqrt{3}}{4} \end{bmatrix}^{-1} = \begin{bmatrix} \frac{3+\sqrt{3}}{4} & -\frac{1+\sqrt{3}}{4} \\ \frac{-1+\sqrt{3}}{4} & \frac{3-\sqrt{3}}{4} \end{bmatrix}$$

and therefore two elementary decimation operators $\tilde{h}^0$ and $\tilde{h}^2$.

For $\lambda = \frac{1}{2}\frac{\sqrt{3}-1}{\sqrt{3}+1}$, the linear combination

$$[\tilde{h}_0 \quad \tilde{h}_1 \quad \tilde{h}_2 \quad \tilde{h}_3] = \lambda[0 \quad 0 \quad \tilde{h}^0_2 \quad \tilde{h}^0_3] + (1-\lambda)[\tilde{h}^2_0 \quad \tilde{h}^2_1 \quad 0 \quad 0]$$

$$= [\frac{1+\sqrt{3}}{8} \quad \frac{3+\sqrt{3}}{8} \quad \frac{3-\sqrt{3}}{8} \quad \frac{1-\sqrt{3}}{8}]$$

provides $\tilde{h} = \frac{1}{\sqrt{2}}h'$. We also get that (29) is verified for $n \leq N - 1 = 1$.

5.3   B-Spline Subdivision

It is well known that the scaling relation satisfied by the B-spline basis functions of order m, $B_m(t) = \sum_k h^m_k B_m(2t - k)$ implies that any spline function $C(t) = \sum_k f^j_k B_m(2^j t - k)$ can also be written as $C(t) = \sum_k f^{j+1}_k B_m(2^{j+1}t - k)$ with

$$f^{j+1}_k = \sum_l h^m_{k-2l} f^j_k \quad k \in \mathbb{Z} \, .$$

From the definition $B_{m+1}(t) = \int_{t-1}^t B_m(\tau)d\tau$ and $B_0(t) = \chi_{[0,1]}$ with χ_ω the characteristic function of the domain ω, it follows that the mask of the B-spline subdivision of degree m is given by

$$h^m_k = \frac{1}{2^m}\binom{m+1}{k}, \quad k = 0, 1, \ldots, m+1 \, . \tag{34}$$

We then have the following

Proposition 10

Given $m \in \mathbb{N}$, the mask of the B-spline subdivision scheme satisfies condition (29) with $L = m$, i.e.

$$\sum_{l=0}^{\lceil m/2 \rceil} h_{2l}(2l)^n = \sum_{l=0}^{\lfloor m/2 \rfloor} h_{2l+1}(2l+1)^n, \quad n = 0, 1, \ldots, m \, . \tag{35}$$

Moreover, it is the only mask $\{h_k\}_{0 \leq k \leq m+1}$ satisfying (29) and (10) with $L = m$.

Proof
It is easy to verify by induction that

$$\sum_{k=0}^{m} \binom{m}{k} k^n (-1)^k = 0, \quad \forall n \le m-1,$$

and splitting the previous sum with respect to even and odd indices leads to (35) where the mask is given by (34).

To prove uniqueness, we rewrite (35) incorporating (10) as

$$A \begin{bmatrix} h_0 \\ -h_1 \\ h_2 \\ -h_3 \\ \vdots \end{bmatrix} = \begin{bmatrix} 2 \\ 0 \\ 0 \\ 0 \\ \vdots \end{bmatrix} \quad with \quad A = \begin{bmatrix} 1 & -1 & 1 & -1 & \cdots \\ 1 & 1 & 1 & 1 & \cdots \\ 0 & 1 & 2 & 3 & \cdots \\ 0 & 1^2 & 2^2 & 3^2 & \cdots \\ \vdots & \vdots & \vdots & \vdots & \end{bmatrix}.$$

After a short calculus, the determinant of A can be written as the sum of strictly positive determinants of Vandermonde matrices. Therefore, $det(A) \ne 0$, that concludes the proof.   □

Example 1
For $m = 3$ we get

$$\begin{cases} f_{2k}^{j+1} = \frac{3}{4} f_k^j + \frac{1}{4} f_{k+1}^j \\ f_{2k+1}^{j+1} = \frac{1}{4} f_k^j + \frac{3}{4} f_{k+1}^j \end{cases}$$

and $M_h = \{h_1, h_0, h_{-1}, h_{-2}\} = \{\frac{1}{4}, \frac{3}{4}, \frac{3}{4}, \frac{1}{4}\}$. We obtain two elementary decimation masks from

$$\tilde{H}_{M_h} = \begin{bmatrix} \tilde{h}_0^0 & \tilde{h}_1^0 \\ \tilde{h}_{-2}^2 & \tilde{h}_{-1}^2 \end{bmatrix} = H_{M_h}^{-1} = \begin{bmatrix} \frac{3}{4} & \frac{1}{4} \\ \frac{1}{4} & \frac{3}{4} \end{bmatrix}^{-1} = \begin{bmatrix} \frac{3}{2} & -\frac{1}{2} \\ -\frac{1}{2} & \frac{3}{2} \end{bmatrix}$$

and other solutions for any $\lambda \in \mathbb{R}$ as

$$\begin{aligned} [\tilde{h}_{-2} \quad \tilde{h}_{-1} \quad \tilde{h}_0 \quad \tilde{h}_1] &= \lambda[0 \quad 0 \quad \tilde{h}_0^0 \quad \tilde{h}_1^0] + (1-\lambda)[\tilde{h}_{-2}^2 \quad \tilde{h}_{-1}^2 \quad 0 \quad 0] \\ &= [-\frac{1}{2}(1-\lambda) \quad \frac{3}{2}(1-\lambda) \quad \frac{3}{2}\lambda \quad -\frac{1}{2}\lambda] \end{aligned}.$$

Remark 4
For splines of order 3, the subdivision given above coincides with the values at 1/4 and 3/4 of the polynomial of degree 1 interpolating $(0, f_k^j), (1, f_{k+1}^j)$. The elementary decimations correspond to the extrapolations from the right (first line of $H_{M_h}^{-1}$) or from the left (second line of $H_{M_h}^{-1}$).

Example 2
For splines of order 5 we get

$$\begin{cases} f_{2k}^{j+1} = \frac{5}{16} f_{k-1}^j + \frac{5}{8} f_k^j + \frac{1}{16} f_{k+1}^j \\ f_{2k+1}^{j+1} = \frac{1}{16} f_{k-1}^j + \frac{5}{8} f_k^j + \frac{5}{16} f_{k+1}^j \end{cases}$$

and $M_h = \{h_3, h_2, h_1, h_0, h_{-1}, h_{-2}\} = \{\frac{1}{16}, \frac{5}{16}, \frac{5}{8}, \frac{5}{8}, \frac{5}{16}, \frac{1}{16}\}$. We obtain four elementary decimation masks from

$$
\tilde{H}_{M_h} = \begin{bmatrix} \tilde{h}_0^0 & \tilde{h}_1^0 & \tilde{h}_2^0 & \tilde{h}_3^0 \\ \tilde{h}_{-2}^2 & \tilde{h}_{-1}^2 & \tilde{h}_0^2 & \tilde{h}_1^2 \\ \tilde{h}_{-4}^4 & \tilde{h}_{-3}^4 & \tilde{h}_{-2}^4 & \tilde{h}_{-1}^4 \\ \tilde{h}_{-6}^6 & \tilde{h}_{-5}^6 & \tilde{h}_{-4}^6 & \tilde{h}_{-3}^6 \end{bmatrix} = H_{M_h}^{-1} = \begin{bmatrix} \frac{5}{16} & \frac{5}{8} & \frac{1}{16} & 0 \\ \frac{1}{16} & \frac{5}{8} & \frac{5}{16} & 0 \\ 0 & \frac{5}{16} & \frac{5}{8} & \frac{1}{16} \\ 0 & \frac{1}{16} & \frac{5}{8} & \frac{5}{16} \end{bmatrix}^{-1} = \begin{bmatrix} \frac{35}{8} & -\frac{47}{8} & \frac{25}{8} & -\frac{5}{8} \\ -\frac{5}{8} & \frac{25}{8} & -\frac{15}{8} & \frac{3}{8} \\ \frac{3}{8} & -\frac{15}{8} & \frac{25}{8} & -\frac{5}{8} \\ -\frac{5}{8} & \frac{25}{8} & -\frac{47}{8} & -\frac{35}{8} \end{bmatrix}.
$$

Other solutions can be constructed following (16) using for instance $\{\lambda_1, \lambda_2, \lambda_3, \lambda_4\}$ as

$$
\begin{aligned}
&[\tilde{h}_{-6} \quad \tilde{h}_{-5} \quad \tilde{h}_{-4} \quad \tilde{h}_{-3} \quad \tilde{h}_{-2} \quad \tilde{h}_{-1} \quad \tilde{h}_0 \quad \tilde{h}_1 \quad \tilde{h}_2 \quad \tilde{h}_3] \\
&= \lambda_1 [0 \quad 0 \quad 0 \quad 0 \quad 0 \quad 0 \quad \tilde{h}_0^0 \quad \tilde{h}_1^0 \quad \tilde{h}_2^0 \quad \tilde{h}_3^0] \\
&+ \lambda_2 [0 \quad 0 \quad 0 \quad 0 \quad \tilde{h}_{-2}^2 \quad \tilde{h}_{-1}^2 \quad \tilde{h}_0^2 \quad \tilde{h}_1^2 \quad 0 \quad 0] \\
&+ \lambda_3 [0 \quad 0 \quad \tilde{h}_{-4}^4 \quad \tilde{h}_{-3}^4 \quad \tilde{h}_{-2}^4 \quad \tilde{h}_{-1}^4 \quad 0 \quad 0 \quad 0 \quad 0] \\
&+ \lambda_4 [\tilde{h}_{-6}^6 \quad \tilde{h}_{-5}^6 \quad \tilde{h}_{-4}^6 \quad \tilde{h}_{-3}^6 \quad 0 \quad 0 \quad 0 \quad 0 \quad 0 \quad 0].
\end{aligned}
$$

Remark 5
The values $\{\lambda_1, \lambda_2, \lambda_3, \lambda_4\} = \{\frac{1}{100}, \frac{47}{300}, \frac{47}{60}, \frac{1}{20}\}$ minimize $\sum_{i=-6}^{3} |\tilde{h}_i|$ to the value $\frac{163}{40}$. Following Remark 2, the corresponding decimation operator gets a smaller stability constant than any of the elementary decimation for which the stability constants are $(14, 6, 6, 14)$.

5.4 Penalized Lagrange Subdivision

We finally consider a non-stationary (i.e. depending on the scale j) subdivision scheme recently introduced in [14] and focus in the sequel on the associated consistent decimation masks generated by our approach.

Using the notations of [14], the scheme is here constructed from a polynomial $P_j(x) = 100(2^{-2j})x^2 - 2^{-4j}x^4$ and a vector of penalization $C = (0, 2, 0, 0)$.

Denoting $M_{h^{(j)}} = \{h_3^{(j)}, h_2^{(j)}, h_1^{(j)}, h_0^{(j)}, h_{-1}^{(j)}, h_{-2}^{(j)}, h_{-3}^{(j)}, h_{-4}^{(j)}\}$, it first comes out that

$$
\lim_{j \to -\infty} M_{h^{(j)}} = \{-\frac{1}{16}, 0, \frac{9}{16}, 1, \frac{9}{16}, 0, -\frac{1}{16}, 0\}, \tag{36}
$$

and

$$
\lim_{j \to +\infty} M_{h^{(j)}} = \{\frac{1}{3}, \frac{1}{8}, 0, 0, 1, \frac{9}{8}, -\frac{1}{3}, -\frac{1}{4}\}. \tag{37}
$$

Therefore, according to the scale j, the subdivision evolves from the classical interpolatory Lagrange subdivision (36) to a non-interpolatory one of Lagrange-type. Indeed, the coefficients in (37) are the point values at $x = 0$ or $x = \frac{1}{2}$ of the Lagrange functions associated with the stencil $\{-1, 1, 2\}$.

According to Proposition 4, it is then possible to generate for each $j \in \mathbb{Z}$ the matrix of associated consistent elementary decimation masks. As an example, Fig. 1 displays the evolution of the third row of $\tilde{H}_{M_h}$ for $j \in [-10, 10]$.

It appears that the decimation mask quickly converges towards its asymptotical limit which is a sub-sampling ($\{0,0,0,0,1,0\}$) when $j \to -\infty$ and $\{-\frac{54}{107}, \frac{144}{107}, -\frac{9}{107}, \frac{24}{107}, -\frac{6}{107}, \frac{8}{107}\}$ when $j \to +\infty$. It is also interesting to notice that, as expected, these decimations are consistent with the asymptotical subdivision schemes associated with the masks (36) and (37) respectively.

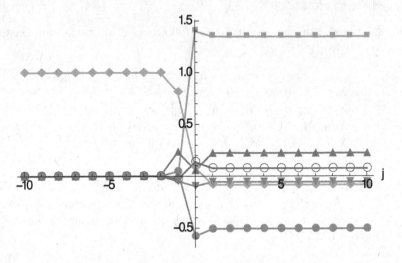

Fig. 1. Six coefficients of the mask of a decimation operator consistent with the penalized Lagrange subdivision scheme for scale $-10 \leq j \leq 10$.

6   Conclusion

We have shown, in this paper, that a construction of all the decimation operators consistent with a given linear and homogeneous subdivision operator can be performed. The proposed generic approach first leads to the construction of elementary decimation operators by exploiting the consistency condition. Then, all the possible consistent decimation operators can be obtained by linear combinations of translated versions of elementary ones. A theoretical analysis has been performed to provide an error bound for the so-called prediction error. It turned out that the decay rate is controlled by the subdivision scheme only. Moreover, the bound depends on the stability constant of the decimation operator that leads to the proposition of an heuristic strategy for the selection of the decimation mask among the family of consistent ones. Several applications have been performed in order to show the interest of our approach. It first appeared that our method is coherent with previous constructions of decimation operators associated with well-known subdivision schemes. Moreover, it allows to go beyond such constructions by offering a large choice of decimation masks that can be used to improve the stability of multi-scale transforms. Finally, it provides an efficient tool to design decimation operators for non classical subdivision schemes. Some generalizations to position dependent schemes as well as to some non linear subdivision operators are in progress [13].

References

1. Amat, S., Donat, R., Liandrat, J., Trillo, J.: Analysis of a fully nonlinear multiresolution scheme for image processing. Found. Comput. Math. **6**(2), 193–225 (2006)
2. Amat, S., Liandrat, J.: On the stability of PPH nonlinear multiresolution. Appl. Comput. Harmon. Anal. **18**, 198–206 (2005)
3. Arandiga, F., Baccou, J., Doblas, M., Liandrat, J.: Image compression based on a multi-directional map-dependent algorithm. Appl. and Comp. Harm. Anal. **23**(2), 181–197 (2007)
4. Baccou, J., Liandrat, J.: Kriging-based interpolatory subdivision schemes. Appl. Comput. Harmon. Anal. **35**, 228–250 (2013)
5. Cavaretta, A., Dahmen, W., Micchelli, C.: Stationary subdivision. In: Memoirs of the American Mathematics Society, vol. 93, No. 453, Providence, Rhode Island (1991)
6. Cohen, A., Daubechies, I., Feauveau, J.C.: Biorthogonal bases of compactly supported wavelets. CPAM **45**(5), 485–560 (1992)
7. Cohen, A., Dyn, N., Matei, B.: Quasilinear subdivision schemes with applications to ENO interpolation. Appl. Comput. Harmon. Anal. **15**, 89–116 (2003)
8. Daubechies, I.: Ten Lectures on Wavelets. SIAM, Philadelphia (1992)
9. De Rham, G.: Un peu de mathématiques à propos d'une courbe plane. Elem. Math. **2**, 73–76 (1947)
10. Deslauries, G., Dubuc, S.: Interpolation dyadique. In: Fractals, dimensions non entières at applications, pp. 44–55 (1987)
11. Dyn, N.: Subdivision schemes in computer-aided geometric design. In: Light, W. (ed.) Advances in Numerical Analysis II, Wavelets, Subdivision Algorithms and Radial Basis Functions, pp. 36–104. Clarendon Press, Oxford (1992)
12. Harten, A.: Multiresolution representation of data: a general framework. SIAM J. Numer. Anal. **33**(3), 1205–1256 (1996)
13. Kui, Z.: Approximation multiechelle non lineaire et applications en analyse de risques. Ph.D. thesis, Ecole Centrale Marseille, Marseille, France (2018)
14. Si, X., Baccou, J., Liandrat, J.: On four-point penalized lagrange subdivision schemes. Appl. Math. Comput. **281**, 278–299 (2016)
15. Sweldens, W.: The lifting scheme: a custom-design construction of biorthogonal wavelets. Appl. Comput. Harmon. Anal. **3**(2), 186–200 (1996)

Translation Surfaces and Isotropic Transport Nets on Rational Minimal Surfaces

Jan Vršek[1,2]([✉]) and Miroslav Lávička[1,2]

[1] Department of Mathematics, Faculty of Applied Sciences,
University of West Bohemia, Univerzitní 8, 306 14 Plzeň, Czech Republic
{vrsekjan,lavicka}@kma.zcu.cz
[2] NTIS – New Technologies for the Information Society,
Faculty of Applied Sciences, University of West Bohemia,
Univerzitní 8, 306 14 Plzeň, Czech Republic

Abstract. We will deal with the translation surfaces which are the shapes generated by translating one curve along another one. We focus on the geometry of translation surfaces generated by two algebraic curves in space and study their properties, especially those useful for geometric modelling purposes. It is a classical result that each minimal surface may be obtained as a translation surface generated by an isotropic curve and its complex conjugate. Thus, we can study the minimal surfaces as special instances of translation surfaces. All the results about translation surfaces will be directly applied also to minimal surfaces. Finally, we present a construction of rational isotropic curves with a prescribed tangent field which leads to the description of all rational minimal surfaces. A close relation to surfaces with Pythagorean normals will be also discussed.

1 Introduction

In computer aided geometric design (CAGD) basic modelling surfaces, with the property being simple and widely used, are applied to construct complex models, see [1,2]. Typical examples are ruled surfaces, rotational surfaces, canal surfaces, swept surfaces, translation surfaces, etc. Recognition of these surfaces from their equations, investigation of suitable parameterization methods and other related topics became an active research area in the past, see e.g. [3–7].

In this paper, we focus on the translation surfaces that are shapes generated by translating one curve along another one. Hence, they are a simple solution to the task to interpolate a surface going through two prescribed curves. From this reason they are typical modelling shapes often used in industrial design and architecture. An invariant feature of a translation surface is the existence of a conjugate Chebyshef (transport) net, see [2]. We will focus on the geometry of translation surfaces generated by two space algebraic curves and study their properties that may be used later especially for geometric modelling purposes.

Furthermore, we will investigate in more detail the minimal surfaces as special instances of translation surfaces. We recall that a non-zero vector $(a_1, a_2, a_3) \in \mathbb{C}^3$

© Springer International Publishing AG 2017
M. Floater et al. (Eds.): MMCS 2016, LNCS 10521, pp. 186–201, 2017.
https://doi.org/10.1007/978-3-319-67885-6_10

is called isotropic if it satisfies the condition $a_1^2 + a_2^2 + a_3^2 = 0$. Then the isotropic curve is a complex curve whose all tangent directions are isotropic. It is a classical geometric result that each minimal surface may be obtained as a translation surface generated by some isotropic curve and its complex conjugate, i.e., an isotropic net on a minimal surface is a transport net, see [8]. Hence all the results about translation surfaces can be directly applied also to minimal surfaces. Finally, we will present a construction of rational isotropic curves with prescribed tangent field which leads to the description of all rational minimal surfaces. A close relation to surfaces with Pythagorean normals (PN surfaces), i.e., rational surfaces with rational offsets (see [9] for more details), will be also discussed.

2   Translation Surfaces

A translation surface is a simple solution to the problem of interpolating a surface passing through two given input curves. In this section we recall some fundamental properties of these surfaces and investigate in more detail their behaviour at infinity. All the obtained results will be in the next section directly applied to minimal surfaces.

2.1   Definition, Fundamental Properties and Singularities

In what follows we will study surfaces in the affine space $\mathbb{A}_{\mathbb{K}}^3$, where $\mathbb{K}$ is some suitable field, typically $\mathbb{K} = \mathbb{R}, \mathbb{C}$. Next let be given two points $\mathbf{p}, \mathbf{q} \in \mathbb{A}_{\mathbb{K}}^3$ then $\mathbf{p} \oplus \mathbf{q} = \frac{\mathbf{p}+\mathbf{q}}{2}$ denotes their midpoint. For two point sets X, Y we define

$$X \oplus Y = \{\mathbf{p} \oplus \mathbf{q} \mid \mathbf{p} \in X, \ \mathbf{q} \in Y\}. \tag{1}$$

Definition 1. *Let $\mathcal{A}, \ \mathcal{B} \subset \mathbb{A}_{\mathbb{K}}^3$ be two curves. The set $\mathcal{A} \oplus \mathcal{B}$ is called a* translation surface.

The construction is illustrated in Fig. 1. We assume that the input curves are algebraic and irreducible. Then the Zariski closure $\overline{\mathcal{A} \oplus \mathcal{B}}$ is an irreducible algebraic surface (i.e., a subset of $\mathbb{A}_{\mathbb{K}}^3$ defined by a single irreducible polynomial), unless $\mathcal{A}$ and $\mathcal{B}$ are parallel lines in which case $\overline{\mathcal{A} \oplus \mathcal{B}}$ is a line.

Assuming that the input curves are given by their parameterizations $\mathbf{a}(u)$ and $\mathbf{b}(v)$ respectively then the parameterization $\mathbf{x}(u,v) = \frac{1}{2}(\mathbf{a}(u) + \mathbf{b}(v))$ of the translation surface fulfils the differential equation

$$\mathbf{x}_{uv} = 0, \tag{2}$$

where the subscript denotes the mixed partial derivative w.r.t. the variables u and v. Conversely, any solution of this differential equation can be written as the sum of two functions in variables u and v, respectively. Hence we arrived at:

Lemma 1. *A surface $\mathcal{X}$ is translational if and only if it can be (locally) given by a parameterization $\mathbf{x}(u,v)$ fulfilling $\mathbf{x}_{uv} = 0$.*

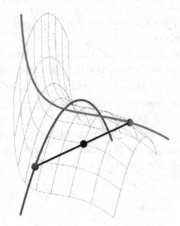

Fig. 1. Construction of a translation surface from two given input curves (red and blue) as the set of all the midpoints $\mathbf{p} \oplus \mathbf{q}$ (black). (Color figure online)

We recall that the translation surfaces are usually defined in an alternative way. Assume $\mathbf{o} \in \mathcal{A}, \mathcal{B}$ then we can construct a surface $\mathcal{X}$ either by translating the curve $\mathcal{A}$ along the curve $\mathcal{B}$, or by translating the curve $\mathcal{B}$ along the curve $\mathcal{A}$, cf. Fig. 2. Of course the obtained shapes are for both approaches the same – they differ only by scaling with factor 2. The approach presented in Fig. 2 immediately demonstrates that $\mathcal{X}$ and thus also $\mathcal{A} \oplus \mathcal{B}$ carries two families of congruent curves; this is the fundamental property of translation surfaces.

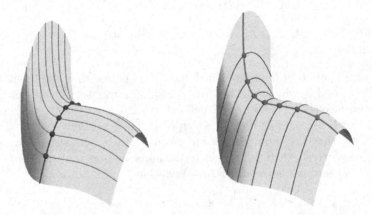

Fig. 2. Construction of a translation surface (yellow) by translating the curve $\mathcal{A}$ along the curve $\mathcal{B}$ (left), or by translating the curve $\mathcal{B}$ along the curve $\mathcal{A}$ (right). (Color figure online)

Clearly, the distinguished property recalled above can help when one wants to recognize the translation surface and reconstruct their generating curves from its given implicit equation. As the curve $\mathcal{A}$ is reproduced via translating along the curve $\mathcal{B}$ (and vice versa) it is efficient to investigate the singular points of $\mathcal{A} \oplus \mathcal{B}$ which must originate from the properties of $\mathcal{A}$ and $\mathcal{B}$.

For almost all input curves (see the discussion after Theorem 1 for the bad cases), a generic point $\mathbf{p} \in \mathcal{A} \oplus \mathcal{B}$ is generated by a unique pair $\mathbf{a} \in \mathcal{A}$ and $\mathbf{b} \in \mathcal{B}$. In this case we can describe the singular locus of $\mathcal{A} \oplus \mathcal{B}$ easily. The singular point set of the variety $\mathcal{X}$ will be denoted by $\mathcal{X}_{Sing}$. First, if $\mathbf{p} \in \mathcal{A}_{Sing}$ then $\mathbf{p} \oplus \mathcal{B}$ is the singular curve on $\mathcal{A} \oplus \mathcal{B}$ (the same holds for $\mathcal{A} \oplus \mathbf{q}$, where $\mathbf{q} \in \mathcal{B}_{Sing}$). Next if $\mathbf{p} \in \mathcal{A}$ and $\mathbf{q} \in \mathcal{B}$ are both regular points (i.e., the points not belonging to $\mathcal{A}_{Sing}$ or $\mathcal{B}_{Sing}$) and the corresponding tangent lines are *not parallel* then the tangent plane at $\mathbf{p} \oplus \mathbf{q}$ is given simply by

$$T_{\mathbf{p} \oplus \mathbf{q}}(\mathcal{A} \oplus \mathcal{B}) = T_{\mathbf{p}}\mathcal{A} \oplus T_{\mathbf{q}}\mathcal{B}. \tag{3}$$

On the other hand, the points with *parallel* tangent lines are singular points of the translation surface and they are contained in its self-intersection curve.

The points of the self-intersection of $\mathcal{A} \oplus \mathcal{B}$ can be expressed as $\mathbf{a}_1 + \mathbf{b}_1 = \mathbf{a}_2 + \mathbf{b}_2$. Let us denote $\mathcal{A} \ominus \mathcal{A} = \mathcal{A} \oplus (-\mathcal{A})$ where $-\mathcal{A}$ is the set centrally symmetric to $\mathcal{A}$ w.r.t. the origin. Then the points of self-intersection correspond to the set $(\mathcal{A} \ominus \mathcal{A}) \cap (\mathcal{B} \ominus \mathcal{B})$. Since any two surfaces always intersect in a curve (at least in the projective space over the algebraic closure of $\mathbb{K}$), we see that $\mathcal{A} \oplus \mathcal{B}$ contains a singular curve originated in the self-intersection of the surface.

Remark 1. Let $\mathcal{A}$ and $\mathcal{B}$ be two complex curves defined by real equations and assume that their real parts $\mathcal{A}_{\mathbb{R}}$ and $\mathcal{B}_{\mathbb{R}}$ are one-dimensional real curves. It may happen that $(\mathcal{A} \oplus \mathcal{B})_{\mathbb{R}}$ differs from $\mathcal{A}_{\mathbb{R}} \oplus \mathcal{B}_{\mathbb{R}}$. We illustrate this on the example. Let $\mathcal{A}_{\mathbb{R}}$ and $\mathcal{B}_{\mathbb{R}}$ be two curves from Fig. 3. Then $\mathcal{A}_{\mathbb{R}} \oplus \mathcal{B}_{\mathbb{R}}$ is the yellow surface from the same figure. However the real part of the surface $\mathcal{A} \oplus \mathcal{B}$ is the union of

Fig. 3. Two generating curves (red and blue) of the translation surface (yellow) from Remark 1 and its self-intersection curve (cyan). (Color figure online)

$\mathcal{A}_{\mathbb{R}} \oplus \mathcal{B}_{\mathbb{R}}$ *and* the self-intersection curve (cyan in the figure). The reason is that this curve is generated by the pairs of complex conjugated points.

2.2   Relation of $\mathcal{A} \oplus \mathcal{B}$ to $\mathcal{A} \times \mathcal{B}$

Since $\mathcal{A} \oplus \mathcal{B}$ is generated by $\mathcal{A}$ and $\mathcal{B}$ it should look like $\mathcal{A} \times \mathcal{B}$ in some sense. Hence, let us investigate this relation in more detail. This will be especially useful when the behaviour of $\mathcal{A} \times \mathcal{B}$ at infinity is thoroughly studied in the next section.

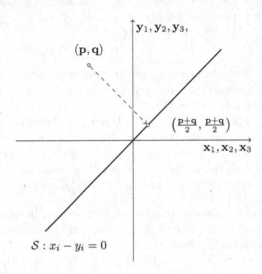

Fig. 4. Projection $\pi : \mathbb{P}_{\mathbb{C}}^6 \dashrightarrow \mathcal{S}$ with the centre $\mathcal{S}^\perp : w_0 = 0, x_i + y_i = 0$ relating $\mathcal{A} \times \mathcal{B}$ with $\mathcal{A} \oplus \mathcal{B}$

Consider the projective closure $\mathbb{P}_{\mathbb{K}}^6$ of $\mathbb{A}_{\mathbb{K}}^3 \times \mathbb{A}_{\mathbb{K}}^3$ and the point coordinates described in the form $(w_0 : x_1 : x_2 : x_3 : y_1 : y_2 : y_3)$. Let us note that for further considerations it is more convenient to work with $\mathbb{P}_{\mathbb{K}}^6$ instead of more usual closure $\mathbb{P}_{\mathbb{K}}^3 \times \mathbb{P}_{\mathbb{K}}^3$. Next let the subspace $\mathcal{S}$ be given by the equations $x_i - y_i = 0$, for $i = 1, 2, 3$. Then the projection $\pi : \mathbb{P}_{\mathbb{C}}^6 \dashrightarrow \mathcal{S}$ with the centre $\mathcal{S}^\perp : w_0 = 0, x_i + y_i = 0$ immediately relates $\mathcal{A} \times \mathcal{B}$ with $\mathcal{A} \oplus \mathcal{B}$, see Fig. 4.

Theorem 1. $\pi(\overline{\mathcal{A} \times \mathcal{B}}) = \overline{\mathcal{A} \oplus \mathcal{B}}$.

Clearly, if the projection $\pi : \mathcal{A} \times \mathcal{B} \to \mathcal{A} \oplus \mathcal{B}$ is one-to-one then these surfaces are the same from the birational point of view. Nonetheless there exist situations where π fails to be birational, for instance:

- If $\mathcal{A}$ and $\mathcal{B}$ are plane curves of degree ≥ 2 in parallel planes (Fig. 5).
- If $\mathcal{A}$ is a line and $\mathcal{B}$ is a curve (then $\mathcal{A} \oplus \mathcal{B}$ is a cylinder); if $\mathcal{B}$ intersects each ruling k times them π is $k : 1$ mapping.
- If $\mathcal{A} \ominus \mathcal{A} = \mathcal{B} \ominus \mathcal{B}$.

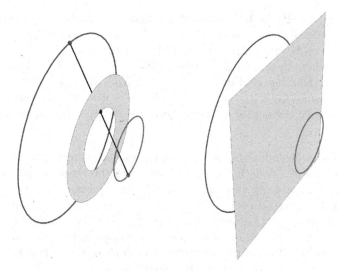

Fig. 5. An example when $(\mathcal{A} \oplus \mathcal{B})_{\mathbb{R}}$ (right) is strictly bigger then $\mathcal{A}_{\mathbb{R}} \oplus \mathcal{B}_{\mathbb{R}}$ (left) when $\mathcal{A}, \mathcal{B}$ are two circles lying in parallel planes.

The last case is closely related to the so called *multitranslation surfaces* which are surfaces obtained as translation surfaces by more than only one way. We would like to recall some classical result describing a property of these surfaces. If $\mathbf{p} \in \mathcal{A}$ is a smooth point on the curve then there exists the well defined tangent direction $\mathbf{v_p}$ (up to multiplication by a scalar) and one can consider a rational map $g_{\mathcal{A}} : \mathcal{A} \to \mathbb{P}^2_{\mathbb{K}}$. The closure $\mathcal{G}_{\mathcal{A}}$ of $g_{\mathcal{A}}(\mathcal{A})$ is an algebraic curve. It holds:

Theorem 2 (S. Lie 1882). *If $\mathcal{A} \oplus \mathcal{B} = \mathcal{C} \oplus \mathcal{D}$ then $\mathcal{G}_{\mathcal{A}} \cup \mathcal{G}_{\mathcal{B}} \cup \mathcal{G}_{\mathcal{C}} \cup \mathcal{G}_{\mathcal{D}}$ is an algebraic curve of degree four.*

When dealing with algebraic curves and surfaces it is more convenient to work over $\mathbb{C}$ then over $\mathbb{R}$. On the other hand, geometric modeling works over the field of real numbers and thus some techniques and results must be reconsidered when used in applications. This means that one has to be very careful before formulating some conclusions. We present it on the case of translation surfaces and show that birationality of π plays a significant role.

If not stated otherwise then assume that $\mathcal{A}$ and $\mathcal{B}$ are curves in complex space but defined by real equations and their real dimensions are one. As mentioned in Remark 1 the surface $(\mathcal{A} \oplus \mathcal{B})_{\mathbb{R}}$ can be strictly bigger then $\mathcal{A}_{\mathbb{R}} \oplus \mathcal{B}_{\mathbb{R}}$. If the projection π is not birational then their difference may become more dramatic.

Theorem 3. *If $\dim_{\mathbb{R}}(\mathcal{A} \oplus \mathcal{B})_{\mathbb{R}} \backslash (\mathcal{A}_{\mathbb{R}} \oplus \mathcal{B}_{\mathbb{R}}) = 2$ then the projection π is not birational.*

Proof. There exists a Zariski open set $U \subset \mathcal{A} \oplus \mathcal{B}$ and a natural number k such that for each $\mathbf{p} \in U$ the cardinality of the fiber $\pi^{-1}(\mathbf{p})$ is exactly k. This number k is called the degree of the projection and it is birational if and only if $k = 1$.

Because $\dim_{\mathbb{R}}(\mathcal{A} \oplus \mathcal{B})_{\mathbb{R}} \backslash (\mathcal{A}_{\mathbb{R}} \oplus \mathcal{B}_{\mathbb{R}}) = 2$ it cannot be contained in any closed set as the dimension of closed sets is at most 1. Hence $U \cap (\mathcal{A} \oplus \mathcal{B})_{\mathbb{R}} \backslash (\mathcal{A}_{\mathbb{R}} \oplus \mathcal{B}_{\mathbb{R}})$ is non-empty. Let $\mathbf{p}$ be an arbitrary point in this intersection. Then $\mathbf{p} = (\mathbf{a}_R + i\mathbf{a}_I + \mathbf{b}_R + i\mathbf{b}_I)/2$, where $\mathbf{a}_I$ and $\mathbf{b}_I$ are non-zero. However $\mathcal{A}$ and $\mathcal{B}$ are defined by real equations and thus they contain also the points $\mathbf{a}_R - i\mathbf{a}_I \in \mathcal{A}$ and $\mathbf{b}_R - i\mathbf{b}_I \in \mathcal{B}$. Trivially $\mathbf{p} = (\mathbf{a}_R - i\mathbf{a}_I + \mathbf{b}_R - i\mathbf{b}_I)/2$, which means that π^{-1} consists of at most two points. Hence the degree of the projection is at least two and it cannot be birational. $\qquad\square$

Next, we can formulate further consequences being implied by the assumption of birationality. Nonetheless, before we state the theorem we recall some fundamental facts dealing with the rationality of algebraic varieties. Let $\mathcal{X}$ be a variety of dimension d over a field $\mathbb{K}$. Then $\mathcal{X}$ is said to be *unirational*, or *parametric*, if there exists a rational map $\mathbf{p} : \mathbb{K}^d \to \mathcal{X}$ defined over $\mathbb{K}$ such that $\mathbf{p}(\mathbb{K}^d)$ is dense in $\mathcal{X}$. We speak about a *(rational) parameterization* $\mathbf{p}(t_1, \ldots, t_d)$ of $\mathcal{X}$. Furthermore, if $\mathbf{p}$ defines a birational map then $\mathcal{X}$ is called *rational*, and we say that $\mathbf{p}(t_1, \ldots, t_d)$ is a *proper parameterization*.

For the sake of brevity, let us mention results at least for algebraic curves and surfaces. By a theorem of Lüroth, a curve has a parameterization iff it has a proper parameterization iff its genus (see [10] for the definition of this notion) vanishes. Hence, for planar curves the notions of rationality and unirationality are equivalent for any field. On contrary, in the surface case unirationality implies rationality over algebraically closed fields only. Hence e.g. there exists a real surface possessing only real non-proper parameterizations.

Theorem 4. *If the projection π is birational then the following statements are equivalent:*

1. $(\mathcal{A} \oplus \mathcal{B})_{\mathbb{C}}$ *is rational,*
2. $(\mathcal{A} \oplus \mathcal{B})_{\mathbb{R}}$ *is rational,*
3. $\mathrm{g}(\mathcal{A}) = \mathrm{g}(\mathcal{B}) = 0$.

Proof. $(1) \Rightarrow (3)$ It is a standard fact in the theory of surfaces that $\mathcal{A} \times \mathcal{B}$ can be rational only if the genus of both curves vanishes. Moreover the birationality of the projection implies that $\mathcal{A} \times \mathcal{B}$ is rational if and only if $\mathcal{A} \oplus \mathcal{B}$ is rational. $(3) \Rightarrow (2)$ Unlike in the surface case the rationality of the curve $\mathcal{A}$ implies the existence of a birational parameterization of $\mathcal{A}_{\mathbb{R}}$ as well. Hence the proper parameterizations $\mathbf{a}(u)$ and $\mathbf{b}(v)$ yield the proper parameterization of $\mathcal{A}_{\mathbb{R}} \times \mathcal{B}_{\mathbb{R}}$ which propagates to the birational parameterization of $\mathcal{A}_{\mathbb{R}} \oplus \mathcal{B}_{\mathbb{R}}$. Finally, by Theorem 3 this is a birational parameterization of $(\mathcal{A} \oplus \mathcal{B})_{\mathbb{R}}$ as well and thus it is rational. $(2) \Rightarrow (1)$ This is evident. $\qquad\square$

Let us emphasize that without the assumption of birationality it may happen that a rational surface is generated by non-rational curves (in this case no parameterization reflecting the kinematic construction of the surface as $\mathcal{A} \oplus \mathcal{B}$ exists). Next, it would also hold $(1) \not\Rightarrow (2)$ in general, i.e., there exist surfaces with $\mathcal{X}_{\mathbb{C}}$ rational but $\mathcal{X}_{\mathbb{R}}$ unirational only.

2.3 Behaviour of Translation Surfaces at Infinity

Recognition of translation surfaces from their implicit equations is a challenging and interesting problem, cf. [5]. In this section, a necessary condition for deciding if the given surface is translational or not is presented. It reflects the behaviour of translation surfaces at infinity. Finding sufficient condition remains an open question. On the other hand, this investigation leads to a compact degree formula for translation surfaces.

Let $\overline{\mathcal{A} \oplus \mathcal{B}} \subset \mathbb{P}^3_{\mathbb{K}}$ be the projective closure of $\mathcal{A} \oplus \mathcal{B}$ and $\omega = \mathbb{P}^3_{\mathbb{K}} \backslash \mathbb{A}^3_{\mathbb{K}}$ be the plane at infinity. Since for $\mathbf{a}, \mathbf{b} \notin \omega$ is $\mathbf{a} \oplus \mathbf{b} \notin \omega$ the points at the boundary of $\mathcal{A} \oplus \mathcal{B}$ at infinity must be generated by intersections of $\overline{\mathcal{A}}$ and/or $\overline{\mathcal{B}}$ with ω.

If $\mathbf{a} \in \omega$, $\mathbf{b} \notin \omega$ then $\mathbf{a} \oplus \mathbf{b} = \mathbf{a}$. So we have to study in more detail the case $\mathbf{a}, \mathbf{b} \in \omega$. We recall the projection $\pi : \mathcal{A} \times \mathcal{B} \to \mathcal{A} \oplus \mathcal{B}$ and use $\overline{\mathcal{A} \times \mathcal{B}} \subset \mathbb{P}^6_{\mathbb{K}}$ instead of $\overline{\mathcal{A}} \times \overline{\mathcal{B}} \subset \mathbb{P}^3_{\mathbb{K}} \times \mathbb{P}^3_{\mathbb{K}}$. The image of a pair $(0 : a_1 : a_2 : a_3) \in \overline{\mathcal{A}}$, $(0 : b_1 : b_2 : b_3) \in \overline{\mathcal{B}}$ in $\overline{\mathcal{A} \times \mathcal{B}}$ is the line

$$L_{\mathbf{ab}} : \alpha(0 : a_1 : a_2 : a_3 : 0 : 0 : 0) + \beta(0 : 0 : 0 : 0 : b_1 : b_2 : b_3). \qquad (4)$$

Then for $\mathbf{a} \neq \mathbf{b}$ we obtain $\pi(L_{\mathbf{ab}}) = \{\alpha\mathbf{a} + \beta\mathbf{b}\} \subset \overline{\mathcal{A} \oplus \mathcal{B}}$. It remains to discuss the case $\mathbf{a} = \mathbf{b}$.

Let $\mathbf{c}$ be a common point of $\mathcal{A}$ and $\mathcal{B}$ in the plane ω. For the sake of simplicity we assume that $\mathbf{c}$ is a regular point on both curves and $\pi(L_{\mathbf{cc}}) = \mathbf{c}$. The point $\mathbf{C} = (0 : c_1 : c_2 : c_3 : c_1 : c_2 : c_3) \in L_{\mathbf{cc}}$ has no image in the projection $\pi(\mathbf{C}) = T_{\mathbf{C}}(\overline{\mathcal{A} \times \mathcal{B}}) \cap \mathcal{S}$. We can distinguish:

1. both tangents intersect ω transversaly $\Rightarrow \pi(\mathbf{C})$ is a "finite" line passing through $\mathbf{c}$,
2. the tangent at one point is contained in $\omega \Rightarrow \pi(\mathbf{C})$ is this tangent line,
3. both tangents lie in $\omega \Rightarrow \mathbf{C}$ is not regular on $\overline{\mathcal{A} \times \mathcal{B}} \Rightarrow \pi(\mathbf{C})$ is a union of lines through $\mathbf{c}$.

The observation discussed above, see also Fig. 6, can be summarized as

Proposition 1. *The boundary at infinity of any translation surface $\mathcal{A} \oplus \mathcal{B}$ consists solely of a union of lines.*

Clearly, Proposition 1 brings a necessary (unfortunately not sufficient) condition for recognizing translation surfaces. Moreover, studying the behaviour at infinity enables us to formulate the *degree formula* for translation surfaces.

Theorem 5. *Consider a translation surface $\mathcal{A} \oplus \mathcal{B}$. Then it holds*

$$\deg \mathcal{A} \oplus \mathcal{B} \leq \deg \mathcal{A} \cdot \deg \mathcal{B} - \#\{\overline{\mathcal{A}} \cap \overline{\mathcal{B}} \cap \omega\}. \qquad (5)$$

Proof. We recall the formula $\deg \mathcal{X} - \mathrm{mult}_{\mathbf{p}} \mathcal{X} = \deg \pi_{\mathbf{p}} \cdot \deg \mathcal{Y}$, where $\pi_{\mathbf{p}} : \mathcal{X} \dashrightarrow \mathcal{Y}$ is the projection from the point $\mathbf{p}$ with the multiplicity $\mathrm{mult}_{\mathbf{p}} \mathcal{X}$, see [11]. Next it holds $\deg(\mathcal{A} \times \mathcal{B}) = \deg \mathcal{A} \cdot \deg \mathcal{B}$ and since π is $1 : 1$ we have $\deg \pi = 1$. $\square$

Intersections of $\overline{\mathcal{A} \times \mathcal{B}}$ with the center of projection correspond exactly to common points at infinity of the curves $\mathcal{A}$ and $\mathcal{B}$. Multiplicity of the point is one if the curves behave "nicely", i.e., when the point is regular and the intersection with ω is transversal.

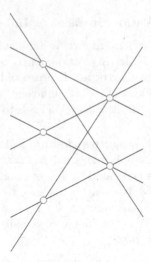

Fig. 6. The boundary at infinity of a translation surface $\mathcal{A} \oplus \mathcal{B}$ consisting of straight lines.

3 Transport Nets on Minimal Surfaces

In the previous sections we investigated translation surfaces and formulated for them several results. In what follows we will focus on minimal surfaces as prominent examples of translation surfaces, which means that all results obtained in the previous parts can be immediately applied also to minimal surfaces.

Minimal surfaces have been studied in various mathematical disciplines and for instance topological problems related to minimal surfaces are still challenging topics for investigations. Due to their applications, they have become an area of intense scientific study also in the fields such as molecular engineering or materials science. In addition, they play a role in general relativity. However, it is beyond the scope of this paper to recall a long history of this class of surfaces and to make a complete overview of all obtained results. We consider them as special instances of translation surfaces and we focus solely on those being algebraic, see also [12] and references therein, and especially on rational ones. Thus, for the sake of brevity, we refer only to the classical and general results on minimal surfaces, see e.g. [13–17] and then, when needed, to the papers/books which are closely related to our point of view. Our plan was not to write a survey paper on minimal surfaces but to present some new results interesting especially for the CAGD community.

3.1 Definition and Weierstrass-Enneper Formula

Minimal surfaces can be introduced in several equivalent ways. This fact and also its long history show how important and interesting they are for many mathematical and other disciplines as e.g. differential geometry, complex analysis and mathematical physics. Let $\mathbb{E}_{\mathbb{R}}^3$ denote the 3-dimensional real Euclidean space in what follows.

Definition 2. *A surface in* $\mathbb{E}_{\mathbb{R}}^3$ *is called* minimal surface *if its mean curvature vanishes identically.*

Minimal surfaces are also often defined as surfaces of the smallest area spanned by a given closed space curve, which is a property that can find some applications e.g. in garment and shoe design (in order to minimize material consumption). Well-known examples of minimal surfaces (although not being algebraic) are catenoids, helicoids, Schwarz minimal surfaces, Riemann's minimal surface, Henneberg surface, etc.

These surfaces can be locally parameterized as $\mathbf{x} : U \to X$, where $U \subset \mathbb{R}^2$, such that $\mathbf{x}$ is *conformal* or *isothermal* (i.e., satisfying the condition $\mathbf{x}_u \cdot \mathbf{x}_v = 0$, $\mathbf{x}_u \cdot \mathbf{x}_u = \mathbf{x}_v \cdot \mathbf{x}_v$) and *harmonic* ($\triangle \mathbf{x} = 0$), which declares a very close connection to complex analysis. In addition, they are the surfaces on which the asymptotic lines form an orthogonal net (i.e., the Dupin indicatrix consists of two conjugate rectangular hyperbolas) and thus they satisfy the distinguishing condition

$$g_{11}h_{22} - 2g_{12}h_{12} + g_{22}h_{22} = 0, \tag{6}$$

where g_{ij} and h_{ij} are the coefficients of the first and second fundamental form, respectively.

A well-known piece of the classical differential geometry is the Weierstrass-Enneper parameterization of minimal surfaces. It holds that any minimal surface defined over a simply-connected parameter domain can be represented by the *Weierstrass-Enneper formula*

$$\mathbf{x}(u,v) = \begin{pmatrix} \mathfrak{Re}\left(\dfrac{1}{2}\displaystyle\int_0^z f(w)(1 - g(w)^2)\,\mathrm{d}w\right) \\[2mm] \mathfrak{Re}\left(\dfrac{i}{2}\displaystyle\int_0^z f(w)(1 + g(w)^2)\,\mathrm{d}w\right) \\[2mm] \mathfrak{Re}\left(\displaystyle\int_0^z f(w)g(w)\,\mathrm{d}w\right) \end{pmatrix}, \quad w = u + iv, \tag{7}$$

where f, g are complex analytic functions defined on $U \subset \mathbb{C}$ and $\mathfrak{Re}$ takes the real part of considered functions, see [8]. The coefficients of the first fundamental form are in this case given as

$$g_{11} = g_{22} = \left(|f|(|g|^2 + 1)\right)^2, \quad g_{12} = 0, \tag{8}$$

i.e., the parameterization is conformal. Investigating (7) in more detail one can see that any choice of f, g as polynomials leads to a polynomial minimal surface $\mathbf{x}(u,v)$. On the other hand, choosing f, g as rational functions does not guarantee that $\mathbf{x}(u,v)$ is rational as integrals of rational functions are not rational, in general. So, it is a question how to generate *all* rational minimal surface. We will show that the approach based on translation surfaces and constructing rational transport nets help.

3.2 Minimal Surfaces as Translation Surfaces

Let us start with recalling the notion of isotropic (or minimal) curve, cf. [8]. An *isotropic curve* is a complex curve in $\mathbb{A}_{\mathbb{C}}^3$ such that a tangent direction $\mathbf{t} = (t_1, t_2, t_3)$ at each regular point satisfies the condition $\mathbf{t} \cdot \mathbf{t} = t_1^2 + t_2^2 + t_3^2 = 0$.

Proposition 2. *The mapping* $\Psi = (\psi_1, \psi_2, \psi_3) : U \to \mathbb{A}_{\mathbb{C}}^3$ *such that*

$$\psi_1(z) = \frac{1}{2} \int_{z_0}^{z} f(w)(1 - g(w)^2) \, dw$$

$$\psi_2(z) = \frac{i}{2} \int_{z_0}^{z} f(w)(1 + g(w)^2) \, dw \tag{9}$$

$$\psi_3(z) = \int_{z_0}^{z} f(w)g(w) \, dw$$

is a parameterization of some isotropic curve.

One can easily confirm that the condition on isotropic curves is satisfied. Formula (9) is called *Weierstrass formula* of the given isotropic curve. Conversely, it is possible to obtain the Weierstrass representation from Ψ by setting simply $f = \psi_1' - i\psi_2'$, $g = \frac{\psi_3'}{\psi_1' - i\psi_2'}$.

It holds that any surface possesses an isotropic parameterization (i.e., the parametric lines are isotropic). In particular, the isotropic parameterization then fulfils

$$g_{11} = g_{22} = 0, \ g_{12} \neq 0. \tag{10}$$

Next for minimal surfaces, assuming the condition for vanishing mean curvature (6) we obtain $h_{12} = 0$. This leads to

$$\Gamma_{12}^1 = \Gamma_{12}^2 = 0, \tag{11}$$

where Γ_{ij}^k are the Christoffel symbols. To sum up, as it holds

$$\mathbf{x}_{uv} = \Gamma_{12}^1 \mathbf{x}_u + \Gamma_{12}^2 \mathbf{x}_v + L\mathbf{N}, \tag{12}$$

where $\mathbf{N}$ is the unit normal vector at $\mathbf{x}$, we can see that condition (2) is satisfied for any isotropic parameterization of a minimal surface. Hence by Lemma 1 we arrive at the classical result (see [8]).

Theorem 6 (S. Lie). *All minimal surfaces are translation surfaces, of which the generating curves are isotropic curves.*

In addition this also relates formulae (7) and (9). Consider an isotropic curve $\mathcal{A} \subset \mathbb{A}_{\mathbb{C}}^3$ parameterized by $\Psi(z)$. Then $\mathbf{x}(u,v) = \mathfrak{Re}(\Psi(z))$ is a minimal surface patch. Denote by $\mathcal{A}^*$ the complex conjugate curve of $\mathcal{A}$. Then $\mathcal{A} \oplus \mathcal{A}^*$ is a complex translation surface in $\mathbb{A}_{\mathbb{C}}^3$. Its real part is a real minimal surface patch described by (7).

We would like to stress that using Theorem 6 all results about translation surfaces hold also for all minimal surfaces, which form their subfamily. For instance, the degree formula (5) formulated for algebraic translation surfaces can be also efficiently used for all algebraic minimal surfaces, too. Furthermore, it holds

Corollary 1. *The boundary at infinity of any algebraic minimal surface consists solely of a union of lines.*

Let us recall that it was already shown by Lie in [18] that the asymptote cone of a real algebraic minimal surface always decays in planes. In addition, Lie refers to the paper [19], in which the intersection of an algebraic minimal surface with the plane at infinity was also identified as consisting of straight lines only.

Remark 2. For the sake of completeness, it must be emphasized that the behaviour of minimal surfaces at infinity is usually associated with its "ends", e.g. ends asymptotic to a plane, see [16,17]. In this paper we are concerned with a different kind of behaviour at infinity. We study the points at infinity of algebraic surfaces which have a canonical extension to projective space. Of course, any planar end yields a straight line at infinity.

Example 1. A planar isotropic curve is a line. Hence, the first non-trivial example are cubics. With some additional work it can be shown that each such a cubic intersects ω with multiplicity 3 and hence it is a polynomial curve. As an example let us consider the isotropic cubic

$$\mathcal{A}: \left(s - \frac{s^3}{3}, \mathrm{i} s + \mathrm{i}\frac{s^3}{3}, s^2 \right) \tag{13}$$

Hence we obtain $\mathcal{A} \cap \omega = (0:1:\mathrm{i}:0)$ and $\mathcal{A}^* \cap \omega = (0:1:-\mathrm{i}:0)$. This gives $(\mathcal{A} \oplus \mathcal{A}^*)_{\mathbb{R}} \cap \omega = \{w = 0, z^9 = 0\}$, which is a 9-tuple line at infinity. Moreover we have $\deg(\mathcal{A} \oplus \mathcal{A}^*) = 9$. The surface $\mathcal{A} \oplus \mathcal{A}^*$ is the Enneper surface, see Fig. 7. Finally it can be proved that any minimal surface generated by cubic isotropic curves is the Enneper surface.

3.3 Rational Isotropic Curves and Rational Minimal Surfaces

As already mentioned before it is an open question how to generate all rational minimal surfaces from Weierstrass-Enneper formula. We show that this problem can be efficiently solved when we consider the minimal surfaces as translation surfaces. This is equivalent to answer (for isotropic curves) the question how to determine a rational function $\varphi(z)$ such that $\int \varphi(z)\mathbf{t}(z)\,\mathrm{d}z$ is again rational.

The approach from [20] gives us a motivation. We recall that the tangent developable surface is the envelope of the osculating planes of the regression curve and the osculating planes are given by the binormal vector field. Let us at least shortly present the main idea of the used approach.

It is known that at each point of a spatial curve $\mathbf{c}(s)$ one can construct the *normal*, *rectifying*, and *osculating planes*, whose points $\mathbf{x} = (x, y, z)^\top$ satisfy

$$\begin{aligned}
\mathbf{t}(s) \cdot \big(\mathbf{x} - \mathbf{c}(s)\big) &= 0, \\
\mathbf{n}(s) \cdot \big(\mathbf{x} - \mathbf{c}(s)\big) &= 0, \\
\mathbf{b}(s) \cdot \big(\mathbf{x} - \mathbf{c}(s)\big) &= 0,
\end{aligned} \tag{14}$$

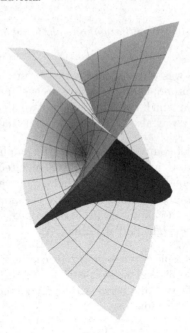

Fig. 7. The minimal Enneper surface as a translation surface with the cubic isotropic transport net.

respectively, where $(\mathbf{t}, \mathbf{n}, \mathbf{b})$ is the *Frenet frame*. The *Frenet-Serret equations* describing the variation of the Frenet frame read

$$\begin{pmatrix} \mathbf{t}' \\ \mathbf{n}' \\ \mathbf{b}' \end{pmatrix} = \sigma \begin{pmatrix} 0 & \kappa & 0 \\ -\kappa & 0 & \tau \\ 0 & -\tau & 0 \end{pmatrix} \begin{pmatrix} \mathbf{t} \\ \mathbf{n} \\ \mathbf{b} \end{pmatrix}, \tag{15}$$

where σ is the speed, κ is the curvature, and τ is the torsion of $\mathbf{c}(s)$. The equation of the osculating plane in (14) has the form

$$\mathbf{b}(s) \cdot \mathbf{x} = h(s), \tag{16}$$

where we define $h(s) := \mathbf{b}(s) \cdot \mathbf{c}(s)$. For a unit vector field $\mathbf{b}(s)$, the function $h(s)$ expresses the distance of the corresponding osculating plane from the origin.

It can be proved that (under some natural assumptions) differentiating (16) and using (15) gives an equation of the rectifying plane and differentiating again yields an equation of the normal plane. It follows that starting from a given rational unit vector $\mathbf{b}(s)$ and a given rational function $h(s)$, the points of a spatial curve $\mathbf{c}(s)$ may be identified as the intersection points of the associated osculating, rectifying and normal planes.

However, $\mathbf{c}(s)$ is not rational in general since the normalized tangent and normal vectors contain square root terms. To resolve this issue, the authors of [20] consider a modified *rational* system of three planes instead, namely

$$\mathbf{b}(s) \cdot \mathbf{x} = h(s), \quad \mathbf{b}'(s) \cdot \mathbf{x} = h'(s), \quad \mathbf{b}''(s) \cdot \mathbf{x} = h''(s). \tag{17}$$

The first plane is still the osculating one, but the other two are not the rectifying and normal planes any more (as $\mathbf{b}'(s)$ is generally not unit and $\mathbf{b}''(s)$ is not orthogonal to $\mathbf{b}'(s)$).

Modifying the system of planes (14), (15), (16) and (17) does not change the main contribution of the introduced approach: we can construct a curve with a given (not necessarily unit) binormal vector field. And moreover in our case when an isotropic curve is considered, $\mathbf{t}(s)$ is both tangent and binormal vector field. This simplifies the considerations and we can immediately formulate:

Lemma 2. *Let* $\mathbf{t}(s)$ *be a rational isotropic vector field and* $h(s)$ *an arbitrary rational function. Then the curve of regression of the tangent developable surface being the envelope of a 1-parametric family of planes* $\mathbf{t}(s) \cdot \mathbf{x} = h(s)$ *is a rational isotropic curve with the tangent field* $\varphi(s)\mathbf{t}(s)$, *where* $\varphi(s)$ *is a rational factor. Conversely any such a curve may be obtained in this way.*

This approach represents a rational alternative to integral formula (9). Furthermore, an arbitrary basepoint-free isotropic vector field can be written in the form

$$\mathbf{t}(s) = \left(1 - g^2(s), \mathrm{i}(1 + g^2(s)), 2g(s)\right). \tag{18}$$

Moreover, considering the rational minimal surfaces as translation surfaces, of which the generating curves are rational isotropic curves, we can formulate

Theorem 7. *All rational minimal surfaces can be generated as translation surfaces* $\mathcal{A} \oplus \mathcal{A}^*$ *where* $\mathcal{A}$ *is a rational isotropic curve from Lemma 2.*

Finally, we recall that the coefficients of the first fundamental form satisfies the condition (8) and the parameterization is thus conformal. This means that for rational minimal surfaces, $g_{11}g_{22} - g_{12}^2$ is a perfect square of some rational function, i.e., the parameterized surface $\mathbf{x}(u, v)$ is a *rational surface with Pythagorean normals* (*PN surface* in short), see [9].

Corollary 2. *All rational minimal surfaces are rational surfaces with Pythagorean normals.*

We recall that rational surfaces with Pythagorean normal vector fields (PN surfaces) were introduced in [9] as a surface analogy to Pythagorean hodograph (PH) curves defined previously in [21]. For a survey of shapes with Pythagorean property see e.g. [22] and references therein. These surfaces provide an elegant and practical solution of various difficult problems occurring in technical applications, in particular in the context of CNC machining. They are distinguished by having rational offsets, i.e., tool paths do not have to be approximated and they are described exactly in NURBS form, which represents currently a universal standard in computer-aided design, see [23]. Recently, it was revealed that all polynomial minimal surfaces are PN, cf. [24]. Now, we can see that this property holds not only for polynomial but for all rational minimal surfaces.

4 Conclusion

In this paper we studied translation surfaces as special instances of shapes important for the technical practice. We dealt especially with the question of their rationality, of their singular locus and it was shown that their boundary at infinity consists of straight lines. Next, a degree formula for translation surfaces was derived. Using a classical result that each minimal surface may be obtained as a translation surface, all obtained results were applied also to minimal surfaces. In addition, we presented a construction of all rational minimal surfaces and discussed their close relation to rational surfaces with Pythagorean normals.

Acknowledgments. The authors were supported by the project LO1506 of the Czech Ministry of Education, Youth and Sports. We thank to all referees for their valuable comments, which helped us to improve the paper.

References

1. Farin, G.: Curves and Surfaces for CAGD: A Practical Guide. Morgan Kaufmann Publishers Inc., San Francisco (2002)
2. Krivoshapko, S., Ivanov, V.: Encyclopedia of Analytical Surfaces. Springer, Cham (2015)
3. Weinert, K., Du, S., Damm, P., Stautner, M.: Swept volume generation for the simulation of machining processes. Int. J. Mach. Tools Manuf. **44**(6), 617–628 (2004)
4. Shen, L.-Y., Pérez-Díaz, S.: Characterization of rational ruled surfaces. J. Symbolic Comput. **63**, 21–45 (2014)
5. Pérez-Díaz, S., Shen, L.-Y.: Parametrization of translational surfaces. In: Proceedings of the 2014 Symposium on Symbolic-Numeric Computation, SNC 2014, New York, NY, USA, pp. 128–129. ACM (2014)
6. Vršek, J., Lávička, M.: Determining surfaces of revolution from their implicit equations. J. Comput. Appl. Math. **290**, 125–135 (2015)
7. Vršek, J., Lávička, M.: Recognizing implicitly given rational canal surfaces. J. Symbolic Comput. **74**, 367–377 (2016)
8. Struik, D.J.: Lectures on Classical Differential Geometry, 2nd edn. Dover Publications, Mineola (1988)
9. Pottmann, H.: Rational curves and surfaces with rational offsets. Comput. Aided Geom. Des. **12**(2), 175–192 (1995)
10. Shafarevich, I.R.: Basic Algebraic Geometry. Springer, Heidelberg (1974)
11. Mumford, D.: Algebraic Geometry I: Complex Projective Varieties. Springer, Heidelberg (1976)
12. Odehnal, B.: On algebraic minimal surfaces. KoG **20**, 61–78 (2016). Scientific and Professional Journal of Croatian Society for Geometry and Graphics
13. Weierstrass, K.: Über die flächen deren mittlere krümmung Überall gleich null ist. Monatsberichte der Berliner Akademie, pp. 612–625 (1866)
14. Schwarz, H.A.: Miscellen aus dem Gebiete der Minimalflächen, pp. 168–169. Springer, Heidelberg (1890)
15. Darboux, G.: Lecons sur la Theorie Generale des Surfaces et les Applications Geometriques Due Calcul Infinitesimal. Chelsea Publishing Series. American Mathematical Society, Providence (2000)

16. Jakob, R., Dierkes, U., Küster, A., Hildebrandt, S., Sauvigny, F.: Minimal Surfaces. Grundlehren der mathematischen Wissenschaften. Springer, Heidelberg (2010)
17. Osserman, R.: A Survey of Minimal Surfaces. Dover Publications, Mineola (2014)
18. Lie, S.: Synthetischanalytische Untersuchungen über Minimalflächen. I. Über reelle algebraische Minimalflächen. Archiv for Mathematik og Naturvidenskab, pp. 157–198 (1877)
19. Geiser, C.F.: Notiz über die algebraischen Minimumsflächen. Math. Ann. 3(4), 530–534 (1871)
20. Farouki, R.T., Šír, Z.: Rational Pythagorean-hodograph space curves. Comput. Aided Geom. Des. 28, 75–88 (2011)
21. Farouki, R., Sakkalis, T.: Pythagorean hodographs. IBM J. Res. Develop. 34(5), 736–752 (1990)
22. Farouki, R.: Pythagorean-Hodograph Curves: Algebra and Geometry Inseparable. Springer, Heidelberg (2008)
23. Piegl, L., Tiller, W.: The NURBS Book. Monographs in Visual Communication, 2nd edn. Springer, New York (1997)
24. Lávička, M., Vršek, J.: On a special class of polynomial surfaces with Pythagorean normal vector fields. In: Boissonnat, J.-D., Chenin, P., Cohen, A., Gout, C., Lyche, T., Mazure, M.-L., Schumaker, L. (eds.) Curves and Surfaces 2010. LNCS, vol. 6920, pp. 431–444. Springer, Heidelberg (2012). doi:10.1007/978-3-642-27413-8_27

Towards Subdivision Surfaces C2 Everywhere

Malcolm Sabin[✉]

Numerical Geometry Ltd., 19 John Amner Close, Ely, Cambs, UK
malcolm.sabin@btinternet.com

Abstract. The conditions for subdivision surfaces which are piecewise polynomial in the regular region to have continuity higher than C1 were identified by Reif [7]. The conditions are ugly and although schemes have been identified and implemented which satisfy them, those schemes have not proved satisfactory from other points of view. This paper explores what can be created using schemes which are not piecewise polynomial in the regular regions, and the picture looks much rosier. The key ideas are (i) use of quasi-interpolation (ii) local evaluation of coefficients in the irregular context. A new method for determining lower bounds on the Hölder continuity of the limit surface is also proposed.

Keywords: Subdivision surfaces · Continuity · Reproduction

1 Motivation

There is a myth among commercial CAD system suppliers that *"Subdivision surfaces are adequate for animation, but they can't be used for serious CAD because they are not C2 at extraordinary points."*

Although this myth cannot be completely discounted (like most myths there is some truth hidden underneath), in this raw form it is false.

1. There are subdivision surface schemes which are C2 at extraordinary vertices (see [9,18,19] and several others). They may all have other problems, but they are C2.
2. Lack of C2 at isolated points does not matter. The ideas of [18] could be applied after a few hundred iterations of a standard method. This would make the limit C2 everywhere, but it would be totally indistinguishable from the raw limit surface. Also, there is a bivariate interpolation technique which minimises the bending energy of the surface. This is widely regarded as good, but the second derivatives are unbounded at the data sites.
3. The 'raw' forms of popular subdivision surfaces [2,4,5] are not merely not C2. They actually have unbounded curvature at extraordinary points, and because the unboundedness accelerates at different rates for positive and negative Gaussian curvatures, this can force regions of the limit surface close to an extraordinary point to have the wrong sense of curvature [15,16]. Such extreme behaviour can be eliminated easily [23,24] by tuning of the coefficients in extraordinary regions, but this does not cure the lack of C2, and is not widely cited.

© Springer International Publishing AG 2017
M. Floater et al. (Eds.): MMCS 2016, LNCS 10521, pp. 202–217, 2017.
https://doi.org/10.1007/978-3-319-67885-6_11

The truth underneath is that the raw forms of the popular schemes, and even tuned versions to a lesser extent, suffer from local distortion of the surface over non-infinitesimal regions. It is these artifacts which are visible and objectionable in the reflection line plots, and it is these which need to be eliminated. The basic cause of these turns out to be the same as the cause of our inability to tune these schemes to perfection: the inability to achieve *quadratic generation* in any simple uniform stationary scheme which gives bi-polynomial pieces in the regular regions. This was understood nearly twenty years ago [7, 9, 10] and that understanding has deepened since [17].

This paper explores the idea of using schemes with fractal limit surfaces, the analog of the quasi-spline curves of [20, 21] to give the kind of behaviour which serious CAD needs. [25] has already explored the quad grid case, and so the focus here is on surfaces where the regular part of the mesh contains triangles.

Any full coverage of this will require many more years of work, and so this paper merely sets out the territory and provides some initial results and some ideas for directions which need to be explored. The exploration here is of surfaces with a quasi-interpolating degree of three.

2   Definitions

A scheme is said to **generate** polynomials of degree d if for any bivariate polynomial of degree d we can find a control polyhedron for which that function is the limit.

A scheme is said to **reproduce** polynomials of degree d if for any bivariate function of degree d we can find a control polyhedron for which that function is the limit, and for which the control points all lie on the function. The limit has to interpolate them.

The distinction between these two was not emphasised in the early papers, but it turns out to be fairly important. Over regular grids, most subdivision schemes can generate polynomials of low degree. Those which are B-spline based do not reproduce them for degrees above 1.

A scheme can reproduce polynomials without being an interpolating scheme for arbitrary data. Such a scheme is called a **quasi-interpolating scheme** of degree d where d is the highest degree of polynomial which it can reproduce.

A scheme is called a **stable generator** of degree d if it generates polynomials of degree d and if the maximum change of dth derivative with respect to addition of perturbations of maximum amplitude ϵ is continuous with respect to ϵ with Hölder exponent greater than 1, 0. The issue of stability is critical. The four-point scheme is a quasi-interpolant of degree 3, but not a stable one, because for any sequence of six data points which do not all lie on a cubic, the second derivatives at dyadic points within the central span diverge at a rate depending on the fourth differences of the local control points.

3 Quadratic and Cubic Reproduction in Functional Subdivision over a Regular Triangulation of the Domain

This section exemplifies an approach. It does not preclude further work on higher quasi-interpolation degrees, on different regular connectivities or on higher arities.

3.1 Stencil Sizes

A quasi-interpolating functional subdivision scheme has the property that if the values at all the old vertices lie on a polynomial, then new vertices will also get vertices on that polynomial. If this property holds at every step, then the limit surface will reproduce that polynomial, which we take as a condition for continuity of the relevant degree, and for avoiding certain artifacts.

In order for a new vertex to have that property, the value has to be chosen to match all polynomials of the target degree, d, which means that unless the scheme is actually interpolating, there is a number n of conditions to be fulfilled equal to the number of independent polynomials of that degree. These are tabulated in Table 1.

Table 1. Stencil sizes

Degree	Number
0	1
1	3
2	6
3	10
4	15
n	$(n+1)(n+2)/2$

Table 2. Number of available vertices

No of rings	Number of vertices in V-vertex stencil	Number of vertices in E-vertex stencil
0	1	
1	7	4
	13	8
2	19	10
	25	14
	31	18
r	$3(r+1)(r+2)+1$	

We get the new value by taking a linear combination of the values at the vertices in the stencil of the new point, and in order to be able to satisfy all these conditions, we need to have at least as many independent coefficients in those linear combinations. This leads to the unwelcome conclusion that high degrees of polynomial reproduction will require very large stencils, involving a number of rings varying at least linearly with the degree. See Table 2. Quadratics and cubics remain (just) within sensible reach.

We can have more coefficients (i.e. more old vertices in the footprint of the stencil) than conditions to satisfy, and in this case the choice of the values of those coefficients is underdetermined. This can be demanded by symmetry, since the number of vertices in a rotationally symmetric stencil goes up with the number of rings included.

However, in the case of the completely regular domain, symmetry also provides a way of resolving the extra freedom.

3.2 Construction of Cubic Quasi-interpolant

We know that each of the stencils must contain at least 10 entries for cubic quasi-interpolation. There are three e-vertex stencils as well as the v-vertex stencil, and so the total number of entries in the mask must be at least 40. We need some understanding in order to be able to construct such a beast.

This understanding comes from two sources: the first is that the v-vertex stencils must have 6-fold rotational symmetry, and that the e-vertex stencils must have two mirror symmetry axes. The second is that we cannot have good behaviour for general data if we do not have good behaviour for extruded data, in which each mesh direction has the same value at all points along each edge in that direction.

If the data is extruded, then the shape of a cross-section limit curve will be given by a univariate refinement scheme whose mask is given by the row-sums of the bivariate scheme. Further, factorisation of the mask is preserved under the taking of row-sums. Because there are three directions for the row sums, the desired symmetries come out in the wash.

We know a univariate scheme with cubic quasi-interpolation. This was described by Hormann and Sabin [20]. Its mask is

$$2 \left(\frac{[1,1]}{2} \right)^6 \frac{[-3, 10, -3]}{4}$$

To map this to a triangular grid we need to take the $[1,1]/2$ factors in pairs, replacing each pair by

$$S = \begin{bmatrix} & 1 & 1 \\ 1 & 2 & 1 \\ & 1 & 1 \end{bmatrix} /8$$

and finding a kernel which, when multiplied by the cube of this, gives the required row sums.

This is easily found to be

$$K = \begin{bmatrix} & -3 & -3 \\ -3 & 26 & -3 \\ & -3 & -3 \end{bmatrix} /8$$

which has the correct row sums to match $[-3, 10, -3]/4$, giving a final mask[1] of $4S^3K$.

$$
\begin{pmatrix}
 & -3 & -12 & -18 & & \\
 & -12 & -28 & -24 & | & \\
-18 & -24 & 66 & 144 & & \\
-12 & -24 & 144 & 468 & | & \\
-3 & -28 & 66 & 468 & 754 & sym \\
 & & & & sym &
\end{pmatrix} /1024
$$

From this we can extract the stencils

$$
\begin{pmatrix}
 & -3 & -18 & -3 & \\
-18 & 66 & 66 & -18 & \\
-3 & 66 & 754 & 66 & -3 \\
-18 & 66 & 66 & -18 & \\
 & -3 & -18 & -3 &
\end{pmatrix} /1024
$$

and

$$
\begin{pmatrix}
 & -12 & -12 & \\
24 & 144 & 24 & \\
-28 & 468 & * & 468 & -28 \\
24 & 144 & 24 & \\
 & -12 & -12 &
\end{pmatrix} /1024
$$

There are enough entries in these stencils (19 and 14 respectively) to give cubic precision, which requires 10 points, but not quite quartic, which requires 15. In any case there is no reason why these coefficients should satisfy the quartic quasi-interpolation conditions. Three of the four cubics and all lower degrees are satisfied because cubic extruded data is matched for three extrusion directions (by symmetry). The fourth cubic condition is based on a linearly varying quadratic, and this is satisfied by mirror symmetry. We therefore assert that this set of stencils satisfies all of the constant, linear, quadratic and cubic quasi-interpolation conditions.

The question remains whether the resulting scheme is stable in the sense that the eigenvalues behave.

---

[1] K here is normalised to sum to 1: the kernel is thus $4K$. This factor of 4 is analogous to the factor of 2 in the expression for the univariate scheme mentioned above.

3.3   The Basis Function

The basis function is the effect on the surface when one control point is moved. If we start the refinement process itself with a single unit valued control point and all the rest with value zero (cardinal data), successive refinements converge towards the basis function. We can get tesselations of the basis function by convolving each of these with the unit row stencil (the row (left) eigenvector of the component with unit eigenvalue and column (right) eigenvector of all 1s) to get the vertices at limit points. These are shown in Fig. 1. The last of these can be compared with the basis functions of the Butterfly [5] and Loop [4] schemes in regular regions in Fig. 2. Our scheme has significantly smaller wriggles than Butterfly, and is narrower·than Loop. In all of these figures, the basis function is shown as far out as the first zero entries. Note that these are not control polyhedra, as the scheme is only quasi-interpolating and the basis function is not a polynomial.

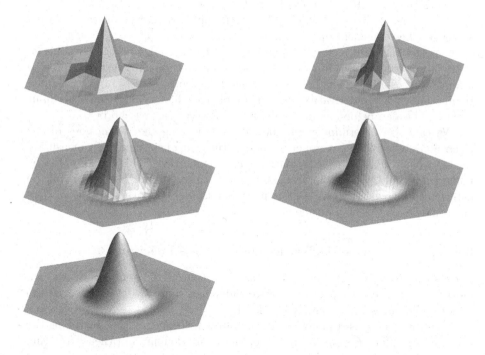

Fig. 1. The first four iterations, in which there is refinement, but no alteration of existing limit points.

3.4   Continuity Analysis

We can address the determination of lower bounds on the Hölder continuity by three routes. The first approach is to apply the bivariate approach of Cavaretta et al. [6], which is capable of showing that whatever the data, a norm of the first

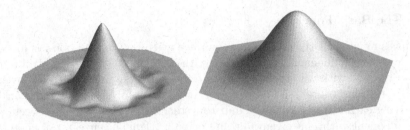

Fig. 2. The basis functions of the butterfly scheme and of Loop in regular regions. (also drawn at four iterations.)

difference of the second or third divided difference contracts with each refinement. This is capable of putting a lower bound on the continuity of the limit surfaces. Unfortunately I have not been able to formulate this approach and have not therefore been able to apply this test.

The second is that of Kobbelt [11], which applies difference operators to the powers of the refinement operation.

I suggest here that a simpler approach may be to check whether the contractivity is true for the basis function. Any other data has its limit surface expressed at any point as a finite weighted sum of basis functions, and so if the basis function is continuous in derivative to a certain degree, then so must any other limit surface. This can be done numerically and is covered in Sect. 3.5 below.

We can also determine upper bounds by checking the magnitudes of the eigenvalues, which can put upper bounds on the level of continuity, by measuring the actual continuity at specific mark points of the surface. This can also be done numerically and is covered in Sect. 3.6 below. Sharper bounds can in principle be determined by eigenanalysis at more mark points. Here only the triangle vertices and triangle centres are considered.

3.5   Contractivity of Differences of the Basis Function

Because we have tesselations of the basis function at different levels, we can, merely by convolving with appropriate difference stencils, determine how the various differences vary from level to level.

Because it is easiest to implement, the tables below are derived by taking the appropriate differences (up to 4th) over the entire domain and then extracting norms. We show here the max-norm and the average absolute value.

Because of the symmetry of the basis, we need only take the first differences in one direction giving Table 3. Similarly the three second differences are just symmetric versions of a single one giving Table 4. We can get the four 3rd differences just by evaluating two of them giving Table 5. The same is true of the 4th differences giving Table 6.

In each table are five columns. The second is the sum of the absolute values of the relevant difference, taken over the tesselations of the basis function shown in Fig. 1. The third is the ratio between each such value and the one above. The fourth column is the maximum absolute value of the relevant difference, and again the fifth holds the ratio between successive refinements.

Table 3. First differences: Because the entries in the second column are sums, the ratios need to be divided by 4 to get the actual ratio of norms. (There are four times as many triangles and therefore four times as many difference entries in each row compared with the previous.) With this correction, both ratios are converging nicely to 1/2, which, because it is less than one indicates that the scheme is convergent, with a continuous limit.

Level	$\Sigma|d|$	Ratio	$|d|_{max}$	Ratio
0	1.954658364122		0.6252769189585	
1	4.4209983756427	2.2617754881314	0.3706062878079	0.5927074494053
2	9.457042987369	2.1391193083155	0.2077661161766	0.5606114170526
3	19.150115219234	2.0249580386608	0.1054197019496	0.5073960272712
4	38.430854767130	2.0068210727280	0.0529147334585	0.5019434932932

Table 4. Differences of first divided differences: The ratios in the third column now need to be divided by 4, but also multiplied by 2, because the raw figures are differences rather than divided differences. Those in the fifth column now need to be multiplied by 2. Again, the ratios are nicely converging to 1/2 which is well below 1, and so we can deduce that the scheme is C1.

Level	$\Sigma|d|$	Ratio	$|d|_{max}$	Ratio
0	1.954658364122		0.625276918958	
1	5.195647796224	2.658084855947	0.509341262301	0.260578150970
2	5.915246892456	1.138500361158	0.148838512720	0.292217661784
3	6.111079357843	1.033106389124	0.039238461919	0.263631107314
4	6.162914058208	1.008482085950	0.009987309486	0.254528567085

3.6 Eigenvalue Analysis

This is very straightforward, as we can work solely with the kernel, which is very simple.

$$4K = \begin{pmatrix} & -3 & -3 \\ -3 & 26 & -3 \\ & -3 & -3 \end{pmatrix}/2$$

Table 5. Differences of second divided differences: We have two different third differences in use here. The ratios in the third column now need to be divided by 4, but also multiplied by 4, because the raw figures are differences rather than divided differences. Those in the fifth column now need to be multiplied by 4. Yet again, the ratios are nicely converging to 1/2 which is well below 1, and so we can deduce that the scheme is C2.

Level	$\Sigma|d|$	Ratio	$|d|_{max}$	Ratio
0	6.3472772348861		1.7968990864628	
1	6.7026946541919	1.0559952568878	0.4211556586767	0.2343791378434
2	4.2440378797963	0.6331838310942	0.0783173770441	0.1859582684705
3	2.2718369140307	0.5353008098362	0.0103408828255	0.1320381659330
4	1.1575101905374	0.5095040860497	0.0013108955912	0.1267682472874
Level	$\Sigma|d|$	Ratio	$|d|_{max}$	Ratio
0	8.457483383656		2.175810559215	
1	19.80725893292	2.341980236248	1.479495404617	0.679974365576
2	18.91360654972	0.954882581874	0.365260442678	0.246881768972
3	11.64307651131	0.615592614804	0.053058514549	0.145262142707
4	6.249221667750	0.536732852496	0.006830399904	0.128733342097

Table 6. Differences of third divided differences: We have three different fourth differences in use here. The ratios in the third column now need to be divided by 4, but also multiplied by 8, because the raw figures are differences rather than divided differences. Those in the fifth column now need to be multiplied by 8. It looks as though we can infer C3.

Level	$\Sigma|d|$	Ratio	$|d|_{max}$	Ratio
0	12.30937074944		3.593798172925	
1	9.351603172048	0.759714152932	0.786811211287	0.218935837080
2	3.573727352791	0.382151304652	0.086012788579	0.109318204095
3	1.044397941094	0.292243318527	0.008115334957	0.094350329662
4	0.280368542187	0.268449918518	0.000710776028	0.087584311946
Level	$\Sigma|d|$	Ratio	$|d|_{max}$	Ratio
0	16.91248082729		4.351621118431	
1	37.84281382575	2.23756728608	2.95899080923	0.679974365576
2	24.47335784142	0.646710837997	0.511077643651	0.172720253829
3	7.817348127387	0.319422785301	0.051984063931	0.101714611423
4	2.094546402655	0.267935669299	0.004790014051	0.092143893521
Level	$\Sigma|d|$	Ratio	$|d|_{max}$	Ratio
0	10.70181694529		2.314131729739	
1	5.6719938630861	0.5300028856857	0.4288936385916	0.1853367434013
2	1.8838035407895	0.3321236916438	0.0439635901700	0.1025046450080
3	0.5026102536695	0.2668060882075	0.0040813827780	0.0928355205346
4	0.1320672373545	0.2627627200007	0.0003559242029	0.0872067684662

The stencils are

$$
\begin{pmatrix}
0 & 0 \\
0 & 26 & 0 \\
0 & 0
\end{pmatrix} /2
\qquad \text{and} \qquad
\begin{pmatrix}
0 \\
-3 * -3 \\
0
\end{pmatrix} /2
$$

The subdivision matrix at vertices for the zero frequency Fourier component becomes

$$
\begin{bmatrix}
26/2 \\
-3/2 & -3/2 \\
0 & -6/2 & 0 \\
0 & 26/2 & 0 & \ddots
\end{bmatrix}
$$

whose only non-zero eigenvalues are $26/2$ and $-3/2$. Those of higher frequency Fourier components are never larger.

We may therefore determine the Hölder continuity at vertices to be

$$
6 - log_2(13) = 6 - 3.7004 = 2.2996,
$$

because each S factor increases the Holder continuity by 2.

The subdivision matrix at triangle centres for the zero frequency Fourier component becomes

$$
\begin{bmatrix}
-6/2 \\
26/2 & 0 \\
-3/2 & -3/2 & 0 \\
-3/2 & 0 & -3/2 & \ddots
\end{bmatrix}
$$

whose only nonzero eigenvalue is -3. Those of higher frequency Fourier components are never larger.

The Hölder continuity here is $6 - log_2(3) = 4.41$

In fact the continuity might be less than 2.2996, because this analysis only tells us about the continuity at the dyadic points, giving upper bounds on the level of continuity. Other mark points might have different (worse) behaviour. There are an infinite number of them and this is why the joint spectral radius calculation is such a lot of work. However, getting two upper bounds on the right side of 2 means that the method has not failed this test.

4   Triangulations with Extraordinary Vertices

The handling of extraordinary vertices requires amazingly little additional analysis. If a scheme has quadratic or cubic precision, it will have linear precision. We can therefore choose a desired natural configuration, for example by taking that of any C2 scheme, and the coefficients computed to give the reproduction properties will automatically make that natural configuration a self-fulfilling prophecy.

We have three types of stencil to determine:

(i) That of the EV itself.

Here use of symmetry resolves the underdeterminedness.

(ii) Those of new vertices[2] lying within the support of the EV: i.e. whose stencils contain the EV in their interior, so that the topology of the stencil is affected asymmetrically by the presence of the EV.

This is where most of the work lies, in that the vertices which need to be taken into account have to be chosen. The EV spoils the regularity of the topology. Calculating the coefficients to give the required reproduction properties applies as in the regular case.

(iii) Those of vertices lying further out, in what we can term the 'far field'. Here the topology of the stencil is regular, but the geometry is influenced by the natural configuration. Thus the actual coefficients will differ from the regular case. This effect propagates unboundedly, although hopefully reducing with distance. However, we can choose to resolve the underdeterminedness by minimising the difference from the regular case. Because of the regress of regularity around the extraordinary vertices, if the stencils converge fast enough towards the regular case, we do not have to worry about continuity at any points other than the extraordinary vertices themselves. This means that eigenanalysis at the EVs is a sufficient tool. Thus most of the really hard work is done in the regular case.

In fact at large distances from an EV we have two possible strategies

(iii.i) to be pure, and insist that all stencils are determined correctly.

In fact, if we are careful, the perturbation of the coefficients from the regular case reduces fast enough with distance that after a certain distance it can be approximated accurately enough (to machine precision) by some simple form, which can be combined when more than one EV is present in the control polyhedron.

(iii.ii) just to revert to the regular stencil for all new vertices whose support does not touch the EV.

This avoids a lot of implementation problems, and will not spoil the continuity at the EV, because only control points within, or on the boundary of, the support of the EV influence the shape of the limit surface in the immediate neighbourhood of the EV. It will not spoil the continuity anywhere else either, because everywhere outside that neighbourhood has its continuity governed by the regular case.

However, we can expect something to go wrong, and an obvious symptom to be expected will be the appearance of artifacts in the region between the two regimes.

A practical resolution might well be to use (iii.ii) to avoid overlapping influences from more than one EV. In regions where more than one EV has a significant effect, who can say what the desired surface should be? As refinement

---

[2] In the case of schemes having multiple kinds of e- and f-vertices (for example, that of [9] or of higher arity) each kind will need individual calculations of this type.

proceeds, the threshold between the two regimes can move in more slowly than the refinement rate, and the artifacts should therefore be spread out and diluted.

4.1   Conformal Characteristic Map

In [13] David Levin conjectured that a stationary subdivision scheme giving the characteristic map

$$u + iv = (s + it)^{6/n}$$

would require a EV mask of unbounded size. Here we have a non-stationary scheme, because the stencils depend on the local layout of control points in the domain, and the corresponding conjecture is that the influence of the presence of the extraordinary point extends unboundedly, even though the EV itself is only accessed by the stencils of a finite number of nearby new vertices.

In the far field we find that the rate of convergence to the regular case turns out to be significantly slower if we enforce cubic reproduction than if we only enforce quadratic, despite the fact that what we are trying to be close to does have cubic precision when the geometry is regular. The Tables 7, 8, 9 and 10 in Annex 2 report $E1 = \Sigma_i \delta_i^2$ and $E2 = max_i |\delta_i|$ for reproduction of degrees 2 and 3, for v-vertices along rays from the EV of valencies 5 and 7.

4.2   Other Characteristic Maps

The choice of the conformal characteristic map is entirely an aesthetic one. It would be equally possible to choose as arbitrary natural configuration that of Loop or of the Butterfly scheme. Anything, in fact, which can be shown to be 1:1 and have its characteristic map join smoothly enough to its scaled instance under the next ring.

4.3   Moving Least Squares

Since presentation of the paper I have seen the recent work of Ivrissimtzis. This uses a combination of a standard subdivision scheme (in this Loop, Butterfly or even the degree 1 box-spline) to determine approximate new vertex positions, which are then refined by projection on to a local least squares fit to some subset of the old vertices. This work is following a line which can be traced back through the work of Boyé, Guennebaud and Schlick [22], Levin [14] and McLain [1]. It looks to be almost equivalent to the above ideas, in that it gives a quasi-interpolant (if enough local points lie on a polynomial, the new point will lie on that polynomial), but goes, apparently, more directly to the new points (projecting rather than first computing coefficients).

5   Directions Still Needing Exploration

5.1   Stronger Proof of Degree of Continuity

The analysis of the regular case above is unsatisfactory, with different approaches giving different opinions. The method of chapter seven in Cavaretta et al. [6] needs to be made to work.

5.2   End and Edge Conditions

Before these ideas can be applied to practical situations, the detail of what to do at the edges of the domain of interest needs to be articulated. Unfortunately we do not yet even have end conditions adequately explored for univariate quasi-splines, and this must obviously come first.

5.3   Extension to Higher Degrees

In principle exactly similar constructions can be made, choosing coefficients to give any desired quasi-interpolation degree. Because the stencils for higher degrees become large quadratically with degree, it is unlikely that higher degrees will be of more than academic interest, but the academic interest is there. It seems likely that the extension in this direction will not expose new problems significantly different from those already seen.

5.4   Extension to Solids and Higher Dimensions

Although regular triangular grids do not have completely regular analogues in higher dimensions, the ideas of using subdivision basis functions which reproduce low degree polynomials without being interpolating for general data must be relevant to IsoGeometric Analysis, overcoming the current disadvantage of subdivision that the lack of polynomial generation gives excessive stiffness because the artifacts contribute spurious energy to the solution.

Acknowledgements. My thanks go to a very diligent referee who bothered to construct a counterexample disproving a conjecture in my first draft of this paper and also pointed out many places where the original text was not clear. I also thank colleagues: Cedric Gerot for helping me understand that counterexample, Leif Kobbelt and Ulrich Reif for prompt replies to my emails requesting clarifications, and to Ioannis Ivrissimtzis for bringing the moving least squares ideas to my attention.

Annex 1: Computation of Stencil Nearest to a Regular Stencil

In the topologically regular but geometrically irregular case, I suggest that the nearest solution to the regular one should be chosen. This gives some measure of continuity with respect to changing layouts in the abscissa plane. The metric for 'nearest' might be chosen by more sophisticated arguments later[3], but for the moment I just use the euclidean distance in coefficient space.

Call the number of coefficients c and the number of quasi-interpolation conditions n.

---

[3] For example, to get better continuity when the set of support points needs to change because of changes in the set of local neighbours.

The quasi-interpolating conditions define a linear subspace of dimension $c-n$ in the space of dimension $c-1$ of sets of coefficients: nearness to the regular coefficients defines a complementary subspace orthogonal to it, and the solution can be found in that subspace by solving a linear system of size $n \times n$.

Let the set of coefficients be a_i, $i \in 1..c$, and let the quasi-interpolation conditions be

$$\forall j \in 1..n \quad \Sigma_i a_i f_j(v_i) = 0$$

If the regular coefficients are $\bar{a}_i$, then we can set up the system

$$\forall i \; a_i = \bar{a}_i + \Sigma_{k \in 1..n} \, \delta_k f_k(v_i)$$

and solve for the δ_j.

$$\begin{aligned}
\forall j \in 1..n \quad 0 &= \Sigma_i a_i f_j(v_i) \\
&= \Sigma_i \left(\bar{a}_i + \Sigma_k \delta_k f_k(v_i) \right) f_j(v_i) \\
&= \Sigma_i \bar{a}_i f_j(v_i) + \Sigma_i \Sigma_k \delta_k f_k(v_i) f_j(v_i) \\
&= \Sigma_i \bar{a}_i f_j(v_i) + \Sigma_k \Sigma_i \delta_k f_k(v_i) f_j(v_i) \\
&= \Sigma_i \bar{a}_i f_j(v_i) + \Sigma_k \delta_k \Sigma_i f_k(v_i) f_j(v_i)
\end{aligned}$$

$$\begin{bmatrix} \ddots & & \\ & \Sigma_i f_k(v_i) f_j(v_i) & \\ & & \ddots \end{bmatrix} \begin{bmatrix} \vdots \\ \delta_k \\ \vdots \end{bmatrix} = - \begin{bmatrix} \vdots \\ \Sigma_i \bar{a}_i f_j(v_i) \\ \vdots \end{bmatrix}$$

The actual coefficients can then be determined from these δ_j values.

$$\forall i \; a_i := \bar{a}_i + \Sigma_k \delta_k f_k(v_i)$$

The rate of convergence with distance from the EV in an interesting natural configuration can be measured by how $\Sigma_j \delta_j^2$ varies with distance.

Annex 2: Convergence for Conformal Characteristic Map

These figures in Tables 7, 8, 9 and 10 ($E1 = \Sigma_i \delta_i^2$ and $E2 = max_i |\delta_i|$) are disappointing in that the convergence is so slow. A conjecture that the rate of convergence would be d^{-6} has been soundly disproved by these calculations.

Table 7. Valency $= 5$, reproduction degree $= 2$

Distance	E1	E2
2	0.0266273481	0.00859447462
4	0.0029902673	0.00098588428
8	0.0004904519	0.00014386747
16	0.000117032	0.00003354449

Table 8. Valency $= 5$, reproduction degree $= 3$

Distance	E1	E2
2	0.043294222	0.0178461598
4	0.035778240	0.0131572241
8	0.020705507	0.0069547056
16	0.01054870	0.0035061301

Table 9. Valency $= 7$, reproduction degree $= 2$

Distance	E1	E2
2	0.03776611323	0.0112682406
4	0.00295127833	0.0010064401
8	0.00022928609	0.0000913945
16	0.0000595678	0.0000175042

Table 10. Valency $= 7$, reproduction degree $= 3$

Distance	E1	E2
2	0.1065766380	0.0439985659
4	0.0327818713	0.0132975787
8	0.0142165239	0.0051765249
16	0.0074573353	0.0025183365

References

1. McLain, D.H.: Two dimensional interpolation from random data. Comput. J. **19**(2), 178–181 (1976)
2. Catmull, E.E., Clark, J.: Recursively generated B-spline surfaces on topological meshes. CAD **10**(6), 350–355 (1978)
3. Doo, D.W.H., Sabin, M.A.: Behaviour of recursive division surfaces near extraordinary points. CAD **10**(6), 356–360 (1978)
4. Loop, C.T.: Smooth subdivision surfaces based on triangles. M.S. Mathematics thesis, University of Utah (1987)
5. Dyn, N., Levin, D., Gregory, J.: A butterfly subdivision scheme for surface interpolation with tension control. ACM ToG **9**(2), 160–169 (1990)
6. Cavaretta, A.S., Dahmen, W., Micchelli, C.A.: Stationary Subdivision. Memoirs of the American Mathematical Society, no. 453. AMS, Providence (1991)
7. Reif, U.: A degree estimate for subdivision surfaces of higher regularity. In: Proceedings of AMS, vol. 124, pp. 2167–2175 (1996)
8. Reif, U.: TURBS - topologically unrestricted rational B-splines. Constr. Approx. **14**(1), 57–77 (1998)
9. Prautzsch, H.: Smoothness of subdivision surfaces at extraordinary points. Adv. Comput. Math. **9**(3–4), 377–389 (1998)
10. Prautzsch, H., Reif, U.: Degree estimates for C^k-piecewise polynomial subdivision surfaces. Adv. Comput. Math. **10**(2), 209–217 (1999)
11. Kobbelt, L.: $\sqrt{3}$-subdivision. In: Proceedings of SIGGRAPH 2000, pp. 103–112 (2000)
12. Velho, L.: Quasi 4-8 subdivision. CAGD **18**(4), 345–357 (2001)
13. Levin, D.: Presentation at Dagstuhl Workshop (2000)
14. Levin, D.: Mesh-independent surface interpolation. In: Brunnett, G., Hamann, B., Müller, H., Linsen, L. (eds.) Geometric Modeling for Scientific Visualization, pp. 37–49. Springer, Heidelberg (2004)
15. Peters, J., Reif, U.: Shape Characterization of subdivision surfaces - basic principles. CAGD **21**, 585–599 (2004)
16. Karciauskas, K., Peters, J., Reif, U.: Shape Characterization of subdivision surfaces - case studies. CAGD **21**, 601–614 (2004)
17. Levin, A.: The importance of polynomial reproduction in piecewise-uniform subdivision. In: Martin, R., Bez, H., Sabin, M. (eds.) IMA 2005. LNCS, vol. 3604, pp. 272–307. Springer, Heidelberg (2005). doi:10.1007/11537908_17. ISBN 3-540-28225-4

18. Levin, A.: Modified subdivision surfaces with continuous curvature. ACM (TOG) **25**(3), 1035–1040 (2006)

19. Zulti, A., Levin, A., Levin, D., Taicher, M.: C2 subdivision over triangulations with one extraordinary point. CAGD **23**(2), 157–178 (2006)

20. Hormann, K., Sabin, M.A.: A family of subdivision schemes with cubic precision. CAGD **25**(1), 41–52 (2008)

21. Dyn, N., Hormann, K., Sabin, M.A., Shen, Z.: Polynomial reproduction by symmetric subdivision schemes. J. Approx. Theor. **155**(1), 28–42 (2008)

22. Boyé, S., Guennebaud, G., Schlick, C.: Least squares subdivision surfaces. Comput. Graph. Forum **29**(7), 2021–2028 (2010)

23. Augsdörfer, U.H., Dodgson, N.A., Sabin, M.A.: Artifact analysis on B-splines, box-splines and other surfaces defined by quadrilateral polyhedra. CAGD **28**(3), 177–197 (2010)

24. Augsdörfer, U.H., Dodgson, N.A., Sabin, M.A.: Artifact analysis on triangular box-splines and subdivision surfaces defined by triangular polyhedra. CAGD **28**(3), 198–211 (2010)

25. Deng, C., Hormann, K.: Pseudo-spline subdivision surfaces. Comput. Graph. Forum **33**(5), 227–236 (2014)

Adaptivity with B-spline Elements

Malcolm Sabin[⊠]

Numerical Geometry Ltd., 19 John Amner Close, Ely, Cambs, UK
malcolm.sabin@btinternet.com

Abstract. This paper takes a stage further the work of Kraft [1] and of Grinspun et al. [2] who used subdivision formulations to show that finite element formulation can be expressed better in terms of the basis functions used to span the space, rather than in terms of the partitioning of the domain into elements. Adaptivity is achieved not by subpartitioning the domain, but by nesting of solution spaces. This paper shows how, with B-spline elements, their approach can be further simplified: a B-spline element of any degree and in any number of dimensions can be refined independently of every other within the basis. This completely avoids the linear dependence problem, and can also give slightly more focussed adaptivity, adding extra freedom only, and exactly, where it is needed, thus reducing the solution times.

Keywords: Finite elements · Adaptivity · Nested spaces

1 Motivation

In the use of the finite element method to solve partial differential equations such as those for thermal analysis and elasticity, an important tool for achieving the best trade-off between accuracy and computing cost is the use of *adaptivity*, whereby the resolution of an initial analysis is improved by using a finer mesh where the solution is varying rapidly, but only there. In places where the solution is already captured to an adequate accuracy, adding extra freedoms adds significant extra cost to the solution without contributing to reducing its overall inaccuracy. The use of B-spline functions as a basis for the analysis turns out to provide a very good way of making an optimal trade-off, which has none of the disadvantages of local remeshing. The key to this is thinking in terms of the basis functions instead of the partitioning of the domain.

2 Prior Work

2.1 Kraft

In [1] Kraft described how a local increase of density in an array of quadratic B-spline functions could be achieved by replacing a single bi-quadratic B-spline function by 16 others. The following Figs. 1, 2, 3 and 4 are my transcriptions of those in his paper.

© Springer International Publishing AG 2017
M. Floater et al. (Eds.): MMCS 2016, LNCS 10521, pp. 218–232, 2017.
https://doi.org/10.1007/978-3-319-67885-6_12

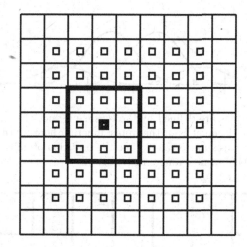

Fig. 1. In this figure the regular grid lines denote knot lines. The small squares denote the centres of biquadratic B-spline functions over the domain in the plane of the paper. The small black square denotes the centre of a specific such function with the boundary of its support highlight. Kraft takes such a function as a candidate for adaptivity.

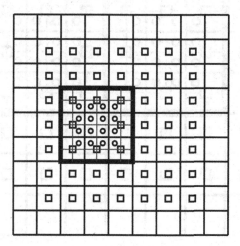

Fig. 2. In this figure the function denoted by the black square of Fig. 1 is replaced by the sixteen functions denoted here by the small circles. Those sixteen functions have knot lines which include the extra lines inside the support of the original function. Note that the remaining functions whose centres are at the squares do not see these extra knots. This process can be viewed as having the new functions defined by knot insertion into the old one (just the old one: no others) or as having the new knots appear as a side-effect of the replacement of the old function by the new ones.

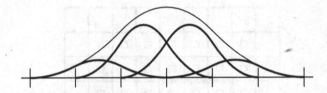

Fig. 3. The new functions are defined by the tensor product of two univariate refinements. In each the replacement is of the upper function here by the four lower ones. Because the new functions sum to the old one the partition of unity property is maintained. The total number of new functions is 16.

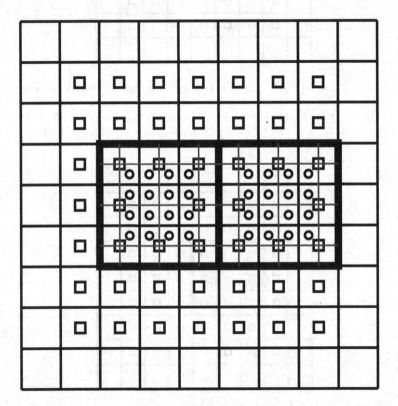

Fig. 4. Kraft was very careful to avoid possible linear dependence issues by specifying that if one old function was refined, those whose support overlapped that of this one should not be refined. They should be left at the original scale. The set of basis functions when two functions as close as permissible are refined is therefore as shown here. This is not the same as would be created by all functions over the new partitioning of the domain.

2.2   Grinspun et al.

In [2] Grinspun et al. used the Loop subdivision scheme to refine a space defined over a triangulation of a curved surface. They report actual analyses of interesting shell structures with non-trivial physics (Fig. 5).

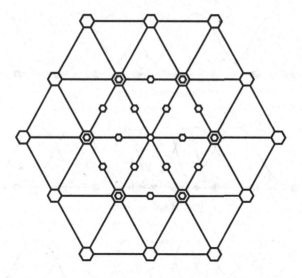

Fig. 5. The method of Grinspun et al. [2]. In this figure (adapted from Figs. 9 and 11 of that paper, with permission) the large hexagons denote centres of basis functions at the coarse level of refinement. They lie on a fairly regular triangular grid. In the centre, a large hexagon is missing but there are small ones denoting the centres of refined basis functions around it. Note that there can be both large and small centred at the same point.

However, their main emphasis, quoted from their abstract, is

...The basic principle of our approach is to refine basis functions, not elements. This removes a number of implementation headaches associated with other approaches and is a general technique independent of domain dimension...

They were able somehow to avoid the linear dependence issue even when supports of refined functions overlapped. If a large region is refined the number of new functions per old one replaced tends down towards four, though for single functions it is three times the valency of the centre of the replaced function plus one.

2.3 Adaptive B-splines

There have been too many papers on this topic to cite them all. These proceedings will contain many papers addressing this topic and references to that literature will be found therein. Two significant contributions are of note, being the Truncated Hierarchical B-splines of Giannelli et al. [3] and the Locally Refined B-splines of Dokken et al. [4]. An excellent comparison of these two with a method like Kraft's but with the functions unscaled by the subdivision ratios, can be found in Johannessen et al. [6].

In the univariate case (illustrated in Fig. 6) both of these recover the partition of unity property which was lost in the original hierarchical work which

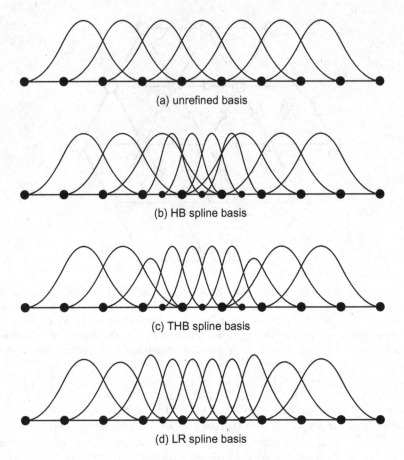

(a) unrefined basis

(b) HB spline basis

(c) THB spline basis

(d) LR spline basis

Fig. 6. In the original hierarchical B-spline (HBspline) basis, one function is split into four by knot insertion at three knots (but cf. Fig. 3). The Truncated Hierarchical Bspline (THB-spline) basis [3] modifies two more of the original functions by subtracting some multiple of the refined functions so that their support is reduced back to three spans in the new knot vector, and so that partition of unity is restored. The Locally Refined B-spline (LR-spline) basis [4] takes this a stage further, truncating all functions to the minimum support. In the univariate case this is exactly the same as the non-uniform B-spline basis over the new knot vector.

they cite. In the bivariate case LR-splines still have a problem with linear dependence, although their paper does address how this may be countered. Both sets of authors regard the problem as being finding a basis spanning all piecewise polynomials of a given degree over a given partitioning of the domain.

3   New Contributions

The main contribution here is to reiterate the emphasis of Grinspun et al. More specifically, we

1. Reduce the number of replacement functions down to two per old one. This is detailed in Sect. 3.1.
2. Provide a proof that this process works for all B-spline degrees. This is detailed in Sect. 3.2.
3. Provide a proof that it works for all dimensions of domain. This is detailed in Sect. 3.3.

We consider various dimensions by starting with the univariate case.

3.1 Adaptivity When the Domain Is Univariate

The B-spline degrees are either even or odd. We take these two cases separately

Even Degrees. These are typified by the quadratic. The argument is made in the captions to Figs. 7, 8, 9 and 10, which should be read in sequence as if they were part of the text. The extension to higher even degrees is implied by the descriptions, though not by the illustrations.

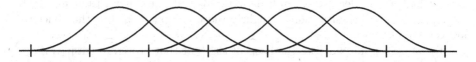

Fig. 7. Suppose that an initial set of quadratic basis functions spanning the solution space is as shown here. This figure shows equal knot intervals, but that is not a precondition. After a number of rounds of selective refinement the spacing of the knots will be seriously uneven.

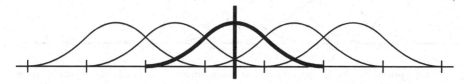

Fig. 8. We can take just one of these functions and insert a single new knot at the midpoint of the central span (all even degree B-splines have an odd number of non-zero spans).

Fig. 9. This knot insertion gives two new functions, which can replace the old. The sum of these two functions is equal to the original, and so the partition of unity property is preserved. In the context where an iterative solver is being used, the coefficients of the new function are both set equal initially to that of the old. This exactly preserves the solution so the refined solution space is a nesting of the old.

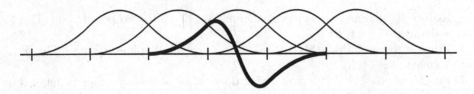

Fig. 10. However, we can 'rotate' the basis within the solution space by rotating the subspace spanned by these two new functions, by taking their sum and difference. This is a non-degenerate transformation which does not alter the rank of the space. The sum of the two new functions is exactly the old one, which has no discontinuity of any derivative at the new knot. Neither do any of the other old functions. The difference function is the only one which has a discontinuity here and it is therefore linearly independent of all the other functions in the extended basis. This is true whether the function being split was part of an initial basis over some coarse partitioning or whether it was the result of many stages of Fig. 9 refinement.

Odd Degrees. These are typified by the cubic. The argument is made in the captions to Figs. 11, 12, 13 and 14, which should be read in sequence. The extension to higher odd degrees is implied by the descriptions there, though not by the illustrations. The situation is slightly more complicated, but the concepts are very similar.

It is a moot point whether the basis of Fig. 9 or that of Fig. 10 (resp., Fig. 13 or Fig. 14) should be used in an actual implementation. The ranks of the Figs. 9 and 10 bases are the same. Each has practical advantages and disadvantages which are not explored further here, except that we note that Figs. 10 and 14

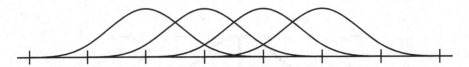

Fig. 11. Suppose that an initial set of cubic basis functions spanning the solution space is as shown here. This figure shows equal knot intervals, but that is not a precondition. After a number of rounds of selective refinement the spacing of the knots will be seriously uneven.

Fig. 12. We can take just two consecutive functions, and insert a single knot at the midpoint of the span between the middle knot of one and the middle knot of the next (all odd degree B-splines have an even number of non-zero spans).

Fig. 13. This knot insertion gives three new functions, which can replace the old two. The sum of the three is equal to the sum of the old two, and so partition of unity is preserved. The detail of the coefficients of the new functions depends on the knot intervals. But the refined solution space is a nesting of the old.

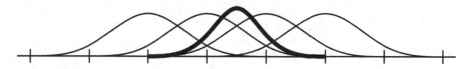

Fig. 14. However, we can again rotate the basis within the solution space by taking appropriate linear combinations of the three new ones, to give the original two, together with one new function which is the only one with a discontinuity at the new knot. It is linearly independent of all other functions in the basis, and the dimension of the space is not altered by the rotation.

no longer preserve partition of unity. This might matter in a modelling context because a coefficient no longer transforms as a point, but as a displacement (see [5]), but in the analysis context it is of no importance whatever. In an iterative solution the coefficient of the new function can be initially set to zero, to match exactly the pre-refinement solution.

The refined space exactly contains the old space as a subspace. This is true whichever of the possible implementations is used.

For even degrees my preference is to use the functions of Fig. 9, though for odd degrees there are more evenly balanced arguments, relating to the necessary infrastructure to steer what refinements can be made.

3.2   Adaptivity When the Domain Is Bivariate

We now assume that all of the original basis functions are tensor products of B-splines in two directions. The argument is made in the captions to Figs. 15, 16, 17, and 18.

The proof of linear independence depends on the fact that a given bivariate function is never split more than once. The splitting introduces a new knot-line and clearly new functions which have different such knot-lines will be linearly independent of each other. If two splits share the same knot-line, but the extents are different in the other direction, then the linear independence of the distribution of discontinuity magnitude in that direction implies linear independence of the new functions. It is useful to notice that, because the split is always of the centre span for even degrees, a given piece of shared knot can only arise at a specific level of the refinement.

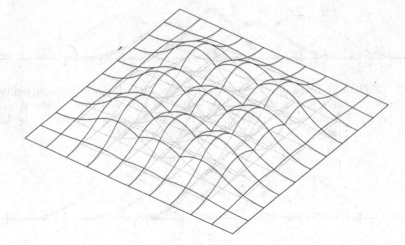

Fig. 15. Suppose that we have an initial set of basis functions over a regular grid. Each basis function is a tensor product of a B-spline in one direction and a B-spline in the other. The two degrees do not have to be equal, although often they will be.

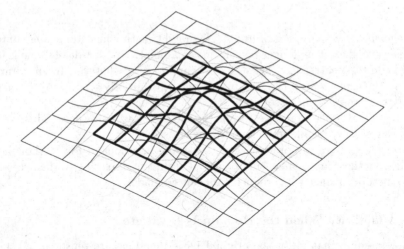

Fig. 16. We can select just one of these basis functions for refinement. In that respect we are following Kraft.

Clearly, in the odd degree case, the two functions split must share the same knots in the other direction. If this is not practical, then the analog of the quadratic case using the span consistently on the same side of the central knot, though less elegant, can be set up.

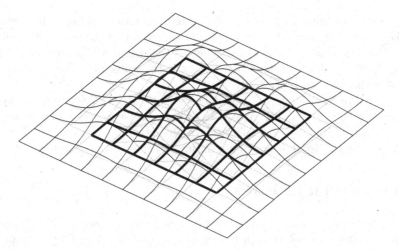

Fig. 17. However, we split in only one direction, not both. We are therefore able to focus the refinement sharply, adding only one dimension to the space, not fifteen. If you want to split in both directions it is necessary to split first in one direction and then both of the new functions in the other. This has the expected symmetry, but because it adds three to the size of the space, takes three refinement operations. The two new functions from a single refinement are given by the tensor product of Fig. 9 (or Fig. 13) with the original B-spline in the other direction.

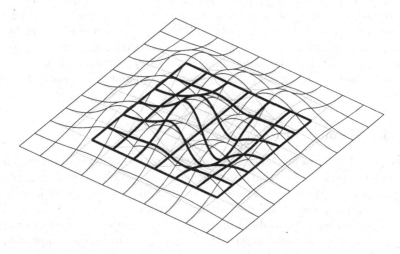

Fig. 18. The rotation in the solution space follows exactly the tensor product of a simple B-spline with the function of Fig. 10.

3.3   Adaptivity When the Domain Has Higher Dimension

Here we take the simple route of specifying that any basis function, which will be a tensor product of as many functions as the dimension of the domain, is split in just one direction, all the other directions keeping their original function. Thus however high the domain dimension becomes, the refinement always increases the size of the solution space by 1.

The proof of linear independence follows that for the bivariate case, except that now the knot insertion is of a piece of knot (hyper)plane. Different hyperplanes cannot be linearly dependent, and different refinements will lead to linearly independent distributions of discontinuity across a common hyperplane.

4   Implementation Issues

An important question is *'How easy is this to incorporate into an existing analysis infrastructure?'*. Even though the scheme is in some ways simpler than the conventional collection of nodes and elements, where the outermost loop during stiffness matrix assembly is through elements, there may be a perceived problem simply because it is different.

4.1   Representation of the Discretisation of the Solution Field

The obvious representation here, assuming isoparametric formulation, is to have a collection of basis function objects, each of which has a number of lists, each as long as 1+degree, one for each independent direction within the domain. These lists contain the domain coordinates of the knots. Thus a triquadratic system over a three-dimensional domain would have a collection of basis functions each represented by three lists of four entries. This has to be supplemented by the map from the parameter domain into the geometric domain, which is used for evaluation of Jacobians. Isogeometric ideas are obviously relevant here, though we are not limited to using the partitioning of the domain which comes from the geometry.

4.2   The Assembly Process

The outermost loops of assembly are now most conveniently taken through pairs of basis functions. Each possible pair corresponds to a single entry in the stiffness matrix, and so using this structure lends itself to effective parallelisation on multi-processor machines with little shared memory.

For each pair we have to do an integration over the intersection of the supports. Here there is a nice surprise: although the domain may appear to be totally splintered by different refinements, the number of polynomial pieces over which the integration must be done is bounded by the number of knots active in the two basis functions forming the pair. They don't see knots which are only discontinuities in other basis functions.

4.3 Solution

There is no reason why, once the stiffness matrix and right hand sides are formed, the solution process has to be any different. However, because of the exact nesting of the spaces, there will be a lot of coherence between successive solutions. This is clearly evident in the case of iterative solution.

However, it is also possible to exploit this coherence in the case of explicit solution by factorisation. When the total number of refinements is small relative to the total number of freedoms, it is possible merely to update the factors at a cost significantly less than starting again.

4.4 Error Estimation

Here for maximum efficiency, we need information not only about the magnitude of the local error, but also about which direction the refinement should be made in. The obvious rule of thumb is to split in the direction in which the derivative of error is largest. (Think about the S-shaped function in Fig. 10. Adding it to the solution space will obviously be good for correcting the derivative of the error.)

4.5 Result Extraction

Because there are no longer any elements which can act as a scaffolding for display of results, this process becomes more complex. At any point where the result has to be evaluated you need to know which basis functions include that point in their support. Some kind of spatial indexing is likely to be indispensable here. This can also give some benefit during the assembly, when identifying which pairs of basis functions have a non-zero overlap could become a significant task.

5 Discussion

'Is this just the same as any of the hierarchical formulations which have been considered, over the last decade or so, which start with a partitioning of the domain and then work out a linearly independent basis for the space of all piecewise polynomials of given continuity over that partitioning?'

No, it is not.

It starts instead with the process of refinement of the basis and a partitioning is the result. That partitioning as such plays a minor role in the solution of any specific problem. The size of the space which the basis spans may be significantly smaller than the maximal linear independent space, as we have seen in Fig. 4, where the maximal dimension over the refined partitioning is six more than Kraft's basis.

'Are you not concerned that you are not using the full space of functions over the partitioning? What does this do to the approximation power?'

Theoretical asymptotic approximation order is not the critical issue in practical use. What matters is whether the solution is sufficiently accurate (usually errors of the order of 10^{-2} in stresses), and whether it is achieved at minimal cost.

Selecting basis functions among those possible is a strength, not a weakness, because the adding of new functions to the basis is done as a response to where the error needs to be reduced. Every extra freedom increases the cost of solution: if there is evidence that it will improve the solution then it can be added: if there is not, you don't want it. Figure 19 illustrates this.

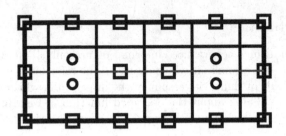

Fig. 19. An example where the full space is wasteful. In this figure just two basis functions in an original 3×6 grid have been refined. The full space over the resulting partitioning of the domain would have had to split all of the functions in the original middle row, making the total number of freedoms four larger. If the error estimator had requested only the two functions to be refined, the extras would have contributed to cost without making the solution any better. This figure may also be compared with Fig. 4.

'This looks OK for the first addition of a new function, but what happens when nearby basis functions have already been split?'

This is very straightforward for even degrees. Think in terms of Fig. 9, where all basis functions are indeed B-splines over their own knot vectors. Any such B-spline can be split independently of anything else around it. Figure 20 shows the univariate case, but there is no extra complication for higher even degrees.

Odd degrees appear more complicated, and indeed they are. However, the new functions in Fig. 14 are still B-splines over their own knot vectors. The central knot is new and its position and extent are not shared by any other functions. This is all that matters. The choice of the other knots is actually arbitrary. In principle we could add a basis function designed to match the local error behaviour: the new knot does not even need to be aligned with the existing mesh directions. That is an exciting prospect far too speculative for this paper. The message here is that almost any choices made in designing software will work. The problem is that there are too many possible effective answers, not the failure of any of them.

'Something that is usually unwanted when dealing with adaptivity is a not bounded number of non-zero basis functions on a certain point of the domain.

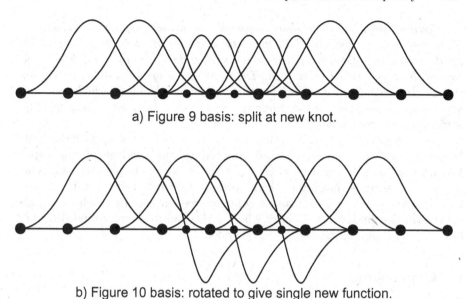

a) Figure 9 basis: split at new knot.

b) Figure 10 basis: rotated to give single new function.

Fig. 20. It is interesting to compare the proposed quadratic functions with those shown in Fig. 6. The upper figure corresponds to Fig. 9 and the lower to Fig. 10.

If one is keeping refining in the same area it seems that the proposed solution does not overcome this problem (while this is partially addressed by the reduced support of THB-spline, or solved by the minimal support of LR B-splines).'

Each entry in the stiffness matrix, representing the interaction of two freedoms, has a bounded number of polynomial regions for integration.

The actual depth of refinement depends on the actual errors found by the error estimator. This should only grow significantly if the solution field contains components of significantly high spatial frequency which, for elliptic problems will happen only near the boundary.

If functions share knots in the other direction(s) rotation of the basis can bring it back to a narrower bandwidth.

'In order to have good numerical properties (e.g., condition numbers of related matrices) certain mesh grading is usually preferable. In the multivariate setting, if the knot insertion is always performed in the same direction, elongated elements (or equivalently, elongated B-spline elements if one prefers to not refer to domain partitioning) may be created. This is why it is common practice to perform dyadic refinement on the marked elements (even if also with the other hierarchical construction the refinement can be carried over along only one direction).'

As is the case with the other approaches, when bad aspect ratios are about to be produced, additional refinement in the other direction(s) can be invoked. The question is the monitoring of the aspect ratio, not the availability of a solution.

'*Section 4 presents some short comments on critical aspects of an adaptive framework without entering into the details of any of them. From a practical point of view the effectivity of this construction that adds only one basis function at the time, should be verified. There are many open questions concerning the resulting adaptive scheme (function based error estimation which takes into account derivative information, some refinement strategies) which may effect the performance of the method.*'

Yes, of course, there are a lot more issues to adaptivity than just the basis. However, these are not significantly affected by the message of this paper. I didn't want to dilute that message by details of loosely-coupled aspects. What is worth stressing is that although the presentation here talks about 'one at a time', there is no reason to invoke a complete re-solution after each one. If the error estimator finds a lot of places where additional freedoms are needed, then of course all those functions need to be added.

6   Conclusions

An approach to adaptive refinement has been described with a number of significant advantages over previous methods. It follows the ideas of Kraft and of Grinspun et al., of starting from the basis rather than the partitioning of the domain, but extends these by making the refinement much more selective. Each refinement adds only one freedom to the system to be solved, and (although it is certainly possible to make several refinements before attempting another solution) this keeps the cost of solution as low as possible. Furthermore, because the refined space is exactly nested (as is the case for the other approaches mentioned above), the next solution can be regarded as an update which may be a lot cheaper than a re-solution.

References

1. Kraft, R.: Adaptive and linearly independent multilevel B-splines. In: Le Méhauté, A., Rabut, C., Schumaker, L. (eds.) Surface Fitting and Multiresolution Methods, pp. 209–218. Vanderbilt University Press (1997)
2. Grinspun, E., Krysl, P., Schröder, P.: A simple framework for adaptive simulation. In: Proceedings of SIGGRAPH 2002 (ACM Transactions on Graphics), pp. 281–290 (2002)
3. Giannelli, C., Jüttler, B., Speleers, H.: THB-splines: the truncated basis for hierarchical splines. CAGD **29**(7), 485–496 (2012)
4. Dokken, T., Lyche, T., Pettersen, K.F.: Polynomial splines over locally refined box-partitions. CAGD **30**(3), 331–356 (2013)
5. Kosinka, J., Sabin, M.A., Dodgson, N.A.: Control vectors for splines. CAD **58**, 173–178 (2015)
6. Johannessen, K.A., Remonato, F., Kvamsdal, T.: On the similarities and differences between classical hierarchical, truncated hierarchical and LR B-splines. CMAME **291**, 64–101 (2015)

Reconstructing Sparse Exponential Polynomials from Samples: Difference Operators, Stirling Numbers and Hermite Interpolation

Tomas Sauer[(✉)]

Lehrstuhl für Mathematik mit Schwerpunkt Digitale Bildverarbeitung & FORWISS,
University of Passau, 94032 Passau, Germany
Tomas.Sauer@uni-passau.de
http://www.fim.uni-passau.de/digitale-bildverarbeitung

Abstract. Prony's method, in its various concrete algorithmic realizations, is concerned with the reconstruction of a sparse exponential sum from integer samples. In several variables, the reconstruction is based on finding the variety for a zero dimensional radical ideal. If one replaces the coefficients in the representation by polynomials, i.e., tries to recover sparse exponential polynomials, the zeros associated to the ideal have multiplicities attached to them. The precise relationship between the coefficients in the exponential polynomial and the multiplicity spaces are pointed out in this paper.

1 Introduction

In this paper we consider an extension of what is known as *Prony's problem*, namely the reconstruction of a function

$$f(x) = \sum_{\omega \in \Omega} f_\omega(x)\, e^{\omega^T x}, \qquad 0 \neq f_\omega \in \Pi, \qquad \omega \in \Omega \subset (\mathbb{R} + i\mathbb{T})^s, \qquad (1)$$

from multiinteger samples, i.e., from samples $f(\Lambda)$ of f on a subgrid Λ of $\mathbb{Z}^s$. Here, the function f in (1) is assumed to be a sparse *exponential polynomial* and the original version of Prony's problem, stated in one variable in [17], is the case where all f_ω are *constants*. Here "sparsity" refers to the fact that the cardinality of Ω is small and that the frequencies are either too unstructured or too irregularly spread to be analyzed, for example, by means of Fourier transforms.

Exponential polynomials appear quite frequently in various fields of mathematics, for example they are known to be exactly the homogeneous solutions of partial differential equations [9,10] or partial difference equations [18] with constant coefficients.

I learned about the generalized problem (1) from a very interesting talk of Bernard Mourrain at the 2016 MAIA conference, September 2016. Mourrain [16] studies extended and generalized Prony problems, especially of the form (1), but also for log-polynomials, by means of sophisticated algebraic techniques like

© Springer International Publishing AG 2017
M. Floater et al. (Eds.): MMCS 2016, LNCS 10521, pp. 233–250, 2017.
https://doi.org/10.1007/978-3-319-67885-6_13

Gorenstein rings and truncated Hankel operators. Much of it is based on the classical duality between series and polynomials that was used in the definition of *least interpolation* [4]. He also gives a recovery algorithm based on finding separating lines, a property that has to defined *a posteriori* and that can lead to severe numerical problems.

This paper here approaches the problem in a different, more direct and elementary way, following the concepts proposed in [19,20], namely by using as a main tool the factorization of a certain Hankel matrix in terms of Vandermonde matrices; this factorization, stated in Theorem 5 has the advantage to give a handy criterion for sampling sets and was a useful tool for understanding Prony's problem in several variables, cf. [20]. Moreover, the approach uses connections to the description of finite dimensional kernels of multivariate convolutions or, equivalently, the homogeneous solutions of systems of partial difference operators.

Before we explore Prony's problem in detail, we show in Sect. 2 that it can also be formulated in terms of kernels of convolution operators or, equivalently, in terms of homogeneous solutions of partial difference equations. From that perspective it is not too surprising that many of the tools used here are very similar to the ones from [18]. The relationship that connects finite differences, the Taylor expansion and the Newton form of interpolation on the multiinteger grid, can be conveniently expressed in terms of multivariate Stirling numbers of the second kind and will be established in Sect. 3. In Sect. 4, this background will be applied to define the crucial "Prony ideal" by means of a generalized Hermite interpolation problem that yields an *ideal projector*. The aforementioned factorization, stated and proved in Sect. 5 then allows us to directly extend the algorithms from [19,20] which generate ideal bases and multiplication tables to the generalized problem without further work. How the eigenvalues of the multiplication tables relate to the common zeros of the ideal in the presence of multiplicities is finally pointed out and discussed in Sect. 6.

The notation used in this paper is as follows. By $\Pi = \mathbb{C}[x_1,\ldots,x_s]$ we denote the ring of polynomials with complex coefficients. For $A \subset \mathbb{N}_0^s$ we denote by $\Pi_A = \operatorname{span}\{(\cdot)^\alpha : \alpha \in A\} \subset \Pi$ the vector space spanned by the monomials with exponents in A, using the fairly common notation $(\cdot)^\alpha$ for the monomial (function) $m \in \Pi$, defined as $m(x) = x^\alpha$. The set of all multiindices $\alpha \in \mathbb{N}_0^s$ of *length* $|\alpha| = \alpha_1 + \cdots + \alpha_s$ is written as $\Gamma_n := \{\alpha \in \mathbb{N}_0^s : |\alpha| \leq n\}$ and defines $\Pi_n = \Pi_{\Gamma_n}$, the vector space of polynomials of total degree at most n.

2   Kernels of Difference Operators

There is a different point of view for Prony's problem (1), namely in terms of *difference operators* and their kernels. Recall that a difference equation can be most easily written as

$$q(\tau)u = v, \qquad u,v : \mathbb{Z}^s \to \mathbb{C}, \quad q \in \Pi, \tag{2}$$

where τ stands for the *shift operator* defined as $\tau_j u := u(\cdot + \epsilon_j)$, $j = 1, \ldots, s$, and $\tau^\alpha := \tau_1^{\alpha_1} \cdots \tau_s^{\alpha_s}$, $\alpha \in \mathbb{N}_0^s$. Then the difference Eq. (2) takes the explicit form

$$v = \left(\sum_{\alpha \in \mathbb{N}_0^s} q_\alpha \tau^\alpha \right) f = \sum_{\alpha \in \mathbb{N}_0^s} q_\alpha \tau^\alpha u = \sum_{\alpha \in \mathbb{N}_0^s} q_\alpha\, u(\cdot + \alpha) = a * u \qquad (3)$$

where a is a finitely supported sequence of filter coefficients with nonzero coefficients $a(-\alpha) = q_\alpha$. This is the well known fact that any difference equation is equivalent to an FIR filter or a convolution or correlation with a finite sequence.

Of particular interest are *kernels* of the convolution operators or, equivalently, homogeneous solutions for the difference Eq. (2), or, more generally of a finite *system* of difference equations

$$q(\tau)u = 0, \qquad q \in Q \subset \Pi, \#Q < \infty. \qquad (4)$$

Indeed, it is easily seen that (4) depends on the space $\mathcal{Q} = \operatorname{span} Q$ and not on the individual basis. But even more is true. Since, for any polynomials $g_q \in \Pi$, $q \in Q$, we also have

$$\sum_{q \in Q} (g_q\, q)(\tau)u = \sum_{q \in Q} g_q(\tau)(q(\tau)u) = \sum_{q \in Q} g_q(\tau)\,0 = 0,$$

the space of homogeneous solutions of (4) depends on the *ideal* $\langle Q \rangle$ generated by Q and any (ideal) basis of this ideal defines a system of difference equations with the same solutions.

The kernel space

$$\ker Q(\tau) := \{u : q(\tau)u = 0, q \in Q\},$$

on the other hand, also has an obvious structure:

$$0 = p(\tau)0 = p(\tau)q(\tau)u = q(\tau)p(\tau)u, \qquad p \in \Pi,$$

tells us that $u \in \ker Q(\tau)$ if and only if $p(\tau) \in \ker Q(\tau)$ for any $p \in \Pi$ so that the homogeneous spaces are closed under translation. This already indicates that relationships between translation invariant polynomial spaces and ideals will play a crucial role.

In one variable, the homogeneous solutions of difference equations are known to be exactly the exponential polynomial sequence, i.e., the sequences obtained by sampling exponential polynomials of the form (1), see [12]. In several variables, the situation is slightly more intricate and studied in [18] where a characterization is given for the case that $\ker Q(\tau)$ is *finite dimensional*. Indeed, the kernel spaces are of the form

$$\bigoplus_{\omega \in \Omega} \mathcal{P}_\omega\, e^{\omega^T \cdot},$$

where $\Omega \subset \mathbb{C}^s$ is a finite set of frequencies and $\mathcal{P}_\omega \subset \Pi$ is a finite dimensional translation invariant subspace of polynomials.

Therefore, we can reformulate the problem of reconstructing a function of the form (1) from integer samples, i.e., from

$$\alpha \mapsto f(\alpha) = \sum_{\omega \in \Omega} f_\omega(\alpha) \, e^{\omega^T \alpha}, \qquad \alpha \in \mathbb{Z}^s, \tag{5}$$

as the problem of finding, for this exponential polynomial sequence, a system Q of partial difference equations such that $Q(\tau)f = 0$. This finding of homogenizing equations is clearly the dual of finding homogeneous solutions of a given system. In fact, we will also study the question of how many elements of the sequence (5) we have to know in order to generate the ideal and how we can finally recover f again from the dual equations. In this respect, we can reformulate the construction from the subsequent sections in the following form.

Theorem 1. *Given any exponential polynomial sequence $f : \mathbb{Z}^s \to \mathbb{C}$ of the form (5), there exists a finite set $Q \subset \Pi$ of polynomials, the so-called* Prony ideal *of f, such that*

$$\mathrm{span}\,\{\tau^\alpha f : \alpha \in \mathbb{Z}^s\} = \ker Q(\tau),$$

and the set Q can be constructed from finitely many values of f.

3   Stirling Numbers and Invariant Spaces of Polynomials

The classical *Stirling numbers* of the second kind, written as $\left\{ {n \atop k} \right\}$ in Karamata's notation, cf. [8, p. 257ff], can be defined as

$$\left\{ {n \atop k} \right\} := \frac{1}{k!} \sum_{j=0}^{k} (-1)^{k-j} \binom{k}{j} j^n. \tag{6}$$

One important property is that they are *differences of zero* [7], which means that

$$\left\{ {n \atop k} \right\} = \frac{1}{k!} \Delta^k 0^n := \frac{1}{k!} \left(\Delta^k (\cdot)^n \right)(0).$$

Since this will turn out to be a very useful property, we define the multivariate Stirling numbers of the second kind for $\nu, \kappa \in \mathbb{Z}^s$ as

$$\left\{ {\nu \atop \kappa} \right\} := \frac{1}{\kappa!} \left(\Delta^\kappa (\cdot)^\nu \right)(0) = \frac{1}{\kappa!} \sum_{\gamma \le \kappa} (-1)^{|\kappa|-|\gamma|} \binom{\kappa}{\gamma} \gamma^\nu, \tag{7}$$

with the convention that $\left\{ {\nu \atop \kappa} \right\} = 0$ if $\kappa \not\le \nu$ where $\alpha \le \beta$ if $\alpha_j \le \beta_j$, $j = 1, \dots, s$. Moreover, we use the usual definition

$$\binom{\kappa}{\gamma} := \prod_{j=1}^{s} \binom{\kappa_j}{\gamma_j}.$$

The identity in (7) follows from the definition of the difference operator

$$\Delta^\kappa := (\tau - I)^\kappa, \qquad \tau p := [p(\cdot + \epsilon_j) : j = 1, \ldots, s], \quad p \in \Pi,$$

by straightforward computations. From [18] we recall the degree preserving operator

$$Lp := \sum_{|\gamma| \le \deg p} \frac{1}{\gamma!} \Delta^\gamma p(0) \, (\cdot)^\gamma, \qquad p \in \Pi, \tag{8}$$

which has a representation in terms of Stirling numbers: if $p = \sum_\alpha p_\alpha (\cdot)^\alpha$, then

$$Lp := \sum_{|\gamma| \le \deg p} (\cdot)^\gamma \sum_{|\alpha| \le \deg p} p_\alpha \frac{1}{\gamma!} (\Delta^\gamma (\cdot)^\alpha) (0) = \sum_{|\gamma| \le \deg p} \left(\sum_{|\alpha| \le \deg p} \left\{ \begin{matrix} \alpha \\ \gamma \end{matrix} \right\} p_\alpha \right) (\cdot)^\gamma,$$

that is,

$$(Lp)_\alpha = \sum_{\beta \in \mathbb{N}_0^s} \left\{ \begin{matrix} \beta \\ \alpha \end{matrix} \right\} p_\beta, \qquad \alpha \in \mathbb{N}_0^s. \tag{9}$$

With the *Pochhammer symbols* or *falling factorials*

$$(\cdot)_\alpha := \prod_{j=1}^s \prod_{k=0}^{\alpha_j - 1} ((\cdot)_j - k), \tag{10}$$

the inverse of L takes the form

$$L^{-1} p := \sum_{|\gamma| \le \deg p} \frac{1}{\gamma!} D^\gamma p(0) \, (\cdot)_\gamma, \tag{11}$$

see again [18]. The *Stirling numbers of first kind*

$$\left[\begin{matrix} \nu \\ \kappa \end{matrix} \right] := \frac{1}{\kappa!} (D^\kappa (\cdot)_\nu) (0), \tag{12}$$

allow us to express the inverse L^{-1} in analogous way for the coefficients of the representation $p = \sum_\alpha \hat{p}_\alpha (\cdot)_\alpha$. Indeed,

$$(L^{-1} p)_\alpha^{\wedge} = \sum_{\beta \in \mathbb{N}_0^s} \left[\begin{matrix} \beta \\ \alpha \end{matrix} \right] \hat{p}_\beta.$$

By the Newton interpolation formula for integer sites, cf. [11,22], and the Taylor formula we then get

$$(\cdot)^\alpha = \sum_{\beta \le \alpha} \frac{1}{\beta!} (\Delta^\beta (\cdot)^\alpha) (0) (\cdot)_\beta = \sum_{\beta \in \mathbb{N}_0^s} \left\{ \begin{matrix} \alpha \\ \beta \end{matrix} \right\} (\cdot)_\beta$$

$$= \sum_{\beta \in \mathbb{N}_0^s} \left\{ \begin{matrix} \alpha \\ \beta \end{matrix} \right\} \sum_{\gamma \le \beta} \frac{1}{\gamma!} (D^\gamma (\cdot)_\beta) (0) (\cdot)^\gamma = \sum_{\beta, \gamma \in \mathbb{N}_0^s} \left\{ \begin{matrix} \alpha \\ \beta \end{matrix} \right\} \left[\begin{matrix} \beta \\ \gamma \end{matrix} \right] (\cdot)^\gamma$$

from which a comparison of coefficients yields the extension of the well-known duality between the Stirling numbers of the two kinds to the multivariate case:

$$\sum_{\beta \in \mathbb{N}_0^s} \left\{ \begin{matrix} \alpha \\ \beta \end{matrix} \right\} \left[\begin{matrix} \beta \\ \gamma \end{matrix} \right] = \delta_{\alpha,\gamma}, \qquad \alpha, \beta \in \mathbb{N}_0^s. \tag{13}$$

Moreover, the multivariate Stirling numbers satisfy a recurrence similar to the univariate case. To that end, note that the Leibniz rule for the forward difference, cf. [2], yields

$$\Delta^\kappa (\cdot)^{\nu+\epsilon_j} = \Delta^\kappa \left((\cdot)^\nu (\cdot)^{\epsilon_j} \right) = \kappa_j \, \Delta^\kappa (\cdot)^\nu + \Delta^{\kappa-\epsilon_j} (\cdot)^\nu,$$

which we substitute into (7) to obtain the recurrence

$$\left\{ \begin{matrix} \nu + \epsilon_j \\ \kappa \end{matrix} \right\} = \frac{1}{\kappa!} \left(\kappa_j \, \Delta^\kappa (\cdot)^\nu + \Delta^{\kappa-\epsilon_j} (\cdot)^\nu \right) (0) = \kappa_j \left\{ \begin{matrix} \nu \\ \kappa \end{matrix} \right\} + \left\{ \begin{matrix} \nu \\ \kappa - \epsilon_j \end{matrix} \right\}. \tag{14}$$

The operator L also can be used to relate structures between polynomial subspaces.

Remark 1. Except [21], which however does not connect to the above, I was not able to find references about *multivariate* Stirling numbers, so the above simple and elementary proofs are added for the sake of completeness. Nevertheless, Gould's statement from [7] may well be true: "... *aber es mag von Interesse sein, daß mindestestens tausend Abhandlungen in der Literatur existieren, die sich mit den Stirlingschen Zahlen beschäftigen. Es ist also sehr schwer, etwas Neues über die Stirlingschen Zahlen zu entdecken.*"

Definition 1. *A subspace $\mathcal{P}$ of Π is called* shift invariant *if*

$$p \in \mathcal{P} \qquad \Leftrightarrow \qquad p(\cdot + \alpha) \in \mathcal{P}, \quad \alpha \in \mathbb{N}_0^s, \tag{15}$$

and it is called D-invariant if

$$p \in \mathcal{P} \qquad \Leftrightarrow \qquad D^\alpha p \in \mathcal{P}, \quad \alpha \in \mathbb{N}_0^s, \tag{16}$$

where $D^\alpha = \frac{\partial^{|\alpha|}}{\partial x^\alpha}$. The principal shift- and D-invariant spaces for a polynomial $p \in \Pi$ are defined as

$$\mathcal{S}(p) := \mathrm{span}\, \{ p(\cdot + \alpha) : \alpha \in \mathbb{N}_0^s \}, \qquad \mathcal{D}(p) := \mathrm{span}\, \{ D^\alpha p : \alpha \in \mathbb{N}_0^s \}, \tag{17}$$

respectively.

Proposition 1. *A subspace $\mathcal{P}$ of Π is shift invariant if and only if $L\mathcal{P}$ is D-invariant.*

Proof. The direction "⇐" has been shown in [18, Lemma 3], so assume that $\mathcal{P}$ is shift invariant and consider, for some $\alpha \in \mathbb{N}_0^s$,

$$D^\alpha Lp = D^\alpha \sum_{|\gamma| \leq \deg p} \frac{1}{\gamma!} \Delta^\gamma p(0) \, (\cdot)^\gamma$$

$$= \sum_{\gamma \geq \alpha} \frac{1}{(\gamma - \alpha)!} \Delta^{\gamma-\alpha} \left(\Delta^\alpha p \right) (0) \, (\cdot)^{\gamma-\alpha} = L\Delta^\alpha p,$$

where $\Delta^\alpha p \in \mathcal{P}$ since the space is shift invariant. Hence $D^\alpha Lp \in L\mathcal{P}$ which proves that this space is indeed D-invariant. □

A simple and well-known consequence of Proposition 1 can be recorded as follows.

Corollary 1. *A subspace $\mathcal{P}$ of Π is invariant under integer shifts if and only if it is invariant under arbitrary shifts.*

Proof. If together with p also all $p(\cdot + \alpha)$ belong to $\mathcal{P}$ then, by Proposition 1, the space $L\mathcal{P}$ is D-invariant from which it follows by [18, Lemma 3] that $p \in \mathcal{P} = L^{-1}L\mathcal{P}$ implies $p(\cdot + y) \in \mathcal{P}$, $y \in \mathbb{C}^s$. □

Proposition 2. *For $q \in \Pi$ we have that $L\mathcal{S}(q) = \mathcal{D}(Lq)$.*

Proof. By Proposition 1, $L\mathcal{S}(q)$ is a D-invariant space that contains Lq, hence $L\mathcal{S}(q) \supseteq \mathcal{D}(Lq)$. On the other hand $L^{-1}\mathcal{D}(Lq)$ is a shift invariant space containing Lq, hence

$$L^{-1}\mathcal{D}(Lq) \supseteq \mathcal{S}(L^{-1}Lq) = \mathcal{S}(q),$$

and applying the invertible operator L to both sides of the inclusion yields that $L\mathcal{S}(q) \subseteq \mathcal{D}(Lq)$ and completes the proof. □

Stirling numbers do not only relate invariant spaces, they also are useful for studying another popular differential operator. To that end, we define the partial differential operators

$$\frac{\hat{\partial}}{\partial x_j} = (\cdot)_j \frac{\partial}{\partial x_j} \quad \text{and} \quad \hat{D}^\alpha := \frac{\hat{\partial}^\alpha}{\partial x^\alpha}, \qquad \alpha \in \mathbb{N}_0^s, \tag{18}$$

also known as θ-*operator* in the univariate case. Recall that the multivariate θ-operator is usually of the form

$$\sum_{|\alpha|=n} \hat{D}^\alpha$$

and its eigenfunctions are the homogeneous polynomials, the associated eigenvalues is their total degree. Here, however, we need the partial θ-operators. To relate differential operators based on $\hat{D}$ to standard differential operators, we use the notation $(\xi D)^\alpha := \xi^\alpha D^\alpha$ for the ξ *scaled* partial derivatives, $\xi \in \mathbb{C}^s$ and use, as common, $\mathbb{C}_* := \mathbb{C} \setminus \{0\}$.

Theorem 2. *For any $q \in \Pi$ and $\xi \in \mathbb{C}^s$ we have that*

$$\left(q(\hat{D})\right)p(\xi) = (Lq(\xi D))\,p(\xi), \qquad p \in \Pi. \tag{19}$$

Proof. We prove by induction that

$$\hat{D}^\alpha = \sum_{\beta \leq \alpha} \left\{ \begin{matrix} \alpha \\ \beta \end{matrix} \right\} (\cdot)^\beta D^\beta, \qquad \alpha \in \mathbb{N}_0^s, \tag{20}$$

which is trivial for $\alpha = 0$. The inductive step uses the Leibniz rule to show that

$$\hat{D}^{\alpha+\epsilon_j} = x_j \frac{\partial}{\partial x_j} \sum_{\beta \leq \alpha} \left\{ \begin{matrix} \alpha \\ \beta \end{matrix} \right\} (\cdot)^\beta D^\beta = \sum_{\beta \leq \alpha} \left\{ \begin{matrix} \alpha \\ \beta \end{matrix} \right\} \left(\beta_j(\cdot)^\beta D^\beta + (\cdot)^{\beta+\epsilon_j} D^{\beta+\epsilon_j} \right)$$

$$= \sum_{\beta \leq \alpha+\epsilon_j} \left(\beta_j \left\{ \begin{matrix} \alpha \\ \beta \end{matrix} \right\} + \left\{ \begin{matrix} \alpha \\ \beta - \epsilon_j \end{matrix} \right\} \right) (\cdot)^\beta D^\beta,$$

from which (20) follows by taking into account the recurrence (14). Thus, by (9),

$$q(\hat{D}) = \sum_{\alpha \in \mathbb{N}_0^s} q_\alpha \hat{D}^\alpha = \sum_{\alpha \in \mathbb{N}_0^s} q_\alpha \sum_{\beta \in \mathbb{N}_0^s} \left\{ \begin{matrix} \alpha \\ \beta \end{matrix} \right\} (\cdot)^\beta D^\beta = \sum_{\beta \in \mathbb{N}_0^s} (Lq)_\beta (\cdot)^\beta D^\beta,$$

and by applying the differential operator to p and evaluating at ξ, we obtain
(19). $\square$

4   Ideals and Hermite Interpolation

A set $I \subseteq \Pi$ of polynomials is called an *ideal* in Π if it is closed under addition
and multiplication with arbitrary polynomials. A projection $P : \Pi \to \Pi$ is
called an *ideal projector*, cf. [3], if $\ker P := \{p \in \Pi : Pp = 0\}$ is an ideal.
Ideal projectors with finite range are *Hermite interpolants*, that is, projections
$H : \Pi \to \Pi$ such that

$$(q(D)Hp)(\xi) = q(D)p(\xi), \qquad q \in \mathcal{Q}_\xi, \quad \xi \in \Xi, \tag{21}$$

where $\mathcal{Q}_\xi$ is a finite dimensional D-invariant subspace of Π and $\Xi \subset \mathbb{C}^s$ is a finite
set, cf. [13]. A polynomial $p \in \ker H$ vanishes at $\xi \in \Xi$ with *multiplicity* $\mathcal{Q}_\xi$, see
[9] for a definition of multiplicity of the common zero of a set of polynomials as
a structured quantity.

A particular case is that $\mathcal{Q}_\xi$ is a *principal D-invariant space* of the form
$\mathcal{Q}_\xi = \mathcal{D}(q_\xi)$ for some $q_\xi \in \Pi$, i.e., the multiplicities are generated by a single polynomial. We say that the respective Hermite interpolation problem and
the associated ideal are of *principal multiplicity* in this case. By means of the
differential operator $\hat{D}$ these ideals are also created by shift invariant spaces.

Theorem 3. *For a finite* $\Xi \subset \mathbb{C}^s$ *and polynomials* $q_\xi \in \Pi$, $\xi \in \Xi$, *the polynomial space*

$$\left\{ p \in \Pi : q(\hat{D})p(\xi) = 0, q \in \mathcal{S}(q_\xi), \xi \in \Xi \right\} \tag{22}$$

is an ideal of principal multiplicity. Conversely, if $\xi \in \mathbb{C}_*^s$ *then any ideal of principal multiplicity can be written in the form (22).*

Proof. For $\xi \in \Xi$ we set $\mathcal{Q}_\xi' = \mathcal{D}(Lq_\xi)$ which equals $L\mathcal{S}(q_\xi)$ by Proposition 2. Then, also

$$\mathcal{Q}_\xi := \left\{ q(\operatorname{diag} \xi \,\cdot\,) : q \in \mathcal{Q}_\xi' \right\}$$

is a D-invariant space generated by $Lq_\xi(\operatorname{diag} \xi \,\cdot\,)$, and by Theorem 2 it follows that

$$q(\hat{D})p(\xi) = 0, \quad q \in \mathcal{S}(q_\xi) \tag{23}$$

if and only if

$$q(D)p(\xi) = 0, \quad q \in \mathcal{Q}_\xi = \mathcal{D}\left(Lq_\xi(\operatorname{diag} \xi \,\cdot\,)\right). \tag{24}$$

This proves the first claim, the second one follows from the observation that the process is reversible provided that $\operatorname{diag} \xi$ is invertible which happens if and only if $\xi \in \mathbb{C}_*^s$. □

The equivalence of (23) and (24) shows that Hermite interpolations can equivalently formulated either in terms of regular differential operators and differentiation invariant spaces or in terms of θ-operators and shift invariant spaces.

The *Hermite interpolation problem* based on Ξ and polynomials q_ξ can now be phrased as follows: given $g \in \Pi$ find a polynomial p (in some prescribed space) such that

$$q(\hat{D})p(\xi) = q(\hat{D})g(\xi), \quad q \in \mathcal{S}(q_\xi), \quad \xi \in \Xi. \tag{25}$$

Clearly, the number of interpolation conditions for this problem is the *total multiplicity*

$$N = \sum_{\xi \in \Xi} \dim \mathcal{S}(q_\xi).$$

The name "multiplicity" is justified here since $\dim \mathcal{Q}_\xi$ is the scalar multiplicity of a common zero of a set of polynomials and N counts the total multiplicity. Note however, that this information is incomplete since problems with the same N can nevertheless be structurally different.

Example 1. Consider $q_\xi(x) = x_1 x_2$ and $q_\xi(x) = x_1^3$. In both cases $\dim \mathcal{S}(q_\xi) = 4$ although, of course, the spaces span $\{1, x_1, x_2, x_1 x_2\}$ and span $\{1, x_1, x_1^2, x_1^3\}$ do not coincide.

A subspace $\mathcal{P} \subset \Pi$ of polynomials is called an *interpolation space* if for any $g \in \Pi$ there exists $p \in \mathcal{P}$ such that (25) is satisfied. A subspace $\mathcal{P}$ is called a *universal interpolation space* of order N if this is possible for *any* choice of Ξ and q_ξ such that

$$\sum_{\xi \in \Xi} \dim \mathcal{S}(q_\xi) \leq N.$$

Using the definition

$$\Upsilon_n := \left\{ \alpha \in \mathbb{N}_0^s : \prod_{j=1}^{s} (1 + \alpha_j) \leq n \right\}, \qquad n \in \mathbb{N},$$

of the *first hyperbolic orthant*, the positive part of the *hyperbolic cross*, we can give the following statement that also tells us that the Hermite interpolation problem is always solvable.

Theorem 4. Π_{Υ_N} *is a universal interpolation space for the interpolation problem (25).*

Proof. Since the interpolant to (25) is an ideal projector by Theorem 3, its kernel, the set of all homogeneous solutions to (25), forms a zero dimensional ideal in Π. This ideal has a Gröbner basis, for example with respect to the graded lexicographical ordering, cf. [6], and the remainders of division by this basis form the space Π_A for some lower set $A \subset \mathbb{N}_0^s$ of cardinality N. Since Υ_N is the union of all lower sets of cardinality $\leq N$, it contains Π_A and therefore Π_{Υ_N} is a universal interpolation space.   □

5   Application to the Generalized Prony Problem

We now use the tools of the preceding sections to investigate the structure of the generalized Prony problem (1) and to show how to reconstruct Ω and the polynomials f_ω from integer samples. As in [19,20] we start by considering for $A, B \subset \mathbb{N}_0^s$ the *Hankel matrix*

$$F_{A,B} = \left[f(\alpha + \beta) : \begin{matrix} \alpha \in A \\ \beta \in B \end{matrix} \right] \tag{26}$$

of samples.

Remark 2. Instead of the Hankel matrix $F_{A,B}$ one might also consider the *Toeplitz matrix*

$$T_{A,B} = \left[f(\alpha - \beta) : \begin{matrix} \alpha \in A \\ \beta \in B \end{matrix} \right], \qquad A, B \subset \mathbb{N}_0^s, \tag{27}$$

which would lead to essentially the same results. The main difference is the set on which f is sampled, especially if A, B are chosen as the total degree index sets $\Gamma_n := \{ \alpha \in \mathbb{N}_0^s : |\alpha| \leq n \}$ for some $n \in \mathbb{N}$.

Remark 3. For a coefficient vector $p = (p_\alpha : \alpha \in A) \in \mathbb{C}^B$, the result of $F_{A,B}p$ is exactly the restriction of the convolution $a * f$ from (3) with $a(-\alpha) = p(\alpha)$, $\alpha \in A$. With the Toeplitz matrix we get the even more direct $T_{A,B}p = (f * p)(A)$.

Given a finite set $\Theta \subset \Pi'$ of linearly independent linear functionals on Π and $A \subset \mathbb{N}_0^s$ the monomial *Vandermonde matrix* for the interpolation problem at Θ is defined as

$$V(\Theta, A) := \left[\theta(\cdot)^\alpha : \begin{array}{l} \theta \in \Theta \\ \alpha \in A \end{array} \right]. \tag{28}$$

It is standard linear algebra to show that the interpolation problem

$$\Theta p = y, \quad y \in \mathbb{C}^\Theta, \qquad \text{i.e.} \qquad \theta p = y_\theta, \quad \theta \in \Theta, \tag{29}$$

has a solution for any data $y = \mathbb{C}^\Theta$ iff $\operatorname{rank} V(\Theta, A) \geq \#\Theta$ and that the solution is unique iff $V(\Theta, A)$ is a nonsingular, hence square, matrix.

For our particular application, we choose Θ in the following way: Let Q_ω be a basis for for the space $\mathcal{S}(f_\omega)$ and set

$$\Theta_\Omega := \bigcup_{\omega \in \Omega} \{ \theta_\omega \, q(\hat{D}) : q \in Q_\omega \}, \qquad \theta_\omega p := p(e^\omega).$$

Since, for any $\omega \in \Omega$,

$$(\Delta^\alpha f_\omega : |\alpha| = \deg f_\omega)$$

is a nonzero vector of complex numbers or constant polynomials, we know that $1 \in \mathcal{S}(f_\omega)$ and will therefore always make the assumption that $1 \in Q_\omega$, $\omega \in \Omega$, which corresponds to $\theta_\omega \in \Theta_\Omega$, $\omega \in \Omega$. Moreover, we request without loss of generality that $f_\omega \in Q_\omega$.

We pattern the Vandermonde matrix conveniently as

$$V(\Theta_\Omega, A) = \left[\left(q(\hat{D})(\cdot)^\alpha \right) (e^\omega) : \begin{array}{l} q \in Q_\omega, \, \omega \in \Omega \\ \alpha \in A \end{array} \right]$$

to obtain the following fundamental factorization of the Hankel matrix.

Theorem 5. *The Hankel matrix $F_{A,B}$ can be factored into*

$$F_{A,B} = V(\Theta_\Omega, A)^T \, F \, V(\Theta_\Omega, B), \tag{30}$$

where F is a nonsingular block diagonal matrix independent of A and B.

Proof. We begin with an idea by Gröbner [9], see also [18], and first note that any $g \in Q_\omega$ can be written as

$$g(x + y) = \sum_{q \in Q_\omega} c_q(y) \, q(x), \qquad c_q : \mathbb{C}^s \to \mathbb{C}.$$

Since $g(x + y)$ also belongs to $\operatorname{span} Q_\omega$ as a function in y for fixed x, we conclude that $c_q(y)$ can also be written in terms of Q_ω and thus have obtained the *linearization formula*

$$g(x + y) = \sum_{q, q' \in Q_\omega} a_{q,q'}(g) \, q(x) \, q'(y), \qquad a_{q,q'}(g) \in \mathbb{C}, \tag{31}$$

from [9]. Now consider

$$(F_{A,B})_{\alpha,\beta} = f(\alpha + \beta) = \sum_{\omega \in \Omega} f_\omega(\alpha + \beta)\, e^{\omega^T(\alpha+\beta)}$$

$$= \sum_{\omega \in \Omega} \sum_{q,q' \in Q_\omega} a_{q,q'}(f_\omega)\, q(\alpha)\, e^{\omega^T \alpha}\, q'(\beta)\, e^{\omega^T \beta}$$

$$= \sum_{\omega \in \Omega} \sum_{q,q' \in Q_\omega} a_{q,q'}(f_\omega) \left(q(\hat{D})(\cdot)^\alpha \right)(e^\omega) \left(q'(\hat{D})(\cdot)^\beta \right)(e^\omega)$$

$$= \left(V(\Theta_\Omega, A)\, \mathrm{diag} \left(\left[a_{q,q'}(f_\omega) : \begin{matrix} q \in Q_\omega \\ q' \in Q_\omega \end{matrix} \right] : \omega \in \Omega \right) V(\Theta_\Omega, B)^T \right)_{\alpha,\beta},$$

which already yields (32) with

$$F := \mathrm{diag} \left(\left[a_{q,q'}(f_\omega) : \begin{matrix} q \in Q_\omega \\ q' \in Q_\omega \end{matrix} \right] : \omega \in \Omega \right) = \mathrm{diag}\left(A_\omega : \omega \in \Omega \right).$$

It remains to prove that the blocks A_ω of the block diagonal matrix F are nonsingular. To that end, we recall that $f_\omega \in Q_\omega$, hence, by (31),

$$f_\omega = f_\omega(\cdot + 0) = \sum_{q,q' \in Q_\omega} a_{q,q'}(f_\omega)\, q(x)\, q'(0),$$

that is, by linear independence of the elements of Q_ω,

$$\sum_{q' \in Q_\omega} a_{q,q'}(f_\omega)\, q'(0) = \delta_{q,f_\omega}, \qquad q \in Q_\omega,$$

which can be written as $A_\omega Q_\omega(0) = e_{f_\omega}$ where Q_ω also stands for the polynomial vector formed by the basis elements. Since Q_ω is a basis for $\mathcal{S}(f_\omega)$, there exist finitely supported sequences $c_q : \mathbb{N}_0^s \to \mathbb{C}$, $q \in Q_\omega$, such that

$$q = \sum_{\alpha \in \mathbb{N}_0^s} c_q(\alpha)\, f(\cdot + \alpha), = \sum_{q',q'' \in Q_\omega} a_{q',q''}(f_\omega) \left(\sum_{\alpha \in \mathbb{N}_0^s} c_q(\alpha)\, q''(\alpha) \right) q'$$

from which a comparison of coefficients allows us to conclude that

$$A_\omega \sum_{\alpha \in \mathbb{N}_0^s} c_q(\alpha) Q_\omega(\alpha) = e_q, \qquad q \in Q_\omega,$$

which even gives an "explicit" formula for the columns of A_ω^{-1}.   □

Remark 4. A similar factorization of the Hankel matrix in terms of Vandermonde matrices for slightly different but equivalent Hermite problems has also been given in [16, Proposition 3.18]. However, the invertibility of the "inner matrix" F was concluded there from the invertibility of the Hankel matrix and the assumption that Π_A must be an *interpolation space*, giving unique interpolants for the Hermite problem. Theorem 5, on the other hand, does not need these assumptions, shows that F is *always* nonsingular and therefore extends the one given in [20, (5)] in a natural way.

Remark 5. If $F_{A,B}$ is replaced by the Toeplitz matrix from (27), then the factorization becomes

$$T_{A,B} = W(\Theta_\Omega, A)\, F\, W(\Theta_\Omega, B)^*, \qquad W(\Theta_\Omega, A) := V(\Theta_\Omega, A)^T \quad (32)$$

which has more similarity to a block Schur decomposition since now a Hermitian of the factorizing matrix appears.

Once the factorization (32) is established, the results from [19,20] can be applied literally and extend to the case of exponential polynomial reconstruction directly. In particular, the following observation is relevant for the termination of the algorithms. It says that if the row index set A is "sufficiently rich", then the full information about the ideal $I_\Omega := \ker \Theta_\Omega$ can be extracted from the Hankel matrix $F_{A,B}$.

Theorem 6. *If Π_A is an interpolation space for Θ_Ω, for example if $A = \Upsilon_N$, then*

1. *the function f can be reconstructed from samples $f(A + B)$, $A, B \subset \mathbb{N}_0^s$, if and only if Π_A and Π_B are interpolation spaces for Θ_Ω.*
2. *a vector $p \in \mathbb{C}^B \setminus \{0\}$ satisfies*

$$F_{A,B}p = 0 \qquad \Leftrightarrow \qquad \sum_{\beta \in B} p_\beta\, (\cdot)^\beta \in I_\Omega \cap \Pi_B.$$

3. *the mapping $n \mapsto \operatorname{rank} F_{A,\Gamma_n}$ is the affine Hilbert function for the ideal I_Ω.*

Proof. Theorem 6 is a direct consequence of Theorem 5 by means of elementary linear algebra. The proof of (1) is a literal copy of that of [20, Theorem 3] for (2) we note that, for $p \in \mathbb{C}^B$,

$$\begin{aligned}
F_{A,B}p &= V(\Theta_\Omega, A)^T\, F\, V(\Theta_\Omega, B)p \\
&= V(\Theta_\Omega, A)^T\, F\, \left[\left(q(\hat{D})p \right)(e^\omega) : q \in Q_\omega,\, \omega \in \Omega \right].
\end{aligned}$$

By assumption, $V(\Theta_\Omega, A)^T$ has full rank, F is invertible by Theorem 5, and therefore $F_{A,B}p = 0$ if and only if the polynomial p belongs to I_Ω. Finally, (3) is an immediate consequence of (2).

Theorem 6 suggests the following generic algorithm: use a nested sequence $B_0 \subset B_1 \subset B_2 \subset \cdots$ of index sets in $\mathbb{N}_0^s$ such that there exist $j(n) \in \mathbb{N}$, $n \in \mathbb{N}$, such that $B_{j(n)} = \Gamma_n$. In other words: the subsets progress in a *graded* fashion. Then, for $j = 0, 1, \ldots$

1. Consider the kernel of F_{Υ_N, B_j}, these are the ideal elements in Π_{B_j}.
2. Consider the complement of the kernel, these are elements of the *normal set* and eventually form a basis for an interpolation space.
3. Terminate if $\operatorname{rank} F_{\Upsilon_N, B_{j(n+1)}} = \operatorname{rank} F_{\Upsilon_N, B_{j(n)}}$ for some n.

Observe that this task of computing an ideal basis from nullspaces of matrices is *exactly* the same as in Prony's problem with constant coefficients. The difference lies only in the fact that now the ideal is not *radical* any more, but this is obviously irrelevant for Theorem 6.

Two concrete instances of this approach were presented and discussed earlier: [19] uses $B_j = \Gamma_j$ and Sparse Homogeneous Interpolation Techniques (DNSIN) to compute an orthonormal H-basis and a graded basis for the ideal and the normal space, respectively. Since these computations are based on *orthogonal decompositions*, mainly QR factorizations, it is numerically stable and suitable for finite precision computations in a floating point environment. A symbolic approach where the B_j are generated by adding multiindices according to a graded term order, thus using Sparse Monomial Interpolation with Least Elements (SMILE), was introduced in [20]. This method is more efficient in terms of number of computations and therefore suitable for a symbolic framework with exact rational arithmetic.

Remark 6. The only a priori knowledge these algorithms need to know is an upper estimate for the *multiplicity* N.

It should be mentioned that also [16] gives algorithms to reconstruct frequencies and coefficients by first determining the *Prony ideal* I_Ω; the way how these algorithms work and how they are derived are different, however. It would be worthwhile to study and understand the differences between and the advantages of the methods.

While we will point out in the next section how the frequencies can be determined by generalized eigenvalue methods, we still need to clarify how the coefficients of the polynomials f_ω can be computed once the ideal structure and the frequencies are determined. To that end, we write

$$f_\omega = \sum_{\alpha \in \mathbb{N}_0^s} f_{\omega,\alpha} (\cdot)^\alpha$$

and note that, with $\xi_\omega := e^\omega \in \mathbb{C}_*^s$

$$f(\beta) = \sum_{\omega \in \Omega} f_\omega(\beta) \, e^{\omega^T \beta} = \sum_{\omega \in \Omega} \sum_{\alpha \in \mathbb{N}_0^s} f_{\omega,\alpha} \beta^\alpha \xi_\omega^\beta = \sum_{\omega \in \Omega} \sum_{\alpha \in \mathbb{N}_0^s} f_{\omega,\alpha} \left(\hat{D}^\alpha (\cdot)^\beta \right) (\xi_\omega).$$

In other words, we have for any choice of $A_\omega \subset \mathbb{N}_0^s$, $\omega \in \Omega$ and $B \subset \mathbb{N}_0^s$ that

$$
\begin{aligned}
f(B) &:= [f(\beta) : \beta \in B] \\
&= \left[\left(\hat{D}^\alpha (\cdot)^\beta \right) (\xi_\omega) : \begin{matrix} \beta \in B \\ \alpha \in A_\omega, \omega \in \Omega \end{matrix} \right] [f_{\omega,\alpha} : \alpha \in A_\omega, \omega \in \Omega] \\
&=: G_{A,B} f_\Omega.
\end{aligned}
$$

The matrix $G_{A,B}$ is another Vandermonde matrix for a Hermite-type interpolation problem with the functionals

$$\theta_\omega \hat{D}^\alpha, \qquad \alpha \in A_\omega, \omega \in \Omega. \tag{33}$$

The linear system

$$G_{A,B} f_\Omega = f(B)$$

can thus be used to determine f_Ω: first note that $L\Pi_n = \Pi_n$ and therefore it follows by Theorem 3 that the interpolation problem is a Hermite problem, i.e., its kernel is an ideal. If we set

$$N = \sum_{\omega \in \Omega} \binom{\deg f_\omega + s}{s} - 1$$

then, by Theorem 4, the space Π_{Υ_N} is a universal interpolation space for the interpolation problem (33). Hence, with $A_\omega = \Gamma_{\deg f_\omega}$, the matrix G_{A,Υ_N} contains a nonsingular square matrix of size $\#A \times \#A$ and the coefficient vector f_Ω is the unique solution of the overdetermined interpolation problem.

Remark 7. The a priori information about the multiplicity N of the interpolation points does not allow for an efficient reconstruction of the frequencies as it only says that there are at most N points or points of local multiplicity up to N.

Nevertheless, the degrees $\deg f_\omega$, $\omega \in \Omega$, more precisely, upper bounds for them, can be derived as a by-product of the determination of the frequencies ω by means of multiplication tables. To clarify this relationship, we briefly revise the underlying theory, mostly due to Möller and Stetter [14], in the next section.

6   Multiplication Tables and Multiple Zeros

Having computed a good basis H for the ideal I_Ω and a basis for the normal set Π/I_Ω, the final step consists of finding the common zeros of H. The method of choice is still to use eigenvalues of the multiplication tables, cf. [1, 23], but things become slightly more intricate since we now have to consider the case of zeros with multiplicities, cf. [14].

Let us briefly recall the setup in our particular case. The multiplicity space at $\xi_\omega = e^\omega \in \mathbb{C}_*^s$ is

$$\mathcal{Q}_\omega := \mathcal{D}\left(Lf_\omega(\operatorname{diag} \xi_\omega \cdot)\right)$$

and since this is a D-invariant subspace, it has a graded basis Q_ω where the highest degree element in this basis can be chosen as $g_\omega := Lf_\omega(\operatorname{diag} \xi_\omega \cdot)$. Since $\mathcal{Q}_\omega = \mathcal{D}(g_\omega)$, all other basis elements $q \in Q_\omega$ can be written as $q = g_q(D)g_\omega$, $g_q \in \Pi$, $q \in Q_\Omega$.

Given a basis P of the normal set Π/I_Ω and a normal form operator $\nu : \Pi \to \Pi/I_\Omega = \operatorname{span} P$ modulo I_Ω (which is an ideal projector and can be computed efficiently for Gröbner and H-bases), the multiplication $p \mapsto \nu\left((\cdot)_j p\right)$ is a linear operation on Π/I_Ω for any $j = 1, \ldots, s$. It can be represented with respect to the basis P by means of a matrix M_j which is called jth *multiplication table* and gives the multivariate generalization of the *Frobenius companion matrix*.

Due to the unique solvability of the Hermite interpolation problem in Π/I_Ω, there exists a basis of fundamental polynomials $\ell_{\omega,q}$, $q \in Q_\omega$, $\omega \in \Omega$, such that

$$q'(D)\ell_{\omega,q}(\xi_{\omega'}) = \delta_{\omega,\omega'}\delta_{q,q'}, \qquad q' \in Q_{\omega'}, \quad \omega' \in \Omega. \tag{34}$$

The projection to the normal set, i.e., the interpolant, can now be written for any $p \in \Pi$ as

$$Lp = \sum_{\omega \in \Omega} \sum_{q \in Q_\omega} q(D)p(\xi_\omega)\,\ell_{\omega,q}$$

hence, by the Leibniz rule and the fact that Q_ω is D-invariant

$$
\begin{aligned}
L\left((\cdot)_j\,\ell_{\omega,q}\right) &= \sum_{\omega' \in \Omega} \sum_{q' \in Q_{\omega'}} q'(D)\left((\cdot)_j\,\ell_{\omega,q}\right)(\xi_\omega)\,\ell_{\omega',q'} \\
&= \sum_{\omega' \in \Omega} \sum_{q' \in Q_{\omega'}} \left((\cdot)_j\,q'(D)\ell_{\omega,q}(\xi_\omega) + \frac{\partial q'}{\partial x_j}(D)\ell_{\omega,q}(\xi_\omega)\right)\ell_{\omega',q'} \\
&= (\cdot)_j\,\ell_{\omega,q} + \sum_{q' \in Q_\omega} \sum_{q'' \in Q_\omega} c_j(q',q'')\,q''(D)\ell_{\omega,q}(\xi_\omega)\,\ell_{\omega,q'} \\
&= (\cdot)_j\,\ell_{\omega,q} + \sum_{q' \in Q_\omega} c_j(q,q')\,\ell_{\omega,q'},
\end{aligned}
\tag{35}
$$

where the coefficients $c_j(q,q')$ are defined by the expansion

$$\frac{\partial q}{\partial x_j} = \sum_{q' \in Q_\omega} c_j(q',q)\,q', \qquad q \in Q_\Omega. \tag{36}$$

Note that the coefficients in (36) are zero if $\deg q' \geq \deg q$. Therefore $c_j(q,q') = 0$ in (35) if $\deg q' \geq \deg q$. In particular, since g_ω is the unique element of maximal degree in Q_ω, it we have that

$$L\left((\cdot)_j\,\ell_{\omega,g_\omega}\right) = (\cdot)_j\,\ell_{\omega,g_\omega}, \qquad \omega \in \Omega. \tag{37}$$

This way, we have given a short and simple proof of the following result from [14], restricted to our special case of principal multiplicities.

Theorem 7. *The eigenvalues of the multiplication tables M_j are the components of the zeros $(\xi_\omega)_j$, $\omega \in \Omega$, the associated eigenvectors the polynomials ℓ_{ω,g_ω} and the other fundamental polynomials form an invariant space.*

In view of numerical linear algebra, the eigenvalue problems for ideals with multiplicities become unpleasant as in general the matrices become derogatory, except when g_ω is a power of linear function, i.e., $g_\omega = (v^T \cdot)^{\deg g_\omega}$ for some $v \in \mathbb{R}^s$, but the method by Möller and Tenberg [15] to determine the joint eigenvalues of multiplication tables and their multiplicities also works in this situation.

There is another remedy described in [5, p. 48]: building a matrix from traces of certain multiplication tables, one can construct a basis for the associated *radical* ideal with simple zeros, thus avoiding the hassle with the structure of multiplicities. In addition, this method also gives the dimension of the multiplicity spaces which is sufficient information to recover the polynomial coefficients. Though this approach is surprisingly elementary, we will not go into details here as it is not in the scope of the paper, but refer once more to the recommendable collection [5].

Moreover, the dimension of the respective invariant spaces is an upper bound for $\deg f_\omega$ which can help to set up the parameters in the interpolation problem in Sect. 5.

7   Conclusion

The generalized version of Prony's problem with polynomial coefficients is a straightforward extension of the standard problem with constant coefficients. The main difference is that in (1) *multiplicities* of common zeros in an ideal play a role where the multiplicity spaces are related to the shift invariant space generated by the coefficients via the operator L from (8). This operator which relates the Taylor expansion and interpolation at integer points in the Newton form, has in turn a natural relationship with multivariate Stirling numbers of the second kind. These properties can be used to extend the algorithms from [19, 20] almost without changes to the generalized case, at least as far the construction of a good basis for the Prony ideal is concerned.

The algorithms from [19, 20], numerical or symbolic, can be reused, the only difference lies in multiplication tables with multiplicities, but the tools from [15] are also available in this case and allow to detect zeros *and* their structure.

Implementations, numerical tests and comparison with the algorithms from [16] are straightforward lines of further work and my be a worthwhile waste of time.

References

1. Auzinger, W., Stetter, H.J.: An elimination algorithm for the computation of all zeros of a system of multivariate polynomial equations. In: Numerical mathematics, Singapore 1988, Internationale Schriftenreihe zur Numerischen Mathematik, vol. 86, pp. 11–30. Birkhäuser, Basel (1988)
2. de Boor, C.: Divided differences. Surv. Approximation Theory **1**, 46–69 (2005). http://www.math.technion.ac.il/sat
3. de Boor, C.: Ideal interpolation. In: Chui, C.K., Neamtu, M., Schumaker, L.L. (eds.) Approximation Theory XI, Gaitlinburg 2004, pp. 59–91. Nashboro Press (2005)
4. de Boor, C., Ron, A.: The least solution for the polynomial interpolation problem. Math. Z. **210**, 347–378 (1992)
5. Cohen, A.M., Cuypers, H., Sterk, M. (eds.): Some Tapas of Computer Algebra. Algorithms and Computations in Mathematics, vol. 4. Springer, Heidelberg (1999)
6. Cox, D., Little, J., O'Shea, D.: Ideals, Varieties and Algorithms. Undergraduate Texts in Mathematics, 2nd edn. Springer, New York (1996)
7. Gould, H.W.: Noch einmal die Stirlingschen Zahlen. Jber. Deutsch. Math.-Verein **73**, 149–152 (1971)
8. Graham, R.L., Knuth, D.E., Patashnik, O.: Concrete Mathematics, 2nd edn. Addison-Wesley, Reading (1998)

9. Gröbner, W.: Über das Macaulaysche inverse System und dessen Bedeutung für die Theorie der linearen Differentialgleichungen mit konstanten Koeffizienten. Abh. Math. Sem. Hamburg **12**, 127–132 (1937)
10. Gröbner, W.: Über die algebraischen Eigenschaften der Integrale von linearen Differentialgleichungen mit konstanten Koeffizienten. Monatsh. Math. **47**, 247–284 (1939)
11. Isaacson, E., Keller, H.B.: Analysis of Numerical Methods. Wiley, New York (1966)
12. Jordan, C.: Calculus of Finite Differences, 3rd edn. Chelsea, New York (1965)
13. Marinari, M.G., Möller, H.M., Mora, T.: On multiplicities in polynomial system solving. Trans. Am. Math. Soc. **348**(8), 3283–3321 (1996)
14. Möller, H.M., Stetter, H.J.: Multivariate polynomial equations with multiple zeros solved by matrix eigenproblems. Numer. Math. **70**, 311–329 (1995)
15. Möller, H.M., Tenberg, R.: Multivariate polynomial system solving using intersections of eigenspaces. J. Symbolic Comput. **32**, 513–531 (2001)
16. Mourrain, B.: Polynomial-exponential decomposition from moments (2016). arXiv:1609.05720v1
17. Prony, C.: Essai expérimental et analytique sur les lois de la dilabilité des fluides élastiques, et sur celles de la force expansive de la vapeur de l'eau et de la vapeur de l'alkool, à différentes températures. J. de l'École Polytechnique **2**, 24–77 (1795)
18. Sauer, T.: Kernels of discrete convolutions and subdivision operators. Acta Appl. Math. **145**, 115–131 (2016). arXiv:1403.7724
19. Sauer, T.: Prony's method in several variables. Numer. Math. (2017, to appear). arXiv:1602.02352
20. Sauer, T.: Prony's method in several variables: symbolic solutions by universal interpolation. J. Symbolic Comput. (2017, to appear). arXiv:1603.03944
21. Schreiber, A.: Multivariate Stirling polynomials of the first and second kind. Discrete Math. **338**, 2462–2484 (2015)
22. Steffensen, I.F.: Interpolation. Chelsea Publication, New York (1927)
23. Stetter, H.J.: Matrix eigenproblems at the heart of polynomial system solving. SIGSAM Bull. **30**(4), 22–25 (1995)

Reparameterization and Adaptive Quadrature for the Isogeometric Discontinuous Galerkin Method

Agnes Seiler[1(✉)] and Bert Jüttler[2]

[1] Doctoral Program "Computational Mathematics", Johannes Kepler University
Linz, Altenberger Str. 69, 4040 Linz, Austria
agnes.seiler@dk-compmath.jku.at
[2] Institute of Applied Geometry, Johannes Kepler University Linz,
Altenberger Str. 69, 4040 Linz, Austria
bert.juettler@jku.at

Abstract. We use the Poisson problem with Dirichlet boundary conditions to illustrate the complications that arise from using non-matching interface parameterizations within the framework of Isogeometric Analysis on a multi-patch domain, using discontinuous Galerkin (dG) techniques to couple terms across the interfaces. The dG-based discretization of a partial differential equation is based on a modified variational form, where one introduces additional terms that measure the discontinuity of the values and normal derivatives across the interfaces between patches. Without matching interface parameterizations, firstly, one needs to identify pairs of associated points on the common interface of the two patches for correctly evaluating the additional terms. We will use reparameterizations to perform this task. Secondly, suitable techniques for numerical integration are needed to evaluate the quantities that occur in the discretization with the required level of accuracy. We explore two possible approaches, which are based on subdivision and adaptive refinement, respectively.

1 Introduction

Isogeometric Analysis (IgA) [5,6] uses the same spaces of spline functions for representing the geometry of a physical domain and for performing a discretization in the context of a PDE-based numerical simulation. This method is based on a parameterization of the physical domain, i.e., on a geometry map that relates the physical domain and the parameter domain.

Many approaches rely on tensor product parameterizations, where the domain is a unit square or a unit cube. Consequently, more complex domains have to be divided into several single patches, forming a multi-patch representation. There exist several methods for coupling the discrete discontinuous Galerkin IgA patch wise solution across the interfaces between single patches and

© Springer International Publishing AG 2017
M. Floater et al. (Eds.): MMCS 2016, LNCS 10521, pp. 251–269, 2017.
https://doi.org/10.1007/978-3-319-67885-6_14

enhancing global continuity of the solution. These include Nitsche's [16] method and mortar techniques [2] as well as the discontinuous Galerkin (dG) approach, which is the focus of the present paper.

DG methods discretize the variational form of a partial differential equation taking into account the discontinuity of the finite dimensional discretization spaces. The publications [4,17] provide a general description of dG techniques in the context of finite elements, which have been transferred to the isogeometric setting in [3,12–14].

So far, only matching interface parameterizations have been studied in the context of dG-IgA methods. More precisely, whenever two patches meet in an interface, then the parameterizations restricted to these interfaces are assumed to be identical (possibly after affine transformations of the parameter domains), see [12–14,18]. On the one hand, this limitation provides the advantage that the elements of the patches on both sides of the interface are perfectly matching, which significantly simplifies the implementation of such methods. On the other hand, it complicates substantially the creation of multi-patch parameterizations.

As notable exceptions we mention the recent publications [10,11], where the authors study gaps and overlaps at the interfaces. While the theory presented in these papers does not require any assumptions regarding matching interfaces, such conditions are assumed to be satisfied in all the computational examples. More precisely, the meshes of the considered domains fulfill restrictive correspondence conditions, which are quite similar to the matching case. This is due to the lack of an implementation for the non-matching case [9].

This recent work has motivated us to investigate the effect of non-matching interface parameterizations in the context of dG-IgA in the present paper. We aim to give a complete description of the necessary computational steps for applying the theoretical results of [10–14,18] to the case of non-matching parameterizations at the interfaces. In order to keep the presentation simple, we restrict ourselves to planar two-patch domains and we assume that the interfaces are geometrically matching, thus they have neither overlaps nor gaps. However, it is clear that the results from [10,11] apply to the non-matching case also, as the theory presented there is sufficiently general.

More precisely, the assembly of the local stiffness matrices derived from the dG bilinear form requires the computation of integrals of the type

$$\int_e D\, b_i^k(\boldsymbol{x}) D'\, b_j^\ell(\boldsymbol{x}) d\boldsymbol{x}, \tag{1}$$

where e is an interface between Ω^k and Ω^ℓ in the physical domain, $\boldsymbol{x} \in e$ is a point on the interface, b_i^k, b_j^ℓ are isogeometric basis functions defined on patches Ω^k, $\Omega^\ell \subset \Omega$ of the multi-patch domain $\Omega \subseteq \mathbb{R}$, and D, D' are differential operators. As we shall see, non-matching interface parameterizations give rise to two problems that need to be treated separately.

The first one concerns the evaluation of $b_i^k(\boldsymbol{x})$ and $b_j^\ell(\boldsymbol{x})$ at the same position $\boldsymbol{x}$ on the interface. Due to the use of non-matching parameterizations, a point $\boldsymbol{x}$ will have two possibly different preimages in the parameter domains of the two patches joined by the interface respectively. To identify pairs of corresponding preimages we use reparameterizations of the preimages of the interface. We also investigate the influence of the quality of the reparameterization on the accuracy of the overall result.

The second problem is related to the use of numerical integration methods. We need to find a quadrature method whose exactness does not depend on the smoothness of the integrands. We present different approaches, one resulting from dividing the element on which quadrature is performed and another one making use of automatized element splitting. The performance of both approaches is explored in numerical experiments.

The remainder of this paper is structured as follows: We establish the notation and describe the example problem we will focus on in the next section. We then state the two issues of evaluation and numerical integration, as described above. Section 3 treats the first problem of finding suitable reparameterizations, while Sect. 4 is devoted to the different quadrature techniques. Results of numerical experiments are presented in Sect. 5. Finally we conclude the paper.

2   Preliminaries

We recall the discontinuous Galerkin isogeometric (dG-IgA) discretization of a given model problem and discuss the computation of the stiffness matrix elements in the case of non-matching interface parameterizations. Hereby, we restrict ourselves to the two-patch case shown in Fig. 1 due to better readability. All observations generalize directly to domains with more than two patches.

2.1   The Model Problem and the Multi-patch Discretization

Given a domain $\Omega \subseteq \mathbb{R}^2$, we consider the Poisson problem

$$\text{Find } u : \quad \begin{cases} -\nabla \cdot (\alpha \nabla u) &= f \quad \text{on } \Omega \\ u &= 0 \quad \text{on } \partial\Omega, \end{cases} \tag{2}$$

where f is given and $\alpha > 0$ is the known diffusion coefficient. We allow α to be piecewise constant, i.e. α may take different values on every single patch (see below).

More precisely, we consider a multi-patch domain $\Omega \subseteq \mathbb{R}^2$ that consists of 2 non-overlapping single patches Ω^1, Ω^2 such that $\bar{\Omega}^1 \cup \bar{\Omega}^2 = \bar{\Omega}$. We use upper indices to refer to the number of the patch, and thus α^k denotes the value of the diffusion coefficient on the k-th patch, $k = 1, 2$.

An interface e between the two patches is the intersection $e = \bar{\Omega}^1 \cap \bar{\Omega}^2$. We consider interfaces that are curve segments only and ignore the remaining ones.

Each physical patch Ω^k is parameterized by an associated geometry mapping G^k with parameter domain $\hat{\Omega}^k = [0,1]^2$, $k = 1, 2$. These mappings are tensor product spline functions

$$G^k(\boldsymbol{\xi}) = \sum_{i \in \mathcal{R}^k} P_i^k \beta_i^k(\boldsymbol{\xi}) \, , \, \boldsymbol{\xi} \in \hat{\Omega}^k, \tag{3}$$

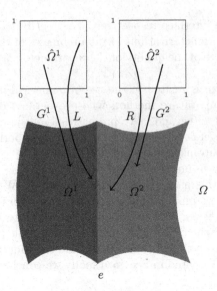

Fig. 1. Multi-patch domain with two patches Ω^1, Ω^2, one interface e and geometry mappings G^1, G^2. The mappings L and R will be introduced later.

which are defined by control points $P_i^k \in \mathbb{R}^2$ and tensor product B-splines β_i^k, where $\mathcal{R}^k$ is the index set of the k-th patch. The lower index i identifies the i-th degree of freedom of the k-th patch.

We do not assume that the knot vectors of the patches are identical. These knot vectors split each parameter domain into elements. We will use open knot vectors which implies that the boundaries of the patches are B-spline curves.

An isogeometric basis function b_i^k on the physical patch Ω^k is the push-forward of a B-spline β^k defined on the parameter domain $\hat{\Omega}^k$,

$$b_i^k(\boldsymbol{x}) = \begin{cases} \left(\beta_i^k \circ \left(G^k \right)^{-1} \right)(\boldsymbol{x}) & \text{if } \boldsymbol{x} \in \Omega^k \\ 0 & \text{otherwise.} \end{cases} \tag{4}$$

These functions span the space which is used to derive the dG-IgA discretization.

For later reference we define the set of all edges

$$\Gamma = \bigcup_{k=1}^{2} \{ G^k([0,1],0),\, G^k([0,1],1),\, G^k(0,[0,1]),\, G^k(1,[0,1]) \} \tag{5}$$

of the multi-patch domain. It is the disjoint union of the set of the interface edges

$$\Gamma_C = \{ e \in \Gamma : e \subseteq \bar{\Omega}^1 \cap \bar{\Omega}^2 \} \tag{6}$$

and the set of boundary edges

$$\Gamma_D = \{ e \in \Gamma : e \subseteq \bar{\Omega}^k \cap \partial\Omega,\, k = 1, 2 \}. \tag{7}$$

2.2 DG-IgA Discretization

The discontinuous Galerkin isogeometric (dG-IgA) discretization space considers the subspace

$$\mathcal{V}_h = \text{span}\,\{b_i^k \,:\, i \in \mathcal{R}^k,\, k = 1, \ldots, n\} \subseteq \prod_{k=1}^{2} \mathcal{H}^1\left(\Omega^k\right), \tag{8}$$

of the broken Sobolev space, see [17]. Functions in $\mathcal{V}_h$ are continuously differentiable on the interior of the single patches but not necessarily smooth across interface edges. This smoothness of the solution is achieved approximately by introducing a penalty term that considers the jump of the solution across the interface. Before stating the final variational formulation we need to define averages and jumps, see [4,17].

For each patch index k, any function $v \in \prod_{k=1}^{2} \mathcal{H}^1\left(\Omega^k\right)$ has a well-defined trace along any edge $e \subset \partial\Omega^k$. Thus, any such function v defines *two* traces on the interface $e \in \Gamma_C$, which we denote as $v|_{\Omega^{1e}}$ and $v|_{\Omega^{2e}}$, respectively. We use them to define the *average*

$$\{v\}^e = \frac{1}{2}\left(v|_{\Omega^{1e}} + v|_{\Omega^{2e}}\right) \tag{9}$$

and the *jump*

$$[v]^e = v|_{\Omega^{1e}} - v|_{\Omega^{2e}} \tag{10}$$

across the interface $e \in \Gamma_C$.

These definitions are further extended to boundary edges $e \in \Gamma_D$,

$$\{v\}^e = v|_{\Omega^k} \text{ and } [v]^e = v|_{\Omega^k},\, k = 1, 2. \tag{11}$$

The dG-IgA discretization

$$\text{Find } u \in \mathcal{V}_h : \quad a(u, v) = F(v) \quad \forall v \in \mathcal{V}_h \tag{12}$$

of the Poisson problem (2) uses the bilinear form

$$a(u, v) = \sum_{k=1}^{2} a_1^k(u, v) - \frac{1}{2} \sum_{e \in \Gamma_C \cup \Gamma_D}\left(a_{2,1}^e(u, v) + a_{2,2}^e(u, v)\right) + \sum_{e \in \Gamma_C \cup \Gamma_D} a_3^e(u, v) \tag{13}$$

with

$$a_1^k(u, v) = \int_{\Omega^k} \alpha^k \nabla u \cdot \nabla v\, d\Omega, \tag{14}$$

$$a_{2,1}^e(u, v) = \int_e \{\nabla u \cdot n\}^e [v]^e de, \qquad a_{2,2}^e(u, v) = \int_e \{\nabla v \cdot n\}^e [u]^e de, \tag{15}$$

$$a_3^e(u, v) = \int_e \frac{\delta}{h} [u]^e [v]^e de \tag{16}$$

and the linear form

$$F(v) = \int_\Omega f v d\Omega. \tag{17}$$

The second group of terms $a_{2,1}^e$ and $a_{2,2}^e$ considers normal vectors $n = n_e$ of the interface e, which need to comply with the chosen orientation of the edges (determined by the patch numbering). The last terms a_3^e in the bilinear form are the penalty terms mentioned before, which involve the sufficiently large parameter δ. They depend on the element size h, i.e. on the length of the knot spans[1].

A detailed derivation of the dG discretization is given in [17]. The adaptation to the isogeometric setting is discussed in the thesis [3], which also comments on the choice of the δ, and in the recent article [12].

The discretization (12) defines the associated *dG norm*

$$\|u\|_{dG}^2 = \sum_{k=1}^2 a_1^k(u, u) + \sum_{e \in \Gamma_C \cup \Gamma_D} a_3^e(u, u), \tag{18}$$

where in $a_1(u, u)$ the gradient of u is restricted to Ω^k, see again [12].

The coefficients u_i^k of the approximate solution

$$u_h = \sum_{k=1}^2 \sum_{i \in \mathcal{R}_k} u_i^k b_i^k \tag{19}$$

are found by solving the linear system $Su = b$ with

$$S = \left(s_{(i,k),(j,\ell)}\right)_{(i,k),(j,\ell)},$$
$$b = \left(b_{(j,\ell)}\right)_{(j,\ell)},$$
$$u = \left(u_i^k\right)_{(i,k)},$$

where

$$s_{(i,k),(j,\ell)} = a\left(b_i^k, b_j^\ell\right), \; i \in \mathcal{R}^k, \; j \in \mathcal{R}^\ell, \; k, \ell = 1, 2, \text{ and}$$
$$b_{(j,\ell)} = F\left(b_j^\ell\right), \; j \in \mathcal{R}^\ell, \; \ell = 1, 2.$$

2.3 Integrals Along Interfaces

Evaluating the forms in (13) involves integrals along interfaces, which pose considerable difficulties. We discuss the evaluation of these quantities in more detail, considering again the domain shown in Fig. 1. As a representative example we shall focus on $a_{2,1}^e$. All observations generalize directly to other terms.

---

[1] For simplicity we consider uniform knots only. If this is not the case then one may consider quasi-uniform knots instead, choosing a parameter that controls the size of all knot spans.

In this situation we obtain

$$a_{2,1}^e(u,v) = \int_e (\nabla u|_{\Omega^1} \cdot n)\, v|_{\Omega^1} + (\nabla u|_{\Omega^2} \cdot n)\, v|_{\Omega^1}$$
$$- (\nabla u|_{\Omega^1} \cdot n)\, v|_{\Omega^2} - (\nabla u|_{\Omega^2} \cdot n)\, v|_{\Omega^2}\mathrm{d}e.$$

The stiffness matrix is a combination of several matrices, each of which is contributed by one of the four forms in (13) defining it. In particular we focus on the contribution of $a_{2,1}^e$.

Taking into account that

$$b_i^2|_{\Omega^1} = 0 \ , \ \nabla b_i^2|_{\Omega^1} = 0 \quad \forall i \in \mathcal{R}^2,$$
$$b_i^1|_{\Omega^2} = 0 \ , \ \nabla b_i^1|_{\Omega^2} = 0 \quad \forall i \in \mathcal{R}^1,$$

we find that only the expressions

$$a_{2,1}^e\left(b_i^k, b_j^\ell\right) = (-1)^{\ell+1} \int_e \left(\nabla b_i^k|_{\Omega^k} \cdot n\right) b_j^\ell|_{\Omega^\ell}\mathrm{d}e \tag{20}$$

contribute to the element $s_{(i,k),(j,\ell)}$ of the stiffness matrix.

In order to compute these values we use an appropriate numerical quadrature rule, which means that we have to evaluate these products on the interface e. This is no major problem if $k = \ell$ since the integral involves only one trace in this case. However, the situation is more complicated if $k \neq \ell$ since the (possibly different) parameterizations of the interface need to be taken into account. In the remainder of this section we discuss the evaluation of $a_{2,1}^e\left(b_i^1, b_j^2\right)$ in more detail.

The interface

$$e = G^1([0,1]^2) \cap G^2([0,1]^2) = G^1(1,[0,1]) = G^2(0,[0,1]) \tag{21}$$

is parameterized by the restrictions

$$L = G^1|_{(G^1)^{-1}(e)} \text{ and } R = G^2|_{(G^2)^{-1}(e)}, \tag{22}$$

see Fig. 1. These two different representations of the same interface are related by the reparameterizations

$$\lambda : [0,1] \to \{1\} \times [0,1] \tag{23}$$

and

$$\varrho : [0,1] \to \{0\} \times [0,1] \tag{24}$$

via

$$L \circ \lambda = R \circ \varrho. \tag{25}$$

The *construction of suitable reparameterizations* λ and ϱ is the first major problem related to the evaluation of this term. We will discuss it in the next section.

These parameterizations will be used to represent the edge as

$$e = (L \circ \lambda)([0,1]) = (G^1 \circ \lambda)([0,1]) = (G^2 \circ \varrho)([0,1]) = (R \circ \varrho)([0,1]). \quad (26)$$

Finally we define $P = L \circ \lambda = R \circ \varrho$ and arrive at

$$
\begin{aligned}
-a_{2,1}^e \left(b_i^1, b_j^2 \right) &= \int_e \left(\nabla b_i^1(\boldsymbol{x})|_{\Omega^1} \cdot n(\boldsymbol{x}) \right) b_j^2(\boldsymbol{x})|_{\Omega^2} \mathrm{d}\boldsymbol{x} \\
&= \int_e \left[\left(\nabla G^1(\boldsymbol{x}) \right)^{-1} \nabla \beta_i^1 \left((G^1)^{-1}(\boldsymbol{x}) \right) |_{\Omega^1} \cdot n(\boldsymbol{x}) \right] \beta_j^2 \left((G^2)^{-1}(\boldsymbol{x}) \right) |_{\Omega^2} \mathrm{d}\boldsymbol{x} \\
&= \int_0^1 \left[\left(\nabla G^1(P(t)) \right)^{-1} \nabla \beta_i^1 \left(L^{-1}(P(t)) \right) \cdot n(P(t)) \right] \beta_j^2 \left(R^{-1}(P(t)) \right) \| \dot{P}(t) \| \mathrm{d}t \\
&= \int_0^1 \left[\left(\nabla G^1(P(t)) \right)^{-1} \nabla \beta_i^1 (\lambda(t)) \cdot n(P(t)) \right] \beta_j^2 (\varrho(t)) \| \dot{P}(t) \| \mathrm{d}t.
\end{aligned}
$$

The integral in the last line is evaluated by a quadrature rule. However, the *choice of the quadrature rule*, which is the second major problem related to the evaluation of this term, is nontrivial and will be discussed further in Sect. 4. In fact, the choice of the rules needs to take the different knots of the functions β_i^1, β_j^2, λ and ϱ into account. While one will generally choose the same knots for λ and ϱ, the knots of β_i^1 and β_j^2 are subject to a non-linear transformation and cannot be assumed to be identical.

3   Finding the Reparameterizations

It is quite common in the literature to assume matching parameterizations or almost matching ones, see [5, p. 4148], [6, p. 87], [12–14,18]. In this situation, the choice of the reparameterizations λ and ϱ is trivial, as they are simply linear parameterizations (possibly reversing the orientation) of the preimages of the interface in the parameter domains. However, the restriction to matching parameterizations poses constraints on the construction of multi-patch parameterizations, making it essentially impossible to parameterize the individual patches separately. This fact motivates us to study the *non-matching* case.

More precisely, we consider situations where the condition (25) cannot be satisfied by considering linear reparameterizations λ and ϱ. Clearly, the condition does not determine λ and ϱ uniquely. We fix one of the mappings, say λ, and compute the remaining one, ϱ. Figure 2 visualizes the relations between the mappings.

The unknown mapping ϱ satisfies $\varrho = R^{-1} \circ L \circ \lambda$. We compute it by least-squares approximation of point samples, as follows:

1. For a given number N of samples, we evaluate

$$\varrho_i = R^{-1} \circ L \circ \lambda \left(\frac{i}{N} \right) \quad (27)$$

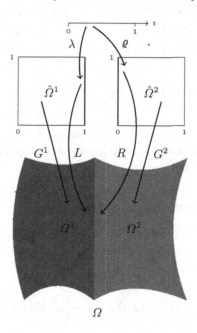

Fig. 2. Multi-patch domain with geometry maps G^1 and G^2, their restrictions L and R to the preimages of the interface and reparameterizations λ and ϱ

by performing the closest point computations

$$\varrho_i = \underset{\boldsymbol{\xi} \in \{0\} \times [0,1]}{\operatorname{argmin}} \left\| L \circ \lambda \left(\frac{i}{N} \right) - R(\boldsymbol{\xi}) \right\|, \ i = 0, \ldots, N, \tag{28}$$

where $\| \cdot \|$ is the Euclidean norm. This formulation also applies to the case of geometrically inexact interfaces (cf. [10,11]).

2. We choose a suitable spline space (e.g. linear, quadric or cubic splines with a few uniformly distributed inner knots) and find the control points $c_j \in \{0\} \times [0,1]$ of the associated B-splines N_j, $j = 1, \ldots, m$, by solving the linear least-squares problem

$$\sum_{i=1}^{N} \left(\sum_{j=1}^{m} c_j N_j \left(\frac{i}{N} \right) - \varrho_i \right)^2 \longrightarrow \min, \tag{29}$$

cf. [7]. The influence of the choice of the spline space for ϱ will be discussed later in Sect. 5. The given reparameterization λ is chosen as a linear polynomial.

We will refer to the case where at least one of the mappings λ and ϱ is different from the identity as non-matching parameterizations at the interface.

4 Numerical Integration

The evaluation of

$$a_{2,1}^e \left(b_i^1, b_j^2 \right) = \int_0^1 \left[\left(\nabla G^1 \left(P(t) \right) \right)^{-1} \nabla \beta_i^1 \left(\lambda(t) \right) \cdot n \left(P(t) \right) \right] \beta_j^2 \left(\varrho(t) \right) \| \dot{P}(t) \| \mathrm{d}t$$

(30)

requires integration with respect to the parameter t, which varies in the parameter domain $[0, 1]$. This is done by applying numerical quadrature and we present several strategies for doing so.

4.1 Gauss Quadrature with Exact Splitting

Gauss quadrature can be applied to segments of analytic functions. Consequently, we split the parameter domain $[0, 1]$ into segments (separated by junctions) where the integrand satisfies this requirement. Three types of junctions arise:

- the inverse images $\lambda^{-1}(\kappa_i^1)$ of the knots κ_i^1 that were used to define the B-splines β_j^1,
- the inverse images $\varrho^{-1}(\kappa_i^2)$ that were used to define the B-splines β_j^2, and
- the knots τ_i that were used to define the B-splines N_j in (29).

These types are visualized in Fig. 3.
Consequently we perform Gauss quadrature with exact splitting by applying the following algorithm:

- Compute all junction points (all three types) in $[0, 1]$,
- sort the junction points, subdivide the domain into segments accordingly,
- subdivide the resulting segments if they are too long,
- apply a Gauss quadrature rule to each segment and sum up the contributions.

As a disadvantage, the inversion of λ and ϱ is costly and has to be done with high accuracy, as the sorting depends on it. Furthermore, the method may result in many segments of varying lengths.

We use Gauss quadrature with $p + 1$ nodes per element (which exactly integrates polynomials of degree $2p + 1$), where p is the degree used for defining the dG-IgA discretization, cf. [15].

4.2 Gauss Quadrature with Uniform Splitting

A computationally simpler approach is to use uniform subdivision, as follows:

- Split the domain $[0, 1]$ uniformly into M segments, where M is a multiple of the number of knot spans used to define the B-splines N_j in (29),
- apply a Gauss quadrature rule to each segment and sum up the contributions.

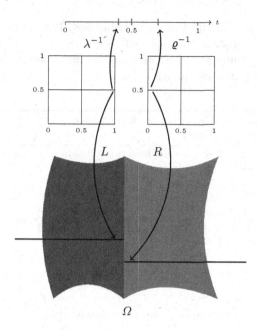

Fig. 3. Exact splitting of a knot span and application of a quadrature rule to each subsegment

As we shall see later, it is mandatory to use large values of M in order to reach the desired level of accuracy. This is due to the fact that the junctions of the first two types listed in the previous section may still be located within the segments obtained by uniform splitting. On the other hand, the use of uniform refinement also creates many small segments that could be merged into larger ones without compromising the accuracy. This can be exploited by using adaptive quadrature.

4.3   Adaptive Gauss Quadrature

We recall the main idea of adaptive quadrature, cf. [8]. In order to evaluate the integral

$$I = \int_a^b f(x)\mathrm{d}x \tag{31}$$

of an integrable function f over an interval $[a, b]$ adaptively one computes two different estimates I_1 and I_2 of I by using two different integration methods. One assumes that one of these estimates, say I_1, is more accurate than the other. Next, one computes the relative distance between I_1 and I_2 taking into account a given (or chosen) tolerance tol, e.g. machine precision. If the difference is small enough, one chooses I_1 as the value of the integral $\int_a^b f(x)\mathrm{d}x$. If this is not the case one splits the interval $[a, b]$ into two subintervals,

$$[a, b] = [a, m] \cup [m, b], \quad \text{where} \quad m = \frac{a+b}{2},$$

and evalues I by summing up the two contributions. This means that one applies the procedure of computing two different estimates and checking their relative difference to both subintervals. Adaptive quadrature is therefore a recursive procedure, which is summarized in Algorithm 1.

Algorithm 1. Adaptive Quadrature: Basic routine.

adaptiveQuadrature(f, a, b, tol)

1: Input: f, a, b, tol where f is an integrable function, a and b are the interval boundaries and tol is a given tolerance
2: Choose knots u_i and weights w_i, $i = 1, \ldots, n$.
3: Compute $I_1 = \sum_{i=1}^{n} w_i f(u_i)$.
4: Choose knots $\tilde{u}_i$ and weights $\tilde{w}_i$, $i = 1, \ldots, m$.
5: Compute $I_2 = \sum_{i=1}^{m} \tilde{w}_i f(\tilde{u}_i)$.
6: **if** $|I_1 - I_2| \leq \text{tol} \cdot |I_1|$ **then**
7:     Return I_1
8: **else**
9:     Return

$$\text{adaptiveQuadrature}\left(f, a, \frac{a+b}{2}, \text{tol}\right) + \text{adaptiveQuadrature}\left(f, \frac{a+b}{2}, b, \text{tol}\right).$$

10: **end if**

Note that the stopping criterion has to be chosen with care and in fact line 6 in the algorithm is a slight oversimplification of it. See [8] for further information.

We apply this procedure to the knot spans that were used to define the B-splines N_j in (29). Therefore we choose I_1 as a Gauss quadrature rule with $p + 1$ quadrature knots, where again p is the degree of the basis functions in the dG-IgA discretization space. For the computation of I_2 we split the interval manually into two halves, apply a Gauss quadrature rule of the same order on both halves, and sum up. The tolerance tol is set to machine precision.

As an advantage, adaptive quadrature can be performed without inverting the reparameterizations. Moreover, it avoids the oversegmentation problem that was present for the previous approach. We observed experimentally that the adaptive procedure accurately detects the junction points and subdivides the domain accordingly. Clearly, the implementation is more costly and requires a recursive algorithm.

5   Numerical Results

We examine the performance of the quadrature methods presented in Sect. 4 as well as the influence of the accuracy of the reparameterization. All experiments were performed using G+Smo[2], an object-oriented C++ IgA library named "Geometry + Simulation Modules".

---

[2] G+Smo: gs.jku.at.

5.1   Reference Results

As a reference we will first show the convergence plot of the solution of the Poisson equation in the case of matching parameterizations, i.e. for $\lambda = \varrho = $ id. In this case we can restrict ourselves to a simple quadrature rule. There is no need for using more elaborate versions of numerical integration. Furthermore, since $\lambda = \varrho = $ id, we do not need to consider the influence of the quality of the reparameterization. More precisely, we consider the two-patch domain with biquadratic matching interface parameterizations shown in Fig. 4, left.

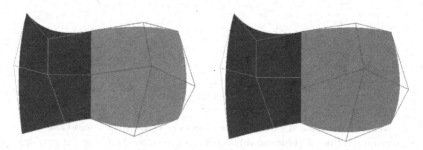

Fig. 4. Patch and its control net. Left: matching parameterizations at the interface. Right: non-matching parameterizations at the interface.

Figure 5 demonstrates the convergence behaviour of the numerical solutions that were obtained for various values of the element size h that was used to define the dG-IgA discretization. We consider a problem with a known solution and measure the error as the difference to it. The quadrature method we used is Gauss quadrature with three quadrature knots. A convergence rate of three for the L2 error and of two for the dG error is clearly visible. This is in accordance with the theoretical predictions, see [1, 6].

5.2   Influence of the Quadrature Rule

We now consider a different parameterization of the same computational domain, with non-matching parameterizations of the interface, see Fig. 4, right. Again we use biquadratic patches. Now we need to use a more complicated integration technique, and we consider the three approaches that were described in Sect. 4.

Figure 6, top and bottom, visualizes the convergence behaviour measured in the L2 and dG norms respectively. Each plot contains four curves, corresponding to four different numerical quadrature techniques. More precisely, we consider Gauss quadrature with exact splitting (yellow), Gauss quadrature with uniform splitting into 10 (blue) and into 30 (red) segments, and adaptive Gauss quadrature (purple). We observe that the first and the last method perform better than the results based on uniform splitting and they achieve the optimal convergence rates (compare with Fig. 5). In particular we note that using uniform quadrature

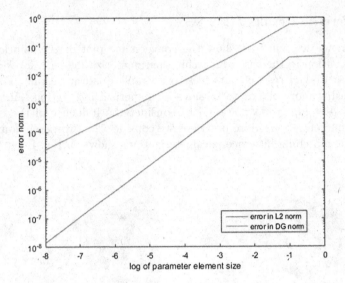

Fig. 5. Matching parameterizations at the interface, convergence behaviour of error in different norms: L2 norm (blue curve), dG norm (red curve). (Color figure online)

leads to a reduced order of convergence for smaller mesh sizes h. Even the use of a very fine but uniform segmentation (30 (red) instead of 10 (blue) segments) does not improve this significantly.

Based on these observations we decided to use solely adaptive and exact Gauss quadrature in the remaining example.

5.3 Influence of the Reparameterization

Next we analyse the influence of the quality of the representation of the reparameterization. Consider again the parameterization of the domain in Fig. 4, right, with non-matching parameterizations of the interface. We compare three different choices of the reparameterizations λ and ϱ.

For the first reparameterization, which generates the results represented by the blue curve in Fig. 7, we choose polynomials λ and ϱ such that the equation $L \circ \lambda = R \circ \varrho$ is exactly satisfied. In this case it was possible to find such polynomials, due to the particular construction of the example. However, this would be impossible in general and it is used here to generate a reference result.

The second and third reparameterizations (red and yellow curves) were obtained using the Algorithm from Sect. 3 to find ϱ, while λ was chosen as a linear polynomial. The second reparameterization uses a linear spline with 8 segments and has an L2 error of $1.3 \cdot 10^{-2}$, and the third reparameterization uses a cubic spline with 4 segments and has an L2 error of $3.1 \cdot 10^{-15}$.

Figure 7 compares the errors in the L2 (top) and dG norms (bottom) for the three reparameterizations. We observe that using a high quality reparameterization is essential for the convergence of the method. Depending on the accuracy of

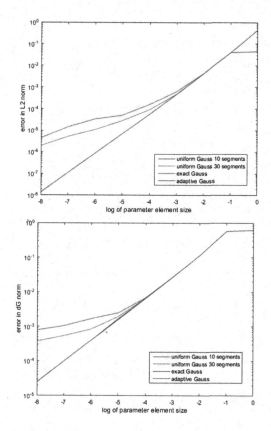

Fig. 6. Influence of the quadrature rule. Top: Convergence behavior of the error in L2 norm. Bottom: Convergence behavior of the error in dG norm. Blue and red curves: 10 and 30 uniform segments per t-knot span. Yellow curves: exact splitting of the knot spans. Purple curves: adaptive quadrature. Note that the yellow curve coincides with the purple one for smaller values of h. Exact representation of the reparameterizations λ and ϱ. (Color figure online)

the reparameterization, h-refinement only works until it reaches a critical mesh size, where further refinement does not have any effect.

The plots show the results obtained by using adaptive Gauss quadrature. The exact method gives virtually identical results.

5.4 Comparison of Exact and Adaptive Quadrature

We perform an experimental comparison of the computational complexity of exact and adaptive quadrature for the domain in Fig. 4, right.

First we demonstrate the effect of using adaptive quadrature, by showing the automatically created splitting points in Fig. 8. We used an accuracy of 10^{-6} instead of machine precision for this picture to obtain a clearer image.

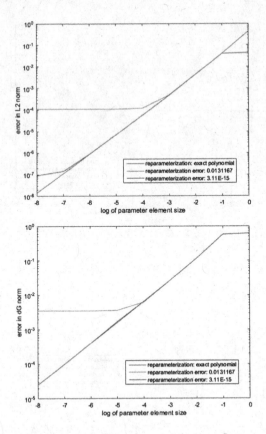

Fig. 7. Influence of the reparameterization. Adaptive quadrature on interface integrals. Top: Convergence behaviour of the error in L2 norm. Bottom: Convergence behaviour of the error in dG norm. Blue curves: Exact representation of λ and ϱ. Red curves: Approximation error of $\varrho \approx 0.0131167$. Yellow curves: Approximation error of $\varrho \approx 3.10616 \cdot 10^{-15}$ (Color figure online)

Both patches were uniformly refined into 4×4 elements by knot insertion. The mappings λ and ϱ are cubic splines on $[0, 1]$ with four knot spans of equal length. Their knots τ_i coincide with the inverse images $\lambda^{-1}(\kappa_i^1)$, as the first mapping is simply the identity. The adaptive quadrature, which is applied to the knot spans $[\tau_i, \tau_{i+1}]$, thus creates additional splitting points around the inverse images $\varrho^{-1}(\kappa_i^2)$, as shown in the Figure. In this particular case, only one splitting point (at 0.5625) is created near $\varrho^{-1}(\kappa_2^2) = 0.5615$ since this suffices to reach the desired accuracy.

These results indicate that, unlike uniform Gauss quadrature, adaptive quadrature avoids over-segmentation of the integration domains. Still, it splits the knot spans more often than exact Gauss quadrature, which also results in a higher number of quadrature knots and thus evaluations.

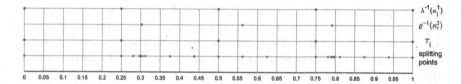

Fig. 8. Splitting points created by adaptive quadrature - see text for details.

In order to analyze this effect, Fig. 9 compares the number of evaluations (i.e., quadrature knots) used by exact and adaptive Gauss quadrature for increasing numbers of elements. In addition, we also show the number of root finding operations (which are more expensive than evaluations) needed to compute the splitting points of exact Gauss quadrature. Clearly, adaptive quadrature requires more evaluations than exact splitting. However, for sufficiently fine discretizations, the number of evaluations in the interior of the patches dominates the total effort.

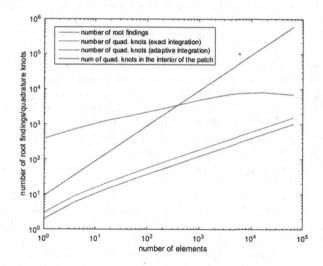

Fig. 9. Number of quadrature knots and root finding operations needed by exact and adaptive quadrature for increasingly finer discretizations.

6   Conclusion

We used a simple model problem to investigate the complications that arise from using non-matching interface parameterizations within the framework of Isogeometric Analysis on a multi-patch domain, using discontinuous Galerkin techniques to couple terms across the interfaces. More precisely, we studied two particular problems. Firstly, we explored the use of reparameterizations to identify pairs of associated points on the common interface. This was found to be useful for correctly evaluating certain terms in the dG discretization. Secondly, we addressed the construction of a suitable procedure for numerical integration. As demonstrated in our numerical experiments, both problems are important for ensuring the optimal rate of convergence for the numerical simulation based on the isogeometric dG discretization.

Future work may be devoted to the extension of the adaptive quadrature-based approach to the three-dimensional case. Moreover, we are currently studying dG-type techniques for performing multi-patch spline surface fitting with approximate geometric smoothness across patch interfaces.

References

1. Bazilevs, Y., de Veiga, L.B., Cottrell, J.A., Hughes, T.J., Sangalli, G.: Isogeometric analysis: approximation, stability and error estimates for h-refined meshes. Math. Models Methods Appl. Sci. **16**, 1031–1090 (2006)
2. Brivadis, E., Buffa, A., Wohlmuth, B., Wunderlich, L.: Isogeometric mortar methods. Comput. Methods Appl. Mech. Eng. **284**, 292–319 (2015)
3. Brunero, F.: Discontinuous Galerkin methods for Isogeometric analysis. Master's thesis, Università degli Studi di Milano (2012)
4. Cockburn, B.: Discontinuous Galerkin methods. J. Appl. Math. Mech. **83**, 731–754 (2003)
5. Cottrell, J.A., Hughes, T.J.R., Bazilevs, Y.: Isogeometric analysis: CAD, finite elements, NURBS, exact geometry and mesh refinement. Comput. Methods Appl. Mech. Eng. **194**, 4135–4195 (2005)
6. Cottrell, J.A., Hughes, T.J.R., Bazilevs, Y.: Isogeometric Analysis. Toward Integration of CAD and FEA. Wiley, Chichester (2009)
7. Dierckx, P.: Curve and Surface Fitting with Splines. Monographs on Numerical Analysis. Oxford Science Publications, Oxford (1995)
8. Gander, W., Gautschi, W.: Adaptive quadrature - revisited. BIT Numer. Math. **40**(1), 84–101 (2000)
9. Hofer, C.: Personal communication
10. Hofer, C., Langer, U., Toulopoulos, I.: Discontinuous Galerkin isogeometric analysis of elliptic diffusion problems on segmentations with gaps. SIAM J. Sci. Comput. (2016). Accepted manuscript, arXiv:1511.05715
11. Hofer, C., Toulopoulos, I.: Discontinuous Galerkin isogeometric analysis of elliptic problems on segmentations with non-matching interfaces. Comput. Math. Appl. **72**, 1811–1827 (2016)
12. Langer, U., Mantzaflaris, A., Moore, S.E., Toulopoulos, I.: Multipatch discontinuous Galerkin isogeometric analysis. In: Jüttler, B., Simeon, B. (eds.) Isogeometric Analysis and Applications 2014. LNCSE, vol. 107, pp. 1–32. Springer, Cham (2015). doi:10.1007/978-3-319-23315-4_1. NFN Technical Report No. 18 at www.gs.jku.at
13. Langer, U., Moore, S.E.: Discontinuous galerkin isogeometric analysis of elliptic pdes on surfaces. In: Dickopf, T., Gander, M.J., Halpern, L., Krause, R., Pavarino, L.F. (eds.) Domain Decomposition Methods in Science and Engineering XXII. LNCSE, vol. 104, pp. 319–326. Springer, Cham (2016). doi:10.1007/978-3-319-18827-0_31. arXiv:1402.1185

14. Langer, U., Toulopoulos, I.: Analysis of multipatch discontinuous Galerkin IgA approximations to elliptic boundary value problems. Comput. Vis. Sci. **17**(5), 217–233 (2016)
15. Mantzaflaris, A., Jüttler, B.: Integration by interpolation and look-up for Galerkin-based isogeometric analysis. Comput. Methods Appl. Mech. Eng. **284**, 373–400 (2015)
16. Nguyen, V.P., Kerfriden, P., Brino, M., Bordas, S.P., Bonisoli, E.: Nitsche's method for two and three dimensional NURBS patch coupling. Comput. Mech. **53**(6), 1163–1182 (2014)
17. Rivière, B.: Discontinuous Galerkin Methods for Solving Elliptic and Parabolic Equations: Theory and Implementation. SIAM (2008)
18. Zhang, F., Xu, Y., Chen, F.: Discontinuous Galerkin methods for isogeometric analysis for elliptic equations on surfaces. Commun. Math. Stat. **2**(3), 431–461 (2014)

Deconfliction and Surface Generation from Bathymetry Data Using LR B-splines

Vibeke Skytt[1]([✉]), Quillon Harpham[2], Tor Dokken[1], and Heidi E.I. Dahl[1]

[1] SINTEF, Forskningsveien 1, 0314 Oslo, Norway
{Vibeke.Skytt,Tor.Dokken,Heidi.Dahl}@sintef.no
[2] HR Wallingford, Howbery Park, Wallingford, Oxfordshire 0x10 8BA, UK
Q.Harpham@hrwallingford.com

Abstract. A set of bathymetry point clouds acquired by different measurement techniques at different times, having different accuracy and varying patterns of points, are approximated by an LR B-spline surface. The aim is to represent the sea bottom with good accuracy and at the same time reduce the data size considerably. In this process the point clouds must be cleaned by selecting the "best" points for surface generation. This cleaning process is called deconfliction, and we use a rough approximation of the combined point clouds as a reference surface to select a consistent set of points. The reference surface is updated using only the selected points to create an accurate approximation. LR B-splines is the selected surface format due to its suitability for adaptive refinement and approximation, and its ability to represent local detail without a global increase in the data size of the surface.

Keywords: Bathymetry · Surface generation · Deconfliction · LR B-splines

1 Introduction

Bathymetry data is usually obtained by single or multi beam sonar or bathymetry LIDAR. Sonar systems acquire data points by collecting information from reflected acoustic signals. Single beam sonar is the traditional technique for acquiring bathymetry data and it collects discrete point data along the path of a vessel equipped with single beam acoustic depth sounders. The equipment is easy to attach to the boat and the acquisition cost is lower than for alternative acquisition methods. The obtained data sets, however, have a scan line like pattern, which gives a highly inhomogeneous point cloud as input to a surface generation application.

Acquisition of bathymetric data with Multi Beam Echo Sounder (MBES) is nowadays of common use. A swath MBES system produces multiple acoustic beams from a single transducer in a wide angle. It generates points in a large band around the vessel on which the equipment is installed. The swath width

M. Floater et al. (Eds.): MMCS 2016, LNCS 10521, pp. 270–295, 2017.
https://doi.org/10.1007/978-3-319-67885-6_15

varies from 3 to 7 times the water depth. In shallow areas, the results of a multi beam sonar degenerates to that of the single beam sonar as the sonar angle is reduced due to a short distance to the sea bottom. Multi beam sonar data acquisition is described in some detail in [10].

LIDAR (light detection and ranging) measures elevation or depth by analyzing the reflections of pulses of laser light from an object. Near shore, especially in shallow areas or in rough waters that are difficult to reach by a sea-borne vessel, data acquisition using bathymetry LIDAR is a good alternative to sonar. Bathymetry LIDAR differs from topography LIDAR by the wavelength of the signals that are used. To be able to penetrate the water, a shorter wavelength is required, so green light is used instead of red. This change reduces the effect of the power used by the laser, and bathymetry LIDAR becomes more costly than the topography equivalent.

Our aim is to represent a specified region with a seamless surface. Some parts of the region are only covered by one survey, while other areas are covered by numerous surveys obtained by different acquisition methods. Where no survey data exists, even vector data created from navigation charts may be taken as input. Collections of bathymetric surveys are a source of potentially "big data" structured as point clouds. Individual surveys vary both spatially and temporally and can overlap with many other similar surveys. Where depth soundings differ greatly between surveys, a strategy needs to be employed to determine how to create an optimal bathymetric surface based on all of the relevant, available data, i.e., select the best data for surface creation.

The digital elevation model (DEM) is the most common format for representing surfaces in geographical information systems (GIS). DEM uses a raster format for storage. Rasters are rectangular arrays of cells (or pixels), each of which stores a value for the part of the surface it covers. A given cell contains a single value, so the amount of detail that can be represented for the surface is limited by the raster cell resolution. The elevation in a cell is frequently estimated using the height values of nearby points. The estimation methods include, but are not restricted to, the inverse weighted interpolation method, also called Shepard's method [22], natural neighbour interpolation, radial basis functions and kriging [7, 16, 20]. Alternatively, one of the existing points lying within the cell can be selected to represent the cell elevation.

Triangulated irregular network (TIN) is used to some extend in GIS context. Sample data points serve as vertices in the triangulation, which normally is computed as a Delaunay triangulation. A triangulated surface can interpolate all points in the point cloud exactly, but for large data sizes an approximate solution is more appropriate. The triangulation data structure is flexible and irregular, and a well-chosen distribution of nodes allows capturing rapid changes in the represented seabed or terrain.

The purpose of trend surfaces is not representation of terrains, but data analytics. These surfaces are described by polynomials of low degree globally approximating the data. Trend surface analysis is used to identify general trends in the data and the input data can be separated into two components: the trend corresponding to the concept of regional features and the residual corresponding to local features. Very often, however, the global polynomial surface becomes too simplistic compared to the data.

In the GIS context, splines are almost entirely understood as regularized splines or splines in tension in the context of radial basis functions. Only in rare instances are splines used for terrain modeling. However, Sulebak et al. [25], use multi-resolution splines in geomorphology and Davydov et al. [3,4], use triangular splines to approximate geographical data partly in combination with radial basis functions. We aim at using polynomial spline surfaces to represent our final result. Moreover, in the process of selecting data surveys for the surface generation, we use spline surfaces as extended trend surfaces. Spline surfaces are able to compactly represent smooth shapes, but our bathymetry data are not likely to describe a globally smooth seabed. Thus, we turn our attention towards locally refineable splines in the form of LR B-spline surfaces.

Section 2 gives a brief overview of the concept of LR B-splines. In Sect. 3, we will present the construction of LR B-spline surfaces and collections of such surfaces approximating point clouds from bathymetry data. The topic of Sect. 4 is the deconfliction process discussed in the context of outliers detection, both for Geo-spatial data and in a more general setting. Finally, we will present a conclusion including plans for further work in Sect. 5.

2   LR B-splines

LR B-spline surfaces are spline surfaces defined on a box partition as visualized in Fig. 1, see [6] for a detailed description of the theory.

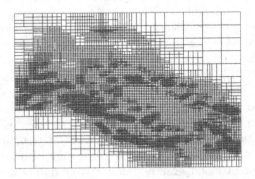

Fig. 1. The polynomial patches in the domain of an LR B-spline surface. This construction will be discussed in some detail in Sect. 3.5.

In contrast to the well-known tensor-product spline surfaces, LR B-spline surfaces posses the property of local refineability. New knot lines, *not* covering the entire domain of the surface, can be added to the surface description. The new knot line must, however, cover the support of at least one B-spline.

The local refinement property implies that models with varying degree of detail can be represented without the drastic increase in model size that would arise in the tensor-product representation. Other approaches addressing the problem of lack of local refinement methods in the tensor-product construction are hierarchical splines [8] and T-splines [21].

An LR-B spline surface F is expressed with respect to parameters u and v as

$$F(u,v) = \sum_{i=1}^{L} s_i P_i N_i^{d_1,d_2}(u,v),$$

where P_i are the surface coefficients, N_i are the associated B-splines and s_i are scaling factors that ensure partition of unity. The B-splines are constructed by taking the tensor-products of univariate B-splines, and are thus defined on a set of knots in both parameter directions. They have polynomial degree d_1 and d_2 in the first and second parameter direction, respectively.

LR B-spline surfaces possess most of the properties of tensor-product spline surfaces, such as non-negative B-spline functions, limited support of B-splines and partition of unity, which ensure numerical stability and modelling accuracy. Linear independence of the B-spline functions is not guaranteed by default. For LR B-spline surfaces of degree two and three, with knot insertion restricted to the middle of knot intervals, no cases of linear dependency are known, but the mathematical proof is still not completed. Actual occurrences of linear dependence can be detected by the peeling algorithm [12] and it can be resolved by a strategy of carefully chosen knot insertions.

3   Surface Generation

We assume the input to be one point cloud where the initial bathymetry data is translated to points represented by their x, y, and z-coordinates. The points can be obtained from one data survey or collected from several surveys. No further preprocessing of the points is performed.

To exploit the local refineability of the LR B-spline surfaces and to optimize the positioning of the degrees of freedom in the surface, we apply an adaptive surface generation approach.

Due to the acquisition methods, bathymetry data is normally projective onto their x and y-coordinates. Thus, it is possible to parameterize the points by these coordinates and approximate the height values (z-coordinates) by a function. In steep areas, however, a parametric surface would be more appropriate. This issue is discussed in [24]. In this paper, we will concentrate on approximation of height values.

The description of the surface generation method in the remainder of this section is partly fetched from [23, 24].

3.1   An Iterative Framework for Approximation with LR-spline Surfaces

The aim of the approximation is to fit an LR-spline surface to a given point cloud within a certain threshold or tolerance. Normally this is achieved for the majority of points in the cloud, and any remaining points that are not within the tolerance after a certain number of iterations can be subject to further investigation. Algorithm 1 outlines the framework of the adaptive surface approximation method.

Data: input point cloud, parameters governing the adaptive procedure:
      tolerance and maximum number of iterations
Result: LR B-spline surface and accuracy information(optionally)
Initiate LR/tensor-product space;
Generate initial surface approximation;
while *there exist out-of-tolerance points or max-levels not reached* **do**
    for *points within each polynomial patch* **do**
        Compute the max. error between points and surface;
        if *max. error is greater than tolerance* **then**
           | Refine LR B-spline surface;
        end
    end
    Perform an iteration of the chosen approximation algorithm;
end
Algorithm 1: The LR B-spline surface generation algorithm

The polynomial bi-degree of the generated LR B-spline surface can be of any degree higher than one, however, in most cases a quadratic (degree two) surface will suffice. Quadratic surfaces ensure C^1-continuity across knot lines with multiplicity one, and as terrains often exhibits rapid variations higher order smoothness may be too restrictive.

The algorithm is initiated by creating a coarse tensor-product spline space. An initial LR B-spline surface is constructed by approximating the point cloud in this spline space. A tensor-product spline space can always be represented by an LR B-spline surface while an LR B-spline surface can be turned into a tensor-product spline surface by extending all knot lines to become global in the parameter domain of the surface.

In each iteration step, a surface approximation is performed. Two approximation methods are used for this purpose, least squares approximation and multi-level B-spline approximation (MBA). Both approximation methods are general algorithms applied to parametric surfaces, which have been adapted for use with LR B-splines. Typically least squares approximation is used for the first iterations as it is a global method with very good approximation properties, while we turn to the MBA method when there is a large variety in the size of the polynomial elements of the surface. The distances between the points in the point cloud and the surface is computed to produce a distance field. In our setting the surface is parameterized by the xy-plane and the computation can be performed by a vertical projection mainly consisting of a surface evaluation.

Next we identify the regions of the domain that do not meet the tolerance requirements and refine the representation in these areas to provide more degrees of freedom for the approximation. Specifically, we identify B-splines whose support contain data points where the accuracy is not satisfied, and introduce new knot lines, in one or two parameter directions depending on the current distance field configuration. The new knot lines must cover the support of at least one B-spline. In each iteration step, many new knot line segments will be inserted in the surface description, giving rise to the splitting of many B-splines. The splitting of one B-spline may imply that an existing knot line segment partly covering its support will now completely cover the support of one of the new B-splines that, in turn, is split by this knot line.

3.2 Least Squares Approximation

Least squares approximation is a global method for surface approximation where the following penalty function is minimized with respect to the coefficients P_i, over the surface domain, Ω:

$$\alpha_1 J(F) + \alpha_2 \sum_{k=1}^{K} (F(x_k, y_k) - z_k)^2. \tag{1}$$

Here $\mathbf{x}_k = (x_k, y_k, z_k), k = 1, \ldots, K$, are the input data points. $J(F)$ is a smoothing term, which is added to the functional to improve the surface quality and ensure a solvable system even if some basis functions lack data points in their support. The approximation is weighted (by the scalars α_1 and α_2, $\alpha_1 + \alpha_2 = 1$) in order to favour either the smoothing term or the least squares approximation, respectively. The smoothing term is given by

$$J(F) = \iint_{\Omega} \int_0^{\pi} \sum_{i=1}^{3} w_i \left(\frac{\partial^i F(x + r \cos \phi, y + r \sin \phi)}{\partial r^i} \bigg|_{r=0} \right) d\phi dx dy. \tag{2}$$

The expression approximates the minimization of a measure involving surface area, curvature and variation in curvature. Using parameter dependent measures, the minimization of the approximation functional is reduced to solving a linear equation system. In most cases $w_1 = 0$ while $w_2 = w_3$. In our case, however, $w_2 = 1$ and $w_3 = 0$ as we utilize 2nd degree polynomials. A number of smoothing terms exist. The one given above is presented in [15]. Other measures can be found in [9, 19] looks into the effect of choosing different smoothing functionals.

In Eq. 2, a directional derivative is defined from the first, second and third derivatives of the surface, and in each point (x, y) in the parameter domain, this derivative is integrated radially. The result is integrated over the parameter domain.

Experience shows that the approximation term must be prioritized in order to achieve a reasonable accuracy in the data points. We use $\alpha_2 = 1.0e^{-3}$ in Eq. 1 and even if the terms have a different magnitude, this greatly favour the least squares term and implies a conflict with the role of the smoothing term as a guarantee for a solvable equation system. Estimated height values in areas sparsely populated by data points, are thus included to stabilize the computations. Some details on the stability of least squares approximation used in this context can be found in [23].

3.3   Locally Refined Multilevel B-spline Approximation (LR-MBA)

Multilevel B-spline approximation (MBA) is a local approximation method [13]. The algorithm is explicit and does not require solving an equation system. It is based on a B-spline approximation technique proposed for image morphing and is explained in [26].

A set of residuals are computed as the difference between the data points and the current approximating surface or, for the initial surface, between the data points and an initial height level which can be selected to be zero. The outline here assumes that the data points are parameterized by their x- and y-coordinates and we approximate the height values, i.e., the z-coordinates. However, the computations can be performed independently in each dimension of the geometry space to create a 3D surface.

A residual surface is computed as follows. Let $\mathbf{x}_c = (x_c, y_c, z_c), c = 1, \ldots, C$, be the data points in the support of a given B-spline N_i, and $r_c = z_c - F(x_c, y_c)$ the corresponding residual. In an interpolative setting the residual surface would satisfy the condition $r_c = \sum_{k=1}^{K} s_k \phi_k N_k(x_c, y_c)$ for all residuals in the support. Here $N_k, k = 1 \ldots K$ are the B-splines overlapping the support of N_i. As the initial point cloud is scattered, there is a large variation in the number of points in the support. If there are no points or if the residuals for all points are smaller than a prescribed tolerance, the coefficient corresponding to N_i is set to zero. Otherwise, we get an under-determined system as we cannot expect interpolation for all residuals in the support. It can be solved for each residual in a least squares sense using the pseudo inverse of the coefficient matrix, giving

$$\phi_c = \frac{s_c N_c(x_c, y_c) r_c}{\sum_{k=1}^{K} (s_k N_k(x_c, y_c))^2}.$$

Since every residual is expected to lead to different values for ϕ, the residual surface coefficient P_i is found by minimizing the error $e(P_i) = \sum_c |P_i N_i(x_c, y_c) - \phi_c N_i(x_c, y_c)|^2$ which leads to the expression

$$P_i = \frac{\sum_c (s_i N_i(x_c, y_c))^2 \phi_c}{\sum_c (s_i N_i(x_c, y_c))^2},$$

In the original setting a number of difference surfaces approximating the distances between the point cloud and the current surface is computed.

The final surface is evaluated by computing the sum of the initial surface and all the difference surfaces. In the LR B-splines setting, the computed difference function is incrementally added to the initial surface at each step giving a unified expression for the surface.

In [23], the approximation accuracy of the LR-MBA algorithm and least squares approximation is compared. It is concluded that LR-MBA does not have the same approximation power as the least squares algorithm, but it is more stable in situations with large variations in height values and when the sizes of the polynomial patches in the surface differ by a large amount. Unlike least squares LR-MBA is an iterative procedure. One update of the coefficient does not lead to the best approximation for a given spline space. The approximation accuracy is improved by applying several coefficient updates between knot insertions.

3.4 Tiling and Stitching

Very large point clouds are unfit for being approximated by one surface due to memory restrictions and high computation times. During surface generation each data point is accessed a number of times, and a tiling approach allows for efficient parallelization over several nodes. Moreover, a large number of points are potentially able to represent a high level of detail, which gives rise to approximating LR B-spline surfaces with higher data size. The surface size should, however, be restricted as the non-regularity of the polynomial patches penalizes data structure traversals when the surface is large (more than 50 000 polynomial patches).

We apply tiling to improve computational efficiency and limit the size of the produced surface, and select a regular tiling approach to enable easy identification of tiles based on the x- and y-coordinates of the points. Figure 2(a) shows a regular tiling based on a dataset with 131 million points, and (b) a set of LR B-spline surfaces approximating the points. The computation is done tile by tile, and applying tiles with small overlaps gives a surface set with overlapping domains. Each surface is then restricted to the corresponding non-overlapping tile yielding very small discontinuities between adjacent surfaces.

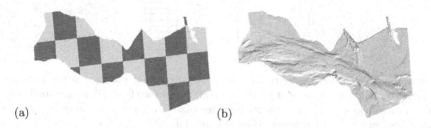

(a)   (b)

Fig. 2. (a) Regular tiling and (b) seamless surface approximating the tiled data points.

To achieve exact C^1-continuity between the surfaces, stitching is applied. The surfaces are refined locally along common boundaries to get sufficient degrees of

freedom to enforce the wanted continuity. For C^0-continuity a common spline
space for the boundary curves enables the enforcement of equality of corresponding coefficients. C^1-continuity is most easily achieved by refining the surface to
get a tensor-product structure locally along the boundary and adapting corresponding pairs of coefficients from two adjacent surfaces along their common
boundary to ensure equality of cross boundary derivatives. C^1-continuity can
always be achieved in the functional setting, for parametric surfaces it may be
necessary to relax the continuity requirement to G^1.

3.5   Examples

Example 1. We will describe the process of creating an LR B-spline surface
from a point cloud with 14.6 million points. The points are stored in a 280 MB
binary file. We apply Algorithm 1 using a combination of the two approximation
methods and examine different stages in the process. Figure 3 shows the point
cloud, thinned with a factor of 32 to be able to distinguish between the points.

Fig. 3. Bathymetry point cloud. Data courtesy HR Wallingford: SeaZone.

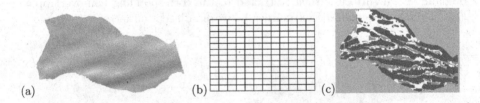

(a)   (b)   (c)

Fig. 4. (a) Initial surface approximation, (b) polynomial patches in the parameter
domain (element structure) and (c) corresponding distance field. White points lie closer
than a threshold of 0.5 m, red points lie more than 0.5 m above the surface and green
points lie more than 0.5 m below. (Color figure online)

The initial surface approximation with a lean tensor-product mesh is shown
in Fig. 4. While the point cloud covers a non-rectangular area the LR B-spline
surface is defined on a regular domain (b), thus the surface (a) is trimmed

with respect to the extent of the point cloud. The last figure (c) shows the points coloured according to the distance to the surface. The surface roughly represents a trend in the point cloud, while the distance field indicates that the points exhibit a wave-like pattern.

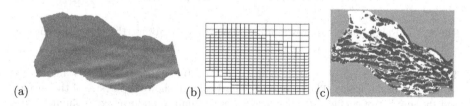

Fig. 5. (a) Surface approximation after one iteration, (b) element structure and (c) corresponding distance field.

Figure 5(a) shows the approximating surface after one iteration, together with (b) the corresponding element structure and (c) the distance field. We see that the domain is refined in the relevant part of the surface. After 4 iterations, it can be seen from Fig. 6 that the surface starts to represent details in the sea floor. We see from the element structure that the surface has been refined more in areas with local detail. The distance field reveals that most of the points are within the 0.5 m threshold.

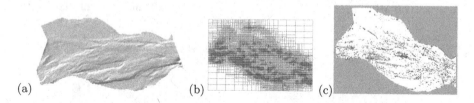

Fig. 6. (a) Surface approximation after four iterations, (b) element structure and (c) corresponding distance field.

After 7 iterations, the surface, Fig. 7(a), represents the shape of the sea floor very well, the corresponding element structure (b) indicates heavy refinement in areas with local details and only a few features in the point cloud fail to be captured by the surface (c). Table 1 shows the evolution of the approximation accuracy throughout the iterative process.

With every iteration, the surfaces size has increased while the average distance between the points and the surface decreased, as did the number of points outside the 0.5 m threshold. The decrease in the maximum distance, however, stopped after 5 iterations. We also find that 2 points have a distance larger than 4 m, while 22 have a distance larger than 2 m. In contrast, the elevation interval is about 50 m. If we look into the details of the last distance field (Fig. 8),

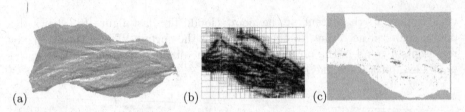

(a)   (b)   (c)

Fig. 7. (a) Final surface approximation after seven iterations, (b) element structure and (c) corresponding distance field.

Table 1. Accuracy related to approximation of a 280 MB point cloud after an increasing number of iterations. The second and third column show the file size of the surface and the number of coefficients. The maximum (column 4) and average (column 5) distance between a point and the surfaces is shown along with the number of points where the distance is larger than 0.5 m (column 6).

Iteration	Surface file size	No. of coefficients	Max. dist.	Average dist.	No. out points
0	26 KB	196	12.8 m	1.42 m	9.9 million
1	46 KB	507	10.5 m	0.83 m	7.3 million
2	99 KB	1336	8.13 m	0.41 m	3.9 million
3	241 KB	3563	6.1 m	0.22 m	1.4 million
4	630 KB	9273	6.0 m	0.17 m	0.68 million
5	1.6 MB	23002	5.3 m	0.12 m	244 850
6	3.7 MB	52595	5.4 m	0.09 m	75 832
7	7.0 MB	99407	5.3 m	0.08 m	20 148

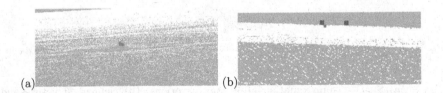

(a)   (b)

Fig. 8. (a) Features not entirely captured by the approximating surface, and (b) outliers in the point set. White points lie closer to the surface than 0.5 m, red and green points have a larger distance. The point size and colour strength are increased with increasing distance. (Color figure online)

we find two categories of large distances: details that have been smoothed out (a) and outliers (b). If, in the first case, a very accurate surface representation is required, a triangulated surface should be applied in the critical areas. Outliers, on the other hand, should be removed from the computation. Still, isolated outliers, as in this case, do not have a large impact on the resulting surface.

Example 2. We approximate a point cloud composed from several data surveys taken from an area in the British channel, and look at the result after four and

Table 2. Approximation accuracy of the point cloud combined from 10 data surveys. The maximum distances below (Max. bel.) and above (Max. av.) and the average distance after 4 and 7 iterations are listed. The elevation range for each data set is given for comparison.

Survey	No. pts.	4 iterations			7 iterations			Elevation
		Max. bel.	Max. ab.	Average	Max. bel.	Max. ab.	Average	
1	71 888	−27.6 m	4.9 m	0.6 m	−26.7 m	2.8 m	0.2 m	35.7 m
2	24 225	−8.3 m	6.7 m	0.6 m	−5.4 m	4.2 m	0.3 m	27.1 m
3	16 248	−10.9 m	12.0 m	0.9 m	−4.1 m	6.0 m	0.3 m	38.4 m
4	483	−1.4 m	6.0 m	0.7 m	−1.5 m	4.1 m	0.4 m	11.3 m
5	7 886	−6.3 m	7.4 m	0.4 m	−4.1 m	5.8 m	0.2 m	33.3 m
6	4 409	−8.3 m	9.2 m	0.5 m	−6.1 m	5.6 m	0.2 m	31.6 m
7	12 240	−7.2 m	8.5 m	0.7 m	−6.8 m	9.0 m	0.5 m	30 m
8	2 910	−6.9 m	7.8 m	1.5 m	−5.5 m	4.4 m	0.7 m	15.4 m
9	1 049 951	−12.7 m	10.5 m	0.4 m	−4.2 m	3.1 m	0.1 m	36.1 m
10	2 047 225	−1.7 m	2.5 m	0.1 m	−1.0 m	1.1 m	0.06 m	11.9 m

seven iterations. 10 partially overlapping surveys contain a total of 3.2 million points. The accuracy threshold is again taken to be 0.5 m. After four iterations, the maximum distance is 27.6 m and the average distance is 0.2 m. After seven iterations, the numbers are 26.9 m and 0.08 m, respectively. The number of points outside the threshold are 367 593 and 38 915, respectively. Although the average approximation error and number of points with a large distance are significantly reduced from the 4th to the 7th iteration, the numbers are clearly poorer than for the previous example. Table 2 gives more detailed information.

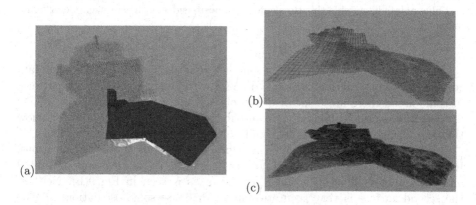

Fig. 9. (a) The combined point cloud, (b) the polynomial patches of the surface approximation after 4 iterations, and (c) after 7 iterations. Data courtesy: SeaZone.

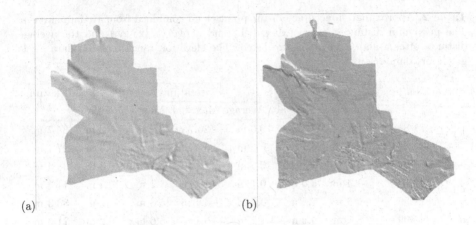

Fig. 10. (a) The surface after 4 iterations, and (b) after 7 iterations.

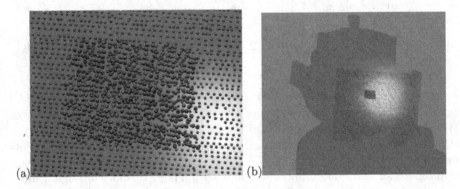

Fig. 11. (a) Detail of the distance field corresponding to the surface after 4 iterations for data surveys 2 and 4 in Table 2, distance threshold 0.5 m, and (b) the detail positioned in the complete surface. Green points lie closer to the surface than 0.5 m, while red and blue points lie outside this threshold on opposite sides of the surface. (Color figure online)

Figure 9 shows the point cloud assembled from the partially overlapping data surveys. This construction leads to a data set with a very heterogeneous pattern, in some areas there are a lot of data points, while in others quite few points describe the sea floor. The polynomial patches of the surface, (b) and (c), show that the surface has been refined significantly during the last 3 iterations.

Figure 10 shows the approximating surfaces after four and seven iterations. In the first case (a), the surface is not very accurate, as we have seen in Table 2 and the polynomial mesh is also quite lean, as is seen in Fig. 9(b). Neither, the second surface is very accurate, but in this case some oscillations can be identified, Fig. 10(b), and the polynomial mesh has become very dense; it is likely that we are attempting to model noise.

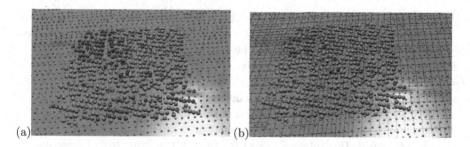

Fig. 12. (a) The same detail as in Fig. 11 corresponding to the surface after 7 iterations, and (b) corresponding distance field with a 2 m threshold.

Figures 11 and 12 zoom into a detail on the surfaces and show the distance fields of two data surveys, number 2 and 4 in Table 2. Data set 2 is shown as small dots and 4 as large dots. In Figs. 11(a) and 12(a), points within the 0.5 m threshold are coloured green while red points and blue points are outside the threshold. Red points lie below the surface and blue points above. We see that points from the two data sets lie on opposite sides of the surface while being geographically close. In Fig. 12(b) the distance threshold is increased to 2 m, and there are still occurrences where close points from the two data sets are placed on opposite sides of the surface. Thus, the vertical distance between these points is at least 4 m. The polynomial elements of the surface included in (b) indicate that a high degree of refinement has taken place in this area. The combined data collection clearly contains inconsistencies, and is a candidate for deconfliction.

4   Deconfliction

Overfitting or fitting to inappropriate data causes oscillations in the surface and unreliable results. Processing the data to remove inconsistencies and selecting the appropriate filtering criteria is a non-trivial task. This filtering process is called deconfliction and is related to outlier detection.

4.1   Outlier Detection

An outlier is an observation that is inconsistent with the remainder of the data set, and as such occurrences can drastically skew the conclusions drawn from a data set. Bathymetry data may contain outliers. Erroneous measurements can be caused by several factors, including air bubbles, complexities in the sea floor and bad weather conditions. These measurements need to be located and excluded from further processing to guarantee that correct results will be generated from the cleaned data. The distinction between outliers and data points describing real features in the sea floor is a challenge. True features should be kept and there are no firm rules saying when outlier removal is appropriate.

Statistical methods [1] for outlier detection have been a topic for a long time. Consider a data set, measurements of discrete points on the sea bottom.

We compare the data points to a trend surface and obtain a set of residuals, and want to test the hyphothesis that a given point belongs to the continuous surface of the real sea floor. Then the corresponding residual should not be unexpectedly large. In statistical terms, the difference surface between the real sea bottom and our trend surface is the population and the residual set is a sample drawn from the population. The sample mean and standard deviation can be used to estimate the population mean. In order to test if a point is an outlier, i.e., not representative of the population, we define a confidence interval. In a perfect world, this interval would relate to the normal distribution having zero mean and a small standard deviation. Other distributions can, however, be more appropriate. For instance, the so called *Student's t distribution* depends on the number of samples and is intended for small sampling sizes.

The confidence interval depends on a confidence level α, and is given by $\left(\tilde{x} - z_{\alpha/2}\frac{S}{\sqrt{n}}, \tilde{x} + z_{\alpha/2}\frac{S}{\sqrt{n}}\right)$. Typically $\alpha \in [0.001, 0.2]$ and the probability that the parameter lies in this interval is $100(1 - \alpha)\%$. The value $z_{\alpha/2}$ denotes the parameter where the integral of a selected distribution to the right of the parameter is equal to $\alpha/2$. It can be computed from the distribution, but tabulated values are also available, see for instance [17] for the Student's t distribution. $\tilde{x}$ is the sample mean and S the sample standard deviation while n is the number of points in the sample.

In the deconfliction setting, we want to test whether the residuals from different data sets can be considered to originate from the same sea floor. I.e., we want to compare two distributions, which requires a slightly different test. To test for equal means of two populations, we can apply the Two-Sample t-Test [18]. To have equal means the value

$$T = \frac{\tilde{x}_1 - \tilde{x}_2}{\sqrt{(s_1^2/N_1 + s_2^2/N_2)}}$$

should lie in an appropriate confidence interval. $\tilde{x}_k$ is the mean of sample $k, k = 1, 2$ and s_k is the standard deviation. N_k is the number of points in the sample. If equal standard deviation is assumed the number of degrees of freedom used to define the confidence interval is $N_1 + N_2 - 1$, otherwise a more complex formula involving the standard deviations is applied to compute the degrees of freedom. This test has, depending on the number of sample points, a thicker tail than the normal distribution, but does still assume some degree of regularity in the data. For instance, the distribution is symmetric. Thus, we need to investigate to what extent the test is applicable for our type of data.

For multi beam sonars, outlier detection is discussed in a number of papers [2,5,10,11]. Traditionally outliers are detected manually by visual inspection. However, due to the size of current bathymetry data surveys, automatic cleaning algorithms are required. The user can define a threshold as a multiple of the computed standard deviation and use statistical methods like confidence intervals or more application specific methods developed from the generic ones to detect outliers. For instance, Grubbs method [11] is based on the Student's t distribution.

Even though computations of statistics for outlier removals may be based on the depth values themselves, residuals with respect to a trend surface are often preferred. The trend surface is typically computed for subsets of the data survey. Selecting the cell size for such subsets is non-trivial. Large cells give larger samples for the computation of statistical criteria, but on the other hand, the cells size must be limited for the trend surface to give a sufficiently adequate representation of the sea floor. In [10] a multi-resolution strategy is applied to get a reasonable level of detail in the model used for outlier detection. Also for proximity based techniques as in k-Nearest Neighbour methods [11], the selection of suitable neighbourhoods is a relevant topic. A problem in trend surface analysis is that the surface tends to be influenced by the outliers. It has been proposed [14] to minimize this influence by using a minimum maximum exchange algorithm (MMEA) to select the data points for creating the trend surface. In [5], the so called M-estimator is utilized for the surface generation.

4.2 Preparing for Deconfliction

Deconfliction becomes relevant when more than one data survey overlap a given area. Two questions arise: are the data surveys consistent, and if not, which survey to choose? The first question is answered by comparing statistical properties of the data surveys. The answer to the second is based on properties of each survey. The data surveys are equipped with metadata information including acquisition method, date of acquisition, number of points and point density. Usually, the most recent survey will be seen as the most reliable, but this can differ depending on the needs of the application, for instance when historical data is requested. In any case, an automated procedure is applied for prioritizing the data surveys resulting in scores that allow, at any sub-area in the region of interest, a sorting of overlapping surveys. We will not go into details about the prioritization algorithm.

In Example 1 in Sect. 3.5, we observed a couple of outliers that could be easily identified by their distance to the surface. Considering outlier data sets, we want to base the identification on residuals to a trend surface, also called reference surface. In [5] low order polynomials approximating hierarchical data partitions defined through an adaptive procedure were used as trend surfaces. We follow a similar approach by choosing an LR B-spline surface as the trend surface and use the framework described in Sect. 3.1 to define a surface roughly approximating the point cloud generated by assembling all data surveys.

The deconfliction algorithm is applied for each polygonal patch in the surface. The size of this patch, or element, has a significant impact on the result. The strategy for adaptive refinement of an LR B-spline surface implies that the surface will be refined in areas where the accuracy is low. Thus, the size of the polynomial elements will vary: in regions where there is a lot of local detail, the element size will be small, while in smooth regions or regions where the point density is too low to represent any detail, the element size is large. The adaptive refinement strategy automatically adjusts the element size to the data configuration. Figures 4, 5, 6 and 7 in Sect. 3.5 shows the element mesh for an LR B-spline surface at different iteration levels. The number of iterations in Algorithm 1 to create the reference surface must be selected to get a good basis for the decisions, see Table 6 for an example.

4.3   The Deconfliction Algorithm

Outliers are data points that appear to be inconsistent with the general trend of the data. Surface generation, even with a careful selection of approximation method, is sensitive to patterns in the data points. Empty regions with significant variation in the height values may lead to unwanted surface artifacts. However, even if one data survey lacks points in an area, another survey may contain this information. Thus, the combination of several surveys can give more complete information than one survey alone, as long as the information from the different surveys is consistent. Our aim is to develop an automatic outlier detection algorithm where the outliers are subsets of data surveys.

Algorithm 2 gives an overview of the deconfliction process. The actual test has to take the configuration of the overlapping point clouds in each element into consideration. The point pattern for the combined point cloud and for each individual cloud may be very non-uniform, and the number of points may differ greatly from element to element.

The algorithm relates to a set of thresholds deduced from the surface generation tolerance and the surveys are considered consistent if the following criteria hold:

– The sample means are within the defined threshold.
– The residual range of the candidate survey does not exceed the range of the higher priority survey with more than a given threshold.
– Most of the residuals of the candidate survey lies within the range of the higher prioritized one.
– The standard deviation computed from the combined data set does not exceed the individual standard deviations with more than a small fraction.

If some of the conditions above do not apply, but the overlap between the surveys is small, the test is repeated on a sub-domain where there is a significant overlap. If the surveys do not overlap, they will be regarded as consistent unless the survey residuals differ significantly.

If the two surveys have the same score and overlap barely or not at all, this probably implies that the surveys originate from the same acquisition, but the point set is split at some stage. This is treated as a special case.

The Two sample t-Test is very strict for this kind of data and the t-Test value becomes very large when the standard deviations of the two samples are small. Thus, this test is not applied directly. However, the t-Test value tends to vary consistently with the other properties. When this tendency is contradicted a closer investigation should be initiated. Also, if the standard deviation of one or both data surveys is large, indicating the existence of outliers within the data sets or a high degree of detail in the sea bottom, or the considerations on the

Data: overlapping data surveys equipped with priority scores, an LR B-spline
reference surface roughly approximating the data set obtained by
combining all individual surveys

Result: a division of the initial surveys into points to use for further processing
and points to reject

for *each data survey* **do**
   for *each polynomial patch in the reference surface (element)* **do**
      Identify the points situated in the element;
      for *each point* **do**
         Compute the residual with respect to the reference surface;
      end
   end
end

for *each element* **do**
   if *more than one survey overlaps the element* **then**
      Compute properties of the highest prioritized sub point cloud: signed
      distance range (residual range) with respect to the reference surface,
      residuals mean, standard deviation of signed residuals, area of overlap
      between the data survey and the element;
      for *each remaining survey in prioritized order* **do**
         Compute properties;
         for *each previously accepted survey* **do**
            Compute characteristics of the combination: standard deviation
            including residuals from both surveys, the Two sample T-test
            value and associated confidence interval;
            Apply deconfliction test;
            if *Possibly conflicting data surveys and small overlap* **then**
               Apply deconfliction test in the overlap area;
            end
            if *Test result is ambiguous* **then**
               Split the element into sub elements and compute modified
               properties;
               for *each sub element* **do**
                  Apply deconfliction test;
               end
               Combine sub element information;
            end
            Mark sub survey as accepted, rejected or uncertain;
         end
      end
   end
end

Post process uncertain sub surveys and include information on adjacent
elements to finalize the decision;

Algorithm 2: Deconfliction algorithm applied on overlapping bathymetry
data sets

residual ranges do not give a clear answer, more testing is beneficial and a sub element investigation is performed.

After deconfliction, the cleaned data surveys are used to update the reference surface to obtain a final surface with better accuracy. This is done by the surface generation algorithm described in Sect. 3 starting the process from the reference surface. Thus, fewer iterations are required to obtain a sufficient accuracy compared to the case when we start from a lean initial surface.

In the following, we will look into a couple of different classes of configurations and discuss them in some detail.

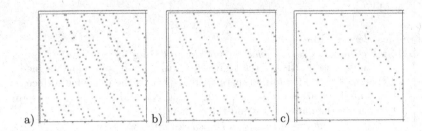

Fig. 13. (a) Pattern of residuals for two surveys, (b) high prioritized survey and (c) survey of lower priority, Element Example 1. Red points lie above the reference surface and green points below. (Color figure online)

Element Example 1. We look at a detail in the test case described in the first example of Sect. 4.4. The element is overlapped by two of the data surveys, and the patterns of the two data surveys are relatively similar as seen in Fig. 13.

Table 3. Characteristic numbers for residuals, the reference surface is created with 3 iteration levels. Element Example 1.

Survey	Score	No. pts.	Range	Mean	Std. dev.	Size
1	0.657	152	−0.232, 0.250	−0.021	0.0088	1863.9
2	0.650	86	−0.155, 0.172	−0.003	0.0046	1823.0

The range of the distance field, the mean distance standard deviation and domain size for the two surveys are given in Table 3. The domain sizes are given as the bounding box of the x- and y- coordinates of the points. The overlap between the surveys has size 1802.3, which imply almost full overlap. The standard deviation computed from the combined point clouds is 0.007. The Two sample t-Test value is 20.5 while the limit with $\alpha = 0.025$ is 1.96. The range and standard deviation for the low priority data surveys is smaller than for the prioritized one. The differences between range extent and mean value for the two surveys are small compared to the threshold of 0.5 and the standard deviation doesn't increase when the two surveys are merged. Thus, the surveys look quite consistent even if the T-test value is high compared to the confidence interval, and this is indeed the conclusion of the test.

Element Example 2. The next example, see Fig. 14, is taken from an area with two overlapping surveys of different patterns. The one with highest score consists of scan lines where the points are close within one scan line, but the distances between the scan lines are large. For the other survey, the points are more sparse, but also more regular. In this configuration, we would prefer to keep most of the points between the scan lines, but only as long as they are consistent with the scan line points.

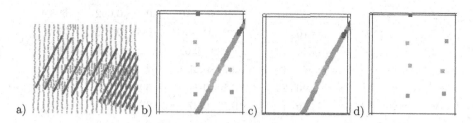

Fig. 14. (a) Overlapping data surveys, (b) residuals pattern for both surveys restricted to one element, (c) prioritized survey and (d) survey to be tested, Element Example 2. Red points lie above the reference surface and green points below. (Color figure online)

Table 4. Characteristic numbers for residuals, deconfliction level 3. Element Example 2.

Survey	Score	No. pts.	Range	Mean	Std. dev.	Size
1	0.640	172	−1.05, 0.625	−0.191	0.177	3045.3
2	0.576	7	−0.64, 1.19	−0.028	0.326	2435.9

The mean values of the residuals are quite similar compared to the 0.5 m threshold, see Table 4, but the ranges don't overlap well, which indicates a rejection of the survey with the lower score. However, the individual standard deviations are relatively high, in particular for the second survey. Thus, a more detailed investigation is initiated. In sub-domain 1, the combined standard deviation is 4.75, which is way above the standard deviations for the individual sub surveys. However, the sub surveys don't overlap and after looking into the closest situated points in the two surveys, the conclusion is that the surveys are consistent. In sub-domain 2, the combined standard deviation is 0.537 and there is no overlap between the two sub surveys. The conclusion is consistence for the same reason as for the previous sub-domain. In sub-domain 3, the combined standard deviation is 0.85. The single point from Survey 2 is well within the range of Survey 1, but the standard deviation tells a different story. However, after limiting the domain even more to cover just the neighbourhood of the survey 2 point, the characteristic residual numbers can be seen in Table 5 as sub-domain 3b and the combined standard deviation is 0.003. The survey is accepted also in this domain. In the last sub-domain, Survey 1 has no points and the final conclusion is acceptance.

Table 5. Characteristic numbers for residuals, sub-domains of Element Example 2.

Sub-domain	Survey	Score	No. pts.	Range	Mean	Std. dev.	Size
1	1	0.640	12	−0.96, −0.56	−0.65	0.015	13.1
1	2	0.576	2	−0.64, −0.24	−0.44	0.040	44.8
2	1	0.640	87	−1.05, 0.10	−0.48	0.062	698.4
2	2	0.576	2	−0.54, −0.15	−0.35	0.039	35.9
3	1	0.640	73	−0.26, 0.62	0.22	0.035	597.1
3	2	0.576	1	0.27, 0.27	0.27		
3b	1	0.640	22	0.18, 0.37	0.30	0.004	35.8
3b	2	0.576	1	0.27, 0.27	0.27		

4.4   Deconfliction Examples

Example 1. Our first example is a small region with three overlapping data surveys, Fig. 15a. The red one (survey 1 in Table 6) has priority score 0.675, the green (survey 2) has score 0.65 and the blue (survey 3) 0.097.

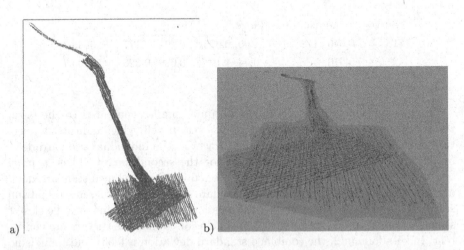

Fig. 15. (a) Three overlapping data surveys and (b) the combined point cloud with the final approximating surface. Data courtesy: SeaZone. (Color figure online)

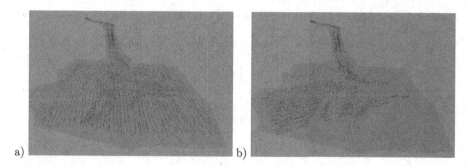

Fig. 16. Surface approximation and (a) the cleaned point set and (b) the points removed by the deconfliction. Green points lie closer to the reference surface than the 0.5 m threshold, red points lie below the surface and blue points lie above, both groups lie outside the threshold. (Color figure online)

The combined data set is approximated by a reference surface using 4 iterations of the adaptive surface generation algorithm. Deconfliction is applied and the surface generation is continued, approximating only the cleaned point set for 3 more iterations. The result can be seen in Fig. 16. About half the points are removed by the deconfliction algorithm and almost all the cleaned points are within the prescribed threshold of 0.5 m of the final surface. The points that have been removed from the computations, are more distant. However, most of them are also close to the surface. In most of the area, the sea floor is quite flat and even if the data surveys are not completely consistent, the threshold is quite large. In the narrow channel at the top of the data set, the shape becomes more steep and the difference between the cleaned and the remaining points becomes larger. Figure 17 shows a detail close to the channel. In Fig. 17(a) two surveys are shown, and the one with large points has highest priority score. For the other one, some points lie outside the 0.5 m threshold (blue points), and we can see that the corresponding scan line has different behaviour vertically than the nearby completely green scan line of the high priority survey.

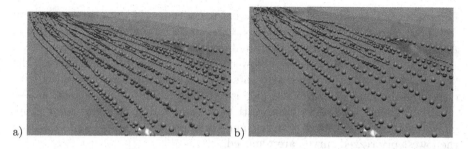

Fig. 17. A detail with data survey nr 2 and 3, (a) both surveys and (b) only the highest prioritized one. (Color figure online)

Table 6. Comparison with different levels of approximation for the reference surface.

Survey		No. pts.	No deconfliction		Deconfliction at level 3			Deconfliction at level 4		
			Range	Mean	Range	Mean	No. pts.	Range	Mean	No. pts.
1	All	6333	−0.83, 0.70	0.12	−0.49, 0.52	0.10		−0.48, 0.56	0.09	
1	Clean				−0.48, 0.52	0.10	6333	−0.48, 0.56	0.09	6333
2	All	3811	−0.64,0.70	0.15	−1.03, 1.75	0.21		−0.89,1.8	0.20	
2	Clean				−0.39, 0.46	0.10	1478	−0.42,0.50	0.10	1546
3	All	11364	−0.55, 0.56	0.10	−1.43, 1.50	0.18		−1.38,1.66	0.18	
3	Clean				−0.6, 0.5	0.10	5209	−0.49, 0.48	0.10	5430

Table 6 shows how the choice of refinement levels for the reference surface influences the accuracy of the final surface, when 3 and 4 iterations for the reference surface is applied. For comparison, the surface approximation is performed also on the combined points set without any deconfliction. The surveys are prioritized according to their number, and the distance range and mean distance to the reference surface is recorded for all computations in addition to the total number of points for each data survey and the number of points in the cleaned survey after deconfliction. All distances are given in meters. In total, for the final surface, the number of iterations is 7 in all cases, but the data size of the final surfaces differ: The surface generated without any deconfliction is of size 329 KB, the surface with deconfliction level 3 is 131 KB while the deconfliction level 4 surface is of size 147 KB. The distances between the final surface and the cleaned point clouds are slightly larger, and some more points are removed when deconfliction is performed at iteration level 3, but the accuracy weighed against surface size is more in favour of this choice of deconfliction level rather than 4. When no deconfliction is applied, we get larger residuals than when the process includes deconfliction (rows marked clean in Table 6). However, if all the distance statistics is computed for all input points even though the ones removed by the deconfliction process are not used for the last iterations of the surface generation, the numbers are higher as expected. The result of this experiment don't clearly favour either deconfliction level 3 or 4. The numbers are roughly comparable, but the smaller surface size for level 3 is preferable.

Example 2. This example is of a different magnitude. 255 data surveys sum up to 1.5 GB. The data set is split into 5 × 3 tiles and are approximated by surfaces. As we can see in Fig. 18, there is limited overlap between the data surveys.

Figure 19 shows overlap zones between three data surveys together with the kept points (a) and the removed points (b). The distances are computed with respect to the reference surface, which is created with deconfliction level 4. The point colours in these zones indicate that the points from different surveys are more than twice the tolerance apart, and consequently the overlap points from the lowest prioritized survey are removed.

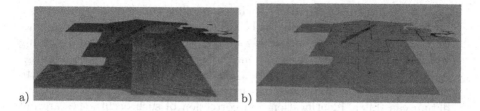

Fig. 18. The reference surface with (a) the points kept by the deconfliction and (b) the points removed. Distances are computed with respect to the reference surface, green points lie closer than 0.5 m, red points lie below and blue points above. Data courtesy: SeaZone. (Color figure online)

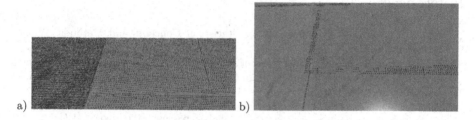

Fig. 19. A detail of the reference surface with (a) the points kept by the deconfliction and (b) the removed points.

5 Conclusion and Further Work

A good data reduction effect has been obtained by approximating bathymetry point clouds with LR B-spline surfaces. The approach handles inhomogeneous point clouds and can be used also for topography data, but is mostly suitable if the data set is to some extent smooth or if we want to extract the trend of the data. Data sets that mainly represent vegetation are less suitable.

We have developed an algorithm for automated deconfliction given a set of overlapping and possibly inconsistent data surveys. The cleaned point sets lead to surfaces with a much smaller risk of oscillations due to noise in the input data. The results so far are promising, but there is still potential for further improvements. Interesting aspects to investigate include:

- Outlier removal in individual data surveys prior to deconfliction.
- Investigation of secondary trend surface approximations based on residuals in situations with many points in an element and small overlaps between the data sets, to detect if there is a systematic behaviour in the approximation errors with respect to the current reference surface.
- Continued investigation of the effect of refinement of the LR B-spline surface to create a suitable reference surface. Aspects to study are number of iterations and a possibility for downwards limitations regarding element size and number of points in an element.

- There is no principal difference in surface modelling and deconfliction between projectable point clouds where the height values can be represented by a function, and when a full 3D surface is required. Still, an investigation regarding which dimensionality to choose in different configurations could be useful.
- A data survey can be subject to a systematic difference with respect to another survey due to differences in registration, for instance the vertical datum can differ. Identification and correction of such occurrences are not covered by the current work. Differences in registration is a global feature of the data set. Indications of it can be detected locally for the reference surface elements, but the determination of an occurrence must be made globally.

References

1. Bhattacharyya, G.K., Johnson, R.A.: Statistical Concepts and Methods. Wiley, New York (1977)
2. Büchenschütz-Nothdurft, O., Pronk, M.J., van Opstal, L.H.: Latest Developments in Bathymetry Data Processing and its Application to Sandwave Detection. Marine Sandwave and River Dune Dynamics, 1–2 April 2004
3. Davydov, O., Zeilfelder, F.: Scattered data fitting by direct extension of local polynomials to bivariate splines. Adv. Comp. Math. **21**, 223–271 (2004)
4. Davydov, O., Morandi, R., Sestini, A.: Local hybrid approximations for scattered data fitting with bivariate splines. CAGD **23**, 703–721 (2006)
5. Debese, N.: Multibeam Echosounder Data Cleaning Through an Adaptive Surface-based Approach. US Hydro 07 Norfolk, May 2007
6. Dokken, T., Pettersen, K.F., Lyche, T.: Polynomial splines over locally refined box-partitions. Comput. Aided Geom. Des. **30**(3), 331–356 (2013)
7. Floater, M.S., Iske, A.: Multistep scattered data interpolation using compactly supported radial basis functions. J. Comput. Appl. Math. **73**, 65–78 (1996)
8. Forsey, D.R., Bartels, R.H.: Surface fitting with hierarchical splines. ACM Trans. Graph. **14**(2), 134–161 (1995)
9. Greiner, G., Hormann, K.: Interpolating and approximating scattered 3D-data with hierarchical tensor product B-splines. In: Le Méhauté, A., Rabut, C., Shumaker, L.L. (eds.) Surface Fitting and Multiresolution Methods, pp. 163–172. Vanderbildt University Press, Nashville (1997)
10. Hennis, N.: Automatic outlier detection in multibeam data. Master thesis, Delft University of Technology, September 2003
11. Hodge, V.J., Austin, J.: A survey of outlier detection methodologies. Artif. Intell. Rev. **22**(2), 85–126 (2004)
12. Johannessen, K.A., Kvamsdal, T., Dokken, T.: Isogeometric analysis using LR B-splines. Comput. Meth. Appl. Mech. Eng. **269**, 471–514 (2013)
13. Lee, S., Wolberg, G., Shin, S.Y.: Scattered data interpolation with multilevel B-splines. IEEE Trans. Visual. Comput. Graph. **3**(3), 229–244 (1997)
14. Lu, D., Li, H., Wei, Y., Zhou, T.: Automatic outlier detection in multibeam bathymetry data using robust LTS estimation. In: 2010 3rd International Congress on Image and Signal Processing (CISP) (2010)
15. Mehlum, E., Skytt, V.: Surface editing. In: Dœhlen, M., Tveito, A. (eds.) Numerical Methods and Software Tools in Industrial Mathematics. Birkhäusser, Boston (1997)

16. Mitas, L., Mitasova, H.: Spatial interpolation. In: Longley, P., Goodchild, M.F., Maguire, D.J., Rhind, D.W. (eds.) Geographic Information Systems - Principles, Techniques, Management, and Applications, pp. 481–498 (2005)

17. NIST/SEMATECH e-Handbook of Statistical Methods (2012). http://www.itl. nist.gov/div898/handbook/eda/section3//eda3672.htm

18. NIST/SEMATECH e-Handbook of Statistical Methods (2012). http://www.itl. nist.gov/div898/handbook/eda/section3//eda353.htm

19. Nowacki, H., Westgaard, G., Heimann, J.: Creation of fair surfaces based on higher order fairness measures with interpolation constraints. In: Nowacki, H., Kaklis, P.D. (eds.) Creating Fair and Shape-Preserving Curves and Surfaces. B.G. Teubner, Stuttgart (1998)

20. Oliver, M.A., Webster, R.: Kriging: a method of interpolation for geographical information system. Int. J. Geogr. Inf. Syst. 4(3), 323–332 (1990)

21. Sederberg, T.W., Zheng, J., Bakenov, A., Nasri, A.: T-splines and T-NURCCs. ACM Trans. Graph. 22(3), 477–484 (2003)

22. Shepard, D.: A two-dimensional interpolation function for irregularly spaced data. In: Proceedings of 23rd National Conference, pp. 517–523. ACM (1968)

23. Skytt, V., Barrowclough, O., Dokken, T.: Locally refined spline surfaces for representation of terrain data. Comput. Graph. 49, 48–58 (2015)

24. Skytt, V., Patané, G., Barrowclough, O., Dokken, T., Spagnuolo, M.: Spatial and environmental data approximation. In: Patané, G., Spagnuolo, M. (eds.) Heterogeneous Spatial Data: Fusion, Modeling and Analysis for GIS Applications. Synthesis Lectures on Visual Computing. Morgan & Claypool Publishers, April 2016

25. Sulebak, J.R., Hjelle, Ø.: Multiresolution spline models and their applications in geomorphology. In: Evans, I.S., Dikau, R., Tokunaga, R., Ohmori, H., Hirano, M. (eds.) Concepts and Modeling in Geomorphology: International Perspectives, pp. 221–237. Terra Publications, Tokyo (2003)

26. Zhang, W., Tang, Z., Li, J.: Adaptive hierachical B-spline surface approximation of large-scale scattered data. In: Sixth Pacific Conference on Computer Graphics and Applications, Pacific Graphics 1998 (1998)

Application of Longest Common Subsequence Algorithms to Meshing of Planar Domains with Quadrilaterals

Petra Surynková[1(⊠)] and Pavel Surynek[2]

[1] Johannes Kepler University Linz, Altenberger Str. 69, 4040 Linz, Austria
petra.surynkova@seznam.cz
[2] Artificial Intelligence Research Center, AIST Tokyo Waterfront, Tokyo, Japan
pavel.surynek@aist.go.jp

Abstract. The problem of mesh matching is addressed in this work. For a given n-sided planar region bounded by one loop of n polylines we are selecting optimal quadrilateral mesh from existing catalogue of meshes. The formulation of matching between planar shape and quadrilateral mesh from the catalogue is based on the problem of finding longest common subsequence (LCS). Theoretical foundation of mesh matching method is provided. Suggested method represents a viable technique for selecting best mesh for planar region and stepping stone for further parametrization of the region.

Keywords: Quadrilaterals · Quadrilateral mesh · Optimal mesh · Longest common subsequence · n-sided planar region

1 Introduction and Motivation

Finding a quadrilateral mesh that matches a given shape - a so called mesh matching problem - represents an important problem in computer aided design, [2,14,20]. The task is specified by user input which describes planar shape to be covered with quadrilateral mesh. In this work we consider only simply connected planar domains which need to be filled by quadrilateral mesh. The planar shape is represented by its boundary drawn in plane typically. In addition to this, the size of mesh, that is, the number of its vertices, is specified by the user as well.

The most important is to have spatially balanced mesh which has crucial effect on computational methods based on quadrilateral decomposition of surfaces of complex shapes because quality of meshes in terms of spatial balance have great impact on accuracy of methods like IGA, [5,12], and FEM, [9]. Computational methods are more efficient and accurate when quadrilateral meshing is well balanced where all quadrilaterals are of similar sizes and angles and even squared, [13,15]. A balance in quadrilateral mesh roughly corresponds to visual attractiveness which is important in computer graphics, presentations, virtual reality, and level of detail modeling.

© Springer International Publishing AG 2017
M. Floater et al. (Eds.): MMCS 2016, LNCS 10521, pp. 296–311, 2017.
https://doi.org/10.1007/978-3-319-67885-6_16

There are many ways how to cover the given shape with a mesh as even single mesh can be matched to the shape in many ways. Moreover there are many non-isomorphic meshes of given number of vertices which makes the problem of finding appropriate mesh for given shape hard.

There exist many previous works dealing with meshing. These methods can be roughly divided into two major directions. Let us call it meshing and mesh matching.

(i) **Meshing.** Meshing is focused on mesh generation for the given shape that is algorithms generate suitable mesh with respect to given shape directly and do not consider alternatives, [7,8], Chaps. 3, 8, and 15 in [10], [17], and [22]. That is in these methods meshes are usually generated by adding points incrementally, more precisely the placement of next point is determined by unambiguous computational process.

(ii) **Mesh matching - our approach.** For given shape we choose suitable mesh from catalogue consisting of a priori existing meshes where in contrast to previous approach many alternative topologically distinct meshes are considered.

Advantage of (i) is that these methods are usually faster but while in (ii) we are able to find topological more appropriate mesh. The idea of mesh selection in style of (ii) is shown in Fig. 1 - an example of two matchings of different quality between user input and quadrilateral meshes.

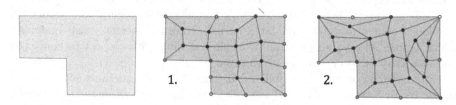

Fig. 1. Two examples of quadrilateral meshes for user input on the left. The first mesh expresses the shape of user input better than the second mesh. First mesh has lower valences of boundary vertices in *convex* vertices of user input and higher valences in *concave* vertices. The various valences of boundary vertices have different color. (Color figure online)

Trying to find a suitable matching between the shape and a mesh in an intuitive way by evaluating all the possible coverings would quickly lead to combinatorial explosion as there is exponential number of such covering. We were trying to mitigate this difficulty by finding an appropriate concept from computer science that can be used for reasoning about the mesh matching problem in a more efficient way. Our finding is that the concept of *longest common subsequence* (LCS) [1,11] can be employed in mesh matching problem to avoid the combinatorial explosion. Given an objective function for measuring quality of matching, the suitable application of the LCS algorithm would automatically prune out coverings of the input shape by mesh that have no chance to maximize the objective function. Hence instead of dealing with the exponential number of coverings the algorithm goes directly to the optimal one in polynomial number of steps.

In this article we describe concept of mesh matching to a given planar shape in the formal mathematical way. Having the precise mathematical formulation of the problem we could describe objective function that corresponds to the quality of matching. We also designed a computationally efficient method for mesh selection which is optimal with respect to the matching quality objective. Our method generalizes the common problem of LCS. In contrast to the original LCS, where only the relation of equality between symbols is considered, we allow more general relations between symbols that reflects various cases that arise in mesh matching problem.

1.1   Contribution

Our main contribution is an introduction of a general framework of application of the LCS problem and associated algorithms in the geometrical problem of mesh matching to a 2D shape. The theoretical properties of LCS algorithms ensures that optimal matching with respect to our notion of quality of matching can be found in polynomial time. This is a significant speedup compared to the baseline methods that exhaustively considers all the possible juxtapositions of a mesh and input 2D shape - the LCS algorithm prunes all the juxtapositions that are not promising in advance. In addition to the general framework we develop a concrete application for a so called corner specific mesh matching which is focused on consideration of mesh matching with respect to convexity and concavity of corners in the given input 2D shape.

The paper is organized as follows. In Sect. 2, we give a short introduction to the quadrilateral meshes and regarding terminology and introduce existing catalogues of meshes of a certain class which we use as the input for our selection methods. Section 3 is devoted to a description of longest common subsequence problem. In Sect. 4 we present an application of LCS to meshing and then we elaborate specific application of LCS to optimal corner sensitive meshing for non-convex shapes. Short summary and future work are given lastly.

2   Background

A quadrilateral mesh, [3, 19, 23] is a triple (V, E, Q) where V is a set of vertices, E is a set of edges, and Q is a set of quadrilaterals. There exists an embedding of (V, E, Q) into 2D plane such that each vertex is assigned a point in the plane and each edge is assigned a curve in the plane, so that curves connect vertices and each quad is represented in the plane as a quadrilateral. In our study we furthermore assume only quadrilateral meshes that form a connected, conforming (i.e. free from T-junctions), orientable *2D manifold with boundary*, [3, 18], i.e. our quadrilateral meshes are defined for segmentation of simply connected planar domains. Let Ω denotes set of quadrilateral meshes.

Regarding the terminology, an edge of the mesh with two incident quads is said to be *internal*, while an edge with just one incident quad is said to be *boundary*. A vertex of a boundary edge is said to be *boundary*, otherwise it is said to be *internal*. The *valence* of a vertex is the number of edges incident to that vertex.

2.1 Existing Catalogues of Meshes

Our work considers the existing context of literature and software in geometry and computer aided design. Extensive works on providing catalogue of meshes of various sizes had been done previously, [4,23]. Our approach in this work is to integrate our mesh selection method with existing mesh catalogue that is kept as a database generated by procedural algorithm - that is, our method will select the optimal mesh out of the catalogue of meshes.

The catalogue of meshes which we use as the input for our selection method consists of quadrilateral meshes of a certain class and was generated in the previous work, [23]. Without constraints, the number of possible meshes is too high. Thus, we consider only valences ≤ 5 for the both boundary and internal vertices which represents standard restriction for numerical applications [16] and what is used in related literature [21,24].

The quadrilateral meshes in the catalogue furthermore satisfy the following invariant:

(I) At least one vertex of each internal edge is internal.

Procedural mesh catalogue generates exponential number of meshes with respect to the number of internal vertices of mesh, i.e. the number of internal vertices is specified as the input. Hence, our mesh selection method should make a consideration about matching of a single mesh very quickly to be able to find suitable mesh for given shape in reasonable time. More precisely, we cannot afford to perform any kind of exponential time search or other time consuming operation.

3 Longest Common Subsequence Problem

The most simplified version of the longest common subsequence problem consists in finding common subsequence within given two sequences of symbols. Informally said, the task is to delete some symbols in given two sequences $x_1, x_2, ..., x_n$ and $y_1, y_2, ..., y_k$ so that the resulting sequences - called common subsequences - will be the same. The objective is to obtain as long as possible sequences at the end (in other words we want to make as few as possible symbol deletions) - that is, longest common subsequences.

Consider a simple example of two strings (set of symbols correspond to Latin alphabet) "`alpha`" and "`aleph`". These two sequences are obviously different but after deleting last 'a' in the first sequence and middle 'e' in the second string we obtain "alph" in both cases which is the longest common subsequence for this example.

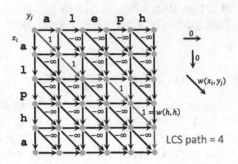

Fig. 2. Graphical scheme of LCS based on weighted directed graphs.

Let us explain LCS on the following graphical scheme based on weighted directed graphs. Let us Σ denote alphabet, let w be a utility function that expresses a match between individual symbols. Formally $w : \Sigma \times \Sigma \to \mathbb{Z} \cup \{-\infty\}$. Usually w is a set so that it returns positive value for a pair of identical symbols and $-\infty$ for a pair of distinct. For example $w(a, a) = 1$ and $w(a, e) = -\infty$. LCS with respect to w can be expressed by following graphical scheme, see Fig. 2. Weight of diagonal edges represents the match between pair of symbols. But there are also vertical and horizontal edges, their weights are typically zeros. More precisely we use zero weight for vertical and horizontal edges if deletion of corresponding symbol is allowed or $-\infty$ if we do not allow deletion. Traversal of a vertical edge corresponds to deletion of a respective symbol in a first string, while horizontal edge represents deletion of a corresponding symbol in a second string. LCS in this scheme is regarded as a directed path from upper-left corner to bottom-right corner in the scheme that has the highest sum of edge's weights. Let us call this an *LCS path*. This LCS path represents longest common subsequences obtained by the deletion of certain symbols. Moreover the LCS path tells us what symbols should be deleted to obtain the resulting sequence. The described schematic approach to LCS is just for didactic purposes, in practice more efficient algorithms based on dynamic programming are used. As the efficient algorithm for LCS are out of scope of our study we refer the reader to relevant literature.

The very positive aspect about the LCS problem is that a variety of efficient (polynomial time) algorithms exist that solve this optimization problem, [25,26].

4   Application of LCS in Shape Matching

Low time complexity makes LCS algorithms good candidates for using them as a basis for consideration about mesh matching. However, mesh matching problem and LCS problem are completely different concepts hence we need to show first what are the similarities between these problems.

When a user specifies his planar shape we can regard his input as an abstract information that can be annotated by a sequence of symbols. Information about the *length* of edges of the boundary, which vertices on the boundary are *convex*, which are *concave* can be read from the input. Such information can be encoded into a sequence of symbols.

At the same time, we need to annotate boundaries of meshes stored in the catalogue using a sequence of symbols in correspondence with annotation of inputs. There is considerable discrepancy between user inputs in the form of planar shapes and representation of meshes in existing catalogues - most catalogues represent meshes in the abstract form as list of quadrilaterals and their interconnections. Moreover we need to reflect a certain level of flexibility of mesh matching with respect to given shape - a mesh may be matched to the shape in a not ideal way while no better matching is possible.

4.1   Formalization of LCS Application in Mesh Matching

We will now introduce formally application of LCS in shape matching. Σ be a set of symbols denoting types of points on the boundary in the 2D shape. Then for each type of point $\sigma \in \Sigma$ we will introduce $w_\sigma : \Omega \times \mathbb{N} \to \mathbb{Z} \cup \{-\infty\}$ a weight function that expresses a match between point of given type and selected vertex in the mesh. Second argument of w_σ represents index of the boundary vertex assumed that boundary vertices are numbered from one.

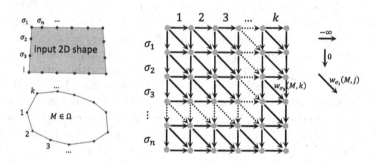

Fig. 3. An example of a 2D shape, a quadrilateral mesh, and graphical scheme of LCS approach.

Then the previously introduced graphical approach to LCS will naturally emerge here within the context of mesh matching. For example, assume a 2D shape where points are denoted $\sigma_1, \sigma_2, ..., \sigma_n$. Next assume a mesh M consisting of k boundary vertices and consider following graphical scheme based on the graphical scheme defining the LCS path, see Fig. 3. Weight of diagonal edges represents the match between point of given type and a vertex in a mesh.

The weights of horizontal edges are $-\infty$ and the weights of vertical edges are zeros, i.e. deletion of a point in a given shape is allowed but deletion from mesh cannot be done ever because the convenience in mesh matching says that all vertices in mesh must be matched somewhere [4].

Mesh matching is now naturally translated to search of LCS path in the described scheme. Moreover as the LCS patch is always optimal which means we have just defined mesh matching that is optimal in a certain sense. From now we will call mesh matching corresponding to LCS path in the defined directed graph scheme and *optimal mesh matching*. Recall again, that this scheme is for pedagogical purpose, in practice we use more efficient approach based on dynamic programming.

4.2   Design of Weight Functions for Corner Specific Shape Matching

We introduce *lattice* of classification of mesh boundary vertices with respect to properties of *convexity*, *concavity*, and *others*. We do not address the positioning of internal vertices. We assume that internal vertices are embedded in regularly distribution. No other topology is considered. Unlike in the case of standard LCS where we compare a pair of symbols whether they are same or not, here we are more flexible. When we compare a vertex from user input shape with a boundary vertex of mesh from the catalogue, the quality of matching between these two vertices is determined by the *lattice* which is a mathematical structure best representing to our needs. For example if a *convex* point from user input is being compared with a vertex from mesh, then *lattice* for classification of *convex* vertices is used to determine the quality of correspondence between the two (the *lattice* thus provide classification from complete match between *convex* and *convex* point to complete mismatch between *convex* and *concave* point).

If we intuitively interpret what our formal definition represent we can say that the operation of comparison between a point from user input and a boundary vertex of a mesh comes as a parameter to an LCS algorithm instead of standard equality between symbols. The output of the algorithm hence will be pair of sequences whose symbols at corresponding positions represent best possible match according to the *lattice*. When this output is interpreted back to the world of geometry we have an optimal matching of a given mesh from catalogue to the given user input. Hence we are able to select optimal mesh from the catalogue provided that consideration about the single mesh is fast enough.

The method works for fixed starting point in a testing quadrilateral mesh which is compared to fixed point in the given user input. For finding an optimal matching between the user input and a mesh from the catalogue we consider all rotation of a testing mesh, i.e. for fixed point in the user input we test all possible starting points in a mesh. This approach is illustrated in Fig. 4.

Our basic mesh matching method assumes a catalogue of meshes represented as a list of interconnected quadrilaterals. The user input given as a planar shape is further processed and is assigned a sequence of boundary points. We distinguish three types of boundary vertices, see Fig. 5:

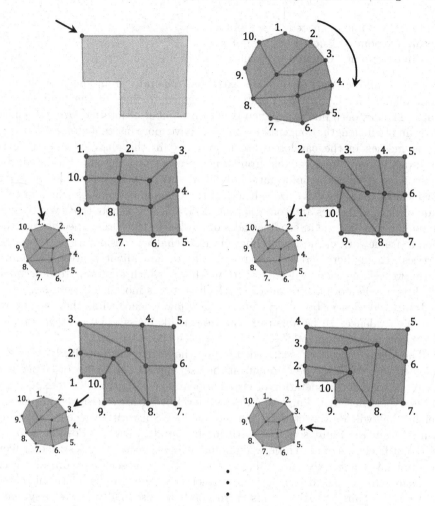

Fig. 4. An example of mesh matching between one quadrilateral mesh from the catalogue and user input. Point in the user input is fixed and the rotations of mesh are considered.

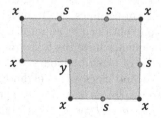

Fig. 5. Three types of boundary vertices in the given user input.

– **straight** point - denoted by symbol s
– **convex** point - denoted by symbol x
– **concave** point - denoted by symbol y

Convex and *concave* points are extracted naturally from the user defined input planar shape. *Straight* points are assigned to lines of the boundary of shape. That is, each line is assigned a certain number of internal *straight* points according to its length. Longer boundary line have more internal *straight* vertices.

As meshes in the catalogue are represented in the abstract way there is not much information available from which a corresponding annotation can be constructed. The usable information about meshes with respect to the property of concavity and convexity are valences of their boundary vertices which in case of quadrilateral meshes are from the range $2, 3, 4, 5$. Let us note that even though we do not consider the internal vertices of meshes in the corner specific case the general framework of mesh matching via LCS enables consideration of internal vertices. To further increase amount of information about a given boundary vertex we also consider valences of its neighbors which allows us to determine which vertex is more likely *convex* or which vertex is more likely *concave*.

Intuitively *convex* boundary vertices have low valence while their neighbors have high valence. To be able to formalize this likeliness to be *straight, convex,* or *concave* point we introduce a lattice for each type of point.

The higher the vertex is classified within the lattice the more likely it can be matched to a certain type of point on the user input. The design of lattice is a matter of careful consideration of visual appearance of matching and experience of the expert designer. Furthermore, each level of lattice is assigned an integer weight that will be reflected by the modified LCS algorithm when comparing symbol from the input sequence with mesh boundary vertex. Positive weights represent likeliness of match between points while negative values stand for likeliness of mismatch. Absolute value of the weight represent measure of match or mismatch. A special weight $-\infty$ is reserved to denote complete mismatch between the pair of vertices (this corresponds to disequality between symbols in the standard LCS algorithm). Figure 7 shows a suggested lattice for *convex* point in the given user input. Some triples of valences of boundary points in the lattice are weeded out because they cannot occur in our certain class of meshes. The lattices for *concave* and *straight* points are designed analogically.

Our modified LCS algorithm assumes two input sequences of symbols - the first from user input; the second obtained by annotating mesh from the catalogue. The symbols of the second sequence have the following format of triples: $[u, v, z]$ where $v \in 2, 3, 4, 5$ is a valence of mesh boundary vertex, and $u, z \in 2, 3, 4, 5$ are valences of counter-clock-wise and clock-wise neighbors of v respectively.

Weights are assigned to triples $[u, v, z]$ by weight functions. Formally we introduce a weight function for each type of point. That is, we have functions:

$$w_s, w_x, w_y : \{2, 3, 4, 5\}^3 \rightarrow \mathbb{Z} \cup \{-\infty\}$$

which assigns triples of valences their weights with respect to the lattice for *straight, convex,* and *concave* point, respectively, see an example in Fig. 6. These weight functions are used by the modified LCS algorithm when it is making comparing a pair of symbols (first one from the first sequence second one from the second sequence).

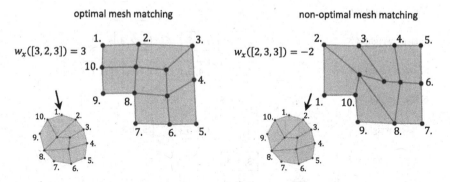

Fig. 6. An example of weights of selected mesh boundary vertex with respect to the fine-grained lattice for *convex* point.

We omit implementation details of the modified LCS algorithm for the sake of brevity. However, important high level specialty of our version of LCS is that it assumes the length of the first sequence to be at least the length of the second sequence. Moreover, deletions of symbols can be made from the first sequence only (from the user input). It is natural assumption as we want to match all the vertices of the catalogue mesh to some point in the user defined shape.

The two sequences of the same length are valuated by a utility which is calculated as the sum of weights of symbols from the second sequence with respect to lattices corresponding to respective symbols from the first sequence. The task of our modified LCS algorithm is to compute longest common subsequence out of given user defined input sequence (the first sequence) with the highest possible utility.

In correspondence to general LCS-like mesh matching framework we define overall utility function of a pair of sequences of symbols discussed above as follows

$$f([\sigma_1, ...\sigma_m], [[u_1, v_1, z_1], ...[u_m, v_m, z_m]]) = \sum_{i=1}^{m} w(\sigma_i, [u_i, v_i, z_i])$$

where

$$w(\sigma_i, [u_i, v_i, z_i]) = \begin{cases} w_s([u_i, v_i, z_i]), & \text{if } \sigma_i = \mathsf{s} \\ w_x([u_i, v_i, z_i]), & \text{if } \sigma_i = \mathsf{x} \\ w_y([u_i, v_i, z_i]) & \text{if } \sigma_i = \mathsf{y} \end{cases}$$

The algorithm finds an LCS path which corresponds to a sequence $[\sigma_1, ...\sigma_m]$ for the input pair of sequences $[\sigma_1', ...\sigma_n']$ and $[[u_1, v_1, z_1], ...[u_m, v_m, z_m]]$ where $m \leq n$ so that $[\sigma_1, ...\sigma_m]$ is a subsequence of $[\sigma_1', ...\sigma_n']$ and its f value is maximum.

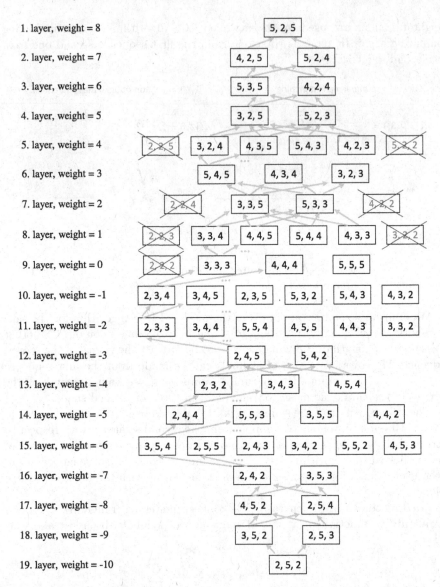

Fig. 7. Suggested lattice for *convex* point in the given user input.

4.3   Design of Lattices

The design of *lattice* is a matter of experience of the expert designer. We sug-gested a lattice for each type of point in the user input. Our lattices are fine-grained so they enable to evaluate mesh matching more precisely than coarse lattices. An example of coarse lattice for *straight* point is shown in Fig. 8,

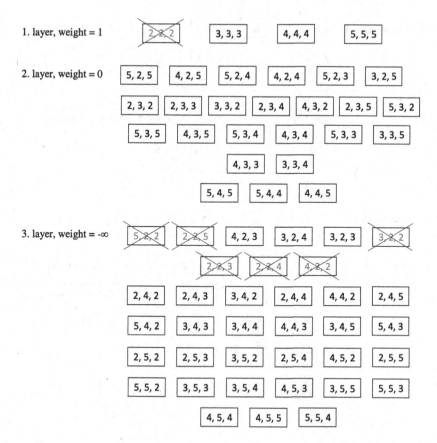

Fig. 8. An example of coarse lattice for *straight* point in the given user input.

an example of coarse lattice for *convex* point is shown in Fig. 9 and an example of coarse lattice for *concave* point is shown in Fig. 10.

We show illustrative examples of lattices constructed according to our intuition for the mesh matching with respect to the corner specific case. These lattices are planned to be evaluated in terms of visual quality of generated mesh matching.

Note that occurrence of $-\infty$ represents strict impossibility to match a point of given type from user input (corresponds to the type of lattice) to a vertex from a mesh. If the LCS algorithm cannot find a correspondence between point from the user input and some mesh vertex so that utility other than $-\infty$ is assigned to that correspondence the point must be treated as deleted by the LCS algorithm. This is the only case when LCS performs deletion from some of its input sequences.

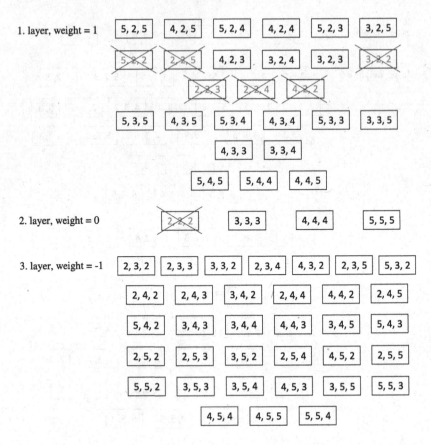

Fig. 9. An example of coarse lattice for *convex* point in the given user input.

4.4   Properties of the Method

Clearly as LCS algorithms are optimal in their nature our modified algorithm enables finding optimal match between used defined shape and mesh from the catalogue with respect to given objective function optimally. Moreover the algorithm requires polynomial time and space which makes it an excellent candidate for selecting a mesh with the best match for the given user input.

The time complexity of the most basic implementation of the LCS-based matching that searches for the best mesh matching through LCS path is $O(|n \times k| \times log_2(n \times k))$. This result comes from the complexity of the standard implementation of the Dijkstra algorithm with the worst case time complexity $O(|E| \times log_2(|V|))$, [6], where V is the set of vertices and E is the set of edges of the underlying graph. In case of LCS path we have a special graph in which both $|V|$ and $|E|$ are $O(n \times k)$ from which we obtain the above result.

The baseline naive algorithm that solves the mesh matching problem by considering all possible coverings of the shape with given mesh needs $O(k \times \binom{k}{n})$

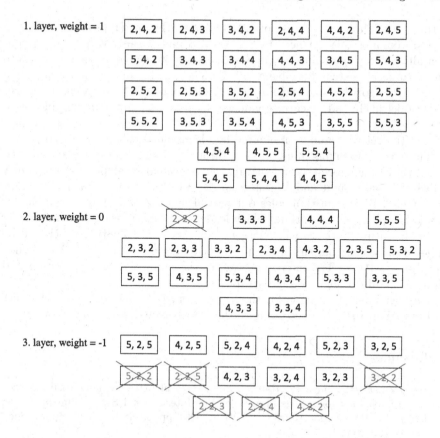

Fig. 10. An example of coarse lattice for *concave* point in the given user input.

mesh to shape comparisons which comes from the fact that there is $\binom{k}{n}$ possible mapping n points of the shape on k vertices of the mesh which needs to be multiplied by $2 \times k$ to account all rotations and orientations of the mesh. If we omit constants in the asymptotic time complexities, we can state that this naive algorithm can find optimal mesh matching with respect to the corner specific case in $2 \times 10 \times \binom{10}{6} = 4200$ steps. This optimal solution is shown in the left part of Fig. 6. This naive method however does not scale for larger meshes and more complex shapes - consider a mesh with 100 boundary vertices and shape with 50 points which is apparently not tractable by the naive algorithm. The LCS-based mesh matching on the other hand can find the optimal solution from Fig. 6 in $10 \times 6 \times log_2(10 \times 6) = 354$ steps and the large case 100 vertices and 50 points becomes tractable as well.

5   Conclusions and Future Work

A new approach for finding a quadrilateral mesh from the catalogue of quadrilateral meshes of a certain class that matches a given user input was described in this work. The method conceptually builds on the known problem of finding longest common subsequence (LCS). As LCS and mesh matching are fundamentally different problems we proposed a series of techniques that allow us to transform mesh matching problem to LCS. These techniques include adaptation of the LCS algorithm and introduction of symbol comparison based on lattices that model likelihood of correspondence between user input points and mesh vertices.

The theoretical foundation of a method is provided. The major contribution is that viewing mesh matching problem through the concept of LCS allows mitigating the combinatorial complexity. Instead of evaluating all possible matchings between the user input and a mesh from the catalogue in the exponential time, the adapted LCS algorithm rules out partial matchings as early as possible if they turn out not to be optimal which leads eventually to the polynomial time.

In the future work we will focus on experimental evaluation which will be targeted on visual comparison of match matching obtained using fine-grained and coarse lattices. The interconnection of the mesh matching algorithm and the procedural catalogue is planned too.

Another interesting topic for the future work is to develop new techniques for evaluation best mesh according to the distribution of internal vertices.

References

1. Bergroth, L., Hakonen, H., Raita, T.: A survey of longest common subsequence algorithms. In: Proceedings of the Seventh International Symposium on String Processing Information Retrieval (SPIRE 2000), pp. 39–48. IEEE Computer Society, Washington, D.C. (2000)
2. Bommes, D., Lévy, B., Pietroni, N., Puppo, E., Silva, C., Tarini, M., Zorin, D.: Quad-mesh generation and processing: a survey. Comput. Graph. Forum **32**(6), 51–76 (2013)
3. Botsch, M., Kobbelt, L., Pauly, M., Alliez, P., Lévy, B.: Polygon Mesh Processing. A K Peters, Natick (2010)
4. Buchegger, F., Jüttler, B.: Planar multi-patch domain parameterization via patch adjacency graphs. Comput. Aided Des. **82**, 2–12 (2017)
5. Cohen, E., Martin, T., Kirby, R., Lyche, T., Riesenfeld, R.: Analysis-aware modeling: understanding quality considerations in modeling for isogeometric analysis. Comput. Meth. Appl. Mech. Eng. **199**(5–8), 334–356 (2010)
6. Cormen, T.H., Stein, C., Rivest, R.L., Leiserson, C.E.: Introduction to Algorithms, 2nd edn. McGraw-Hill Higher Education, London (2001)
7. Díaz-Morcillo, A., Bernal-Ros, A., Nuño, L.: Mesh generation methods over plane and curved surfaces. In: Proceedings of the 7th International Meshing Roundtable, IMR 1998, Dearborn, Michigan, USA, 26–28 October 1998, pp. 397–407 (1998)
8. Edelsbrunner, H.: Geometry and Topology for Mesh Generation. Cambridge University Press, New York (2001)
9. Floater, M., Hormann, K.: Surface Parameterization: A Tutorial and Survey, pp. 157–186. Springer, Heidelberg (2005)
10. Frey, P.J., George, P.L.: Mesh Generation: Application to Finite Elements. Wiley, London (2008)

11. Hirschberg, D.S.: Algorithms for the longest common subsequence problem. J. ACM **24**(4), 664–675 (1977)

12. Hughes, T., Cottrell, J., Bazilevs, Y.: Isogeometric analysis: CAD, finite elements, NURBS, exact geometry and mesh refinement. Comput. Meth. Appl. Mech. Eng. **194**(39–41), 4135–4195 (2005)

13. Jimack, P.K., Mahmood, R., Walkley, M.A., Berzins, M.: A multilevel approach for obtaining locally optimal finite element meshes. Adv. Eng. Softw. **33**(7–10), 403–415 (2002)

14. Kälberer, F., Nieser, M., Polthier, K.: Quadcover - surface parameterization using branched coverings. Comput. Graph. Forum **26**(3), 375–384 (2007)

15. Knupp, P.: Remarks on Mesh Quality, Reno, NV, 7–10 January 2007

16. Liu, Y., Xing, H.L., Guan, Z.: An indirect approach for automatic generation of quadrilateral meshes with arbitrary line constraints. Numer. Meth. Eng. **87**(9), 906–922 (2011)

17. Lo, D.S.H.: Finite Element Mesh Generation. CRC Press, Boca Raton (2015)

18. Luebke, D., Reddy, M., Cohen, J.D., Varshney, A., Watson, B., Huebner, R.: Level of Detail for 3D Graphics. Morgan Kaufmann Publishers, San Francisco (2003)

19. Mitchell, S.A.: A characterization of the quadrilateral meshes of a surface which admit a compatible hexahedral mesh of the enclosed volume. In: Puech, C., Reischuk, R. (eds.) STACS 1996. LNCS, vol. 1046, pp. 465–476. Springer, Heidelberg (1996). doi:10.1007/3-540-60922-9_38

20. Nasri, A., Sabin, M., Yasseen, Z.: Filling n-sided regions by quad meshes for subdivision surfaces. Comput. Graph. Forum **28**(6), 1644–1658 (2009)

21. Peng, C.H., Barton, M., Jiang, C., Wonka, P.: Exploring quadrangulations. ACM Trans. Graph. **33**(1), 1–13 (2014)

22. Ramaswami, S., Siqueira, M., Sundaram, T.A., Gallier, J.H., Gee, J.C.: A new algorithm for generating quadrilateral meshes and its application to FE-based image registration. In: Proceedings of the 12th International Meshing Roundtable, IMR 2003, Santa Fe, New Mexico, USA, 14–17 September 2003, pp. 159–170 (2003)

23. Surynkova, P., Buchegger, F: Enumerating quadrilateral meshes. Comput. Aided Geom. Des. (submitted 2017)

24. Takayama, K., Panozzo, D., Sorkine-Hornung, O.: Pattern-based quadrangulation for N-sided patches. Comput. Graph. Forum **33**(5), 177–184 (2014)

25. Ukkonen, E.: Algorithms for approximate string matching. Inf. Control **64**(1–3), 100–118 (1985)

26. Ullman, J.D., Aho, A.V., Hirschberg, D.S.: Bounds on the complexity of the longest common subsequence problem. J. ACM **23**(1), 1–12 (1976)

Order-Randomized Laplacian Mesh Smoothing

Ying Yang[1,2], Holly Rushmeier[2], and Ioannis Ivrissimtzis[3(✉)]

[1] Fujian Provincial Key Laboratory of Information Processing
and Intelligent Control, Minjiang University, Fuzhou, China
[2] Yale University, New Haven, USA
{ying.yang.yy368,holly.rushmeier}@yale.edu
[3] Durham University, Durham, UK
ioannis.ivrissimtzis@durham.ac.uk

Abstract. In this paper we compare three variants of the graph Laplacian smoothing. The first is the standard synchronous implementation, corresponding to multiplication by the graph Laplacian matrix. The second is a voter process inspired asynchronous implementation, assuming that every vertex is equipped with an independent exponential clock. The third is in-between the first two, with the vertices updated according to a random permutation of them. We review some well-known results on spectral graph theory and on voter processes, and we show that while the convergence of the synchronous Laplacian is graph dependent and, generally, does not converge on bipartite graphs, the asynchronous converges with high probability on all graphs. The differences in the properties of these three approaches are illustrated with examples including both regular grids and irregular meshes.

Keywords: Laplacian smoothing · Graph Laplacian matrix · Voter processes · Triangle meshes · Regular grids · Taubin smoothing

1 Introduction

Laplacian smoothing is used in a variety of applications as a simple yet effective method for data denoising. Assuming that each data point has a well-defined neighborhood, each iterative Laplacian smoothing step updates the data points by weighted averages of their neighborhoods. Since neighborhood relations are naturally described by graphs with their edges connecting neighboring points, Laplacian smoothing is often described as an operator acting on graphs, smoothing the values of a function defined over the graph's vertices.

In graphics applications, Laplacian smoothing, applied either globally or locally, is often used for smoothing discrete surfaces, especially triangle meshes. In cases of global mesh smoothing in particular, higher quality results are obtained by using more sophisticated variants of the fundamental technique, such as the HC-Laplacian smoothing [18], the curvature flow smoothing [5], or the Taubin smoothing [15]. The latter, which will be used in our experiments, is based on the alteration between a smoothing step with a positive weight w_1 and an anti-smoothing step with a negative weight w_2.

© Springer International Publishing AG 2017
M. Floater et al. (Eds.): MMCS 2016, LNCS 10521, pp. 312–323, 2017.
https://doi.org/10.1007/978-3-319-67885-6_17

Since Laplacian smoothing is simple, general and well-understood, it is often the technique of choice when a smoothing or noise suppression algorithm needs to be incorporated into a general mesh processing framework [10], or into a more complex algorithm, as for example the machine learning algorithm for surface reconstruction in [1], or the Boundary Element Method based evolutionary structural optimization in [17].

In image processing applications, weighted neighborhood averaging, usually referred as Gaussian smoothing, is a fundamental technique at the heart of classic edge detection algorithms, or more sophisticated variants of them such as the bilateral edge preserving smoothing in [16], a technique which has been extended to triangle meshes [7]. Bilateral filtering has been studied through the spectrum of the graph Laplacian in [8].

Several curve and surface subdivision algorithms can defined as combined operators with at least one Laplacian smoothing component. The classic Lane-Riesenfeld algorithms [11] apply Laplacian smoothing on a very simple graph, i.e. a graph with the connectivity of a polygon, while in some recent generalizations the smoothing step corresponds to the Laplacians of more dense graphs [3].

In this paper we compare three implementations of Laplacian smoothing, depending on the order in which vertices are updated. In the first implementation we consider, L_{sync}, each step updates all graph vertices synchronously. This implementation, which is ubiquitous in practical applications, has the major advantage that we do not have to define a specific order in which the vertices are updated, thus do not have to arbitrarily impose such an order.

In the second implementation, L_{exp}, every vertex carries an independent exponential clock and is updated when that clock rings. L_{exp} has several similarities with the discrete operators called *voter processes*, which are a major modeling and simulation tool. They have been traditionally employed on regular grid settings [4] to model physical phenomena, and recently on irregular graph settings to model social interactions, e.g. spread of influences on social media, or consumer behavior [12,19]. In the latter cases, the basic voter processes are usually augmented with special features that allow them to capture the complexity of social interactions. As an advantage of the L_{exp} operator, we note that it is more natural than L_{sync} in the sense that it does not assume instant interaction between the parts of the system, here the vertices of the graph, and as a result it can avoid the problem of non-convergence which appears when L_{sync} is applied on bipartite graphs. Moreover, it has a trivial memory efficient implementation since, unlike L_{sync}, it does not require to retain the existing values of the function until the end of the current iteration. One potential disadvantage is that the outcome in non-deterministic, something that in several applications can be unacceptable.

In between L_{sync} and L_{exp}, we also consider an operator L_{perm}, which updates the vertices one by one as in L_{exp}, but following a random permutation of the vertex set rather than using independent exponential clocks. In L_{perm}, no vertex will be updated twice before all vertices have been updated once, allowing for a clear distinction between the iterative steps of the smoothing process, similarly to L_{sync}. This is particularly convenient when Taubin smoothing is applied and thus, we alternate between two distinct Laplacian smoothing steps. Nevertheless, L_{perm} seems to be less interesting than L_{sync} and L_{perm} from a theoretical point of view and here will only be studied experimentally.

Contribution and Limitations: The main contribution of the paper is the demonstration of theoretical and practical shortcomings of the synchronous implementation of the Laplacian smoothing, which have been largely overlooked in the relevant literature. As the main limitation of the paper we note that given the theoretical and practical shortcomings of the two randomized implementations we tested, a conclusive case for their use instead of the ubiquitous synchronous implementation can only be made in the context of a specific application scenario, as for example a specific mesh processing pipeline or a specific grid simulation. The latter however is beyond the scope of this paper.

Overview: In Sect. 2, we describe the three implementations of Laplacian smoothing in detail, and review the theoretical properties of L_{sync} and L_{exp}. In Sect. 3, we test all three implementations on regular height maps and 3D triangle meshes, using Laplacian smoothing or Taubin smoothing as appropriate. We briefly conclude in Sect. 4.

2   Theoretical Properties of L_{sync} and L_{exp}

First, we review the convergence properties of the usual synchronous Laplacian smoothing. While the convergence or not of L_{sync} is a direct consequence of well-known spectral properties of the Laplacian matrix, to the best of our knowledge, bipartite graphs have not been identified in the relevant literature as the class of graphs where L_{sync}, generally, does not converge.

Let $G = (V, E)$ be a connected graph with N vertices and let f_n be a real function defined over the vertex set V, which can also be written as a vector

$$\mathbf{f}_n = (f_n(v_0), f_n(v_1), \ldots, f_n(v_{N-1}))^T. \tag{1}$$

One iterative step of the synchronous Laplacian smoothing L_{sync} updates $\mathbf{f}_n$ by

$$\mathbf{f}_{n+1} = L \cdot \mathbf{f}_n \tag{2}$$

where the Laplacian matrix L, indexed by the vertices of G, is given by

$$L = L(i, j) = \begin{cases} 1/d_i & \text{if } (v_i, v_j) \in E, \\ 0 & \text{otherwise} \end{cases} \tag{3}$$

where d_i is the valence of the vertex v_i. Regarding the spectral properties of L, see for example [2], its eigenvalues satisfy

$$1 = \lambda_0 > \lambda_1 \geq \cdots \geq \lambda_{N-1} \geq -1, \tag{4}$$

that is, all the eigenvalues are in the interval $[1, -1]$ and the largest is equal to 1 and has multiplicity one. Moreover, $\lambda_{N-1} = -1$ if and only if G is bipartite, that is, when its vertices can be split into two subsets, let say the white and the black vertices, such that only vertices in different subsets are connected with an edge. In that case, the spectrum of L is symmetric about the origin and thus λ_{N-1} also has multiplicity one [2].

Let $\mathbf{v}_i$ for $i = 0, \ldots, N-1$ be an orthonormal basis of $\mathbb{R}^N$ consisting of eigenvectors of L. The initial function $\mathbf{f}_0$ of vertex values can be written in that basis as

$$\mathbf{f}_0 = a_0 \mathbf{v}_0 + \cdots + a_{N-1} \mathbf{v}_{N-1} \tag{5}$$

giving

$$\mathbf{f}_n = L^n \mathbf{f}_0 = a_0 \lambda_0^n \mathbf{v}_0 + \cdots + a_{N-1} \lambda_{N-1}^n \mathbf{v}_{N-1} \tag{6}$$

which, on bipartite graphs, for large number of iterations n converges to

$$a_0 \mathbf{v}_0 + a_{N-1}(-1)^n \mathbf{v}_{N-1}. \tag{7}$$

From Eq. 7 we see that L_{sync} does not converge if the graph G is bipartite and $a_{N-1} \neq 0$, i.e., the initial vertex values $\mathbf{f}_0$ contain a non-zero component of $\mathbf{v}_{N-1}$. The eigenvector $\mathbf{v}_{N-1}$ of bipartite graphs has a quite simple form, namely, it has values $+1$ on white vertices and -1 on black. On the other hand, for any graph G, bipartite or not, we have

$$\mathbf{v}_0 = (1, \ldots, 1)^T. \tag{8}$$

and simple computations show that if $\mathbf{f}_n$ converges to a function $\mathbf{f}$ then

$$\mathbf{f} = a_0 \mathbf{v}_0 \tag{9}$$

where a_0 is the coefficient of the eigenvector $\mathbf{v}_0$ when we write $\mathbf{f}_0$ in the basis of the eigenvectors of L.

Relevance: Bipartite graphs appear often in practice. Open polygons are bipartite, as well as regular quadrilateral grids and regular cubic grids. While triangle meshes are not bipartite, regular hexagonal meshes are bipartite. Regarding general polygonal meshes, the graph with vertex set the vertices and the faces of the mesh, and edges from the vertex-vertices to the incident face-vertices and the face-vertices to the incident vertex-faces, is bipartite. Thus, if we smooth attributes defined on both vertices and faces, e.g. normals, colors or material properties, by updating the vertices by the mean of the incident faces and the faces by the mean of the incident vertices, then the process would generally not converge.

2.1 L_{exp} Smoothing and Voter Processes

Since L_{exp} is a probabilistic process we cannot study it through the spectral properties of the Laplacian matrix. Instead, we will study it through its relation to voter processes, which can be seen as the discrete counterparts of L_{exp}. While voter processes are well-understood and have already found numerous applications, to the best of our knowledge their connection with the L_{exp} implementation of the Laplacian smoothing has not been studied. Next we will make this connection explicit, essentially formalizing the following simple intuitive argument; under L_{exp} smoothing, the values of the function f on the graph vertices correspond to opinion expectations in a voter process.

The L_{exp} Smoothing Process: Following the standard terminology on voter processes, each vertex of G carries an independent exponential clock of rate 1. Each time a clock rings the value on its vertex is updated, becoming the mean of the values of its 1-ring neighborhood. In matrix notation, this is equivalent to multiplication by the matrix L_k derived from the identity matrix by substituting its k-th row with the k-th row of the Laplacian matrix. That is,

$$\mathbf{f}_{m+1} = L_k \cdot \mathbf{f}_m \tag{10}$$

where

$$L_k = L_k(i,j) = \begin{cases} L(i,j) & \text{if } i = k \\ 1 & \text{if } i = j \neq k \\ 0 & \text{otherwise} \end{cases} \tag{11}$$

Let $t_0, t_1, \ldots, t_n$ be the sequence of vertex indices in the order they rang during the smoothing process. We have

$$\mathbf{f}_{n+1} = L_{t_n} \cdots L_{t_0} \mathbf{f}_0 \tag{12}$$

which, by writing $\mathbf{f}_0$ in the natural base becomes

$$\mathbf{f}_{n+1} = \sum_{i=0}^{N-1} f_0^i \cdot L_{t_n} \cdot \ldots \cdot L_{t_0} \cdot \mathbf{e}_i \tag{13}$$

where f_0^i is i-th coordinate of $\mathbf{f}_0$ and $\mathbf{e}_i$ is the vector with 0's everywhere and 1 at the i-th coordinate. From Eq. 13 we see that for large values of n the behavior of L_{exp} depends on the limit of

$$L_{t_n} \cdot \ldots \cdot L_{t_0} \cdot \mathbf{e}_i. \tag{14}$$

The Corresponding Voter Process: In its standard form, a voter process is defined on a graph whose vertices carry a value from the set $\{0,1\}$, commonly called the *opinion*. Each vertex carries an independent exponential clock and upon ringing that vertex will choose with uniform random probability one of its neighbors and adopt its opinion. To each vertex i we associate the binary random variable X_i with

$$X_i(0) = 0, \quad X_i(1) = 1 \tag{15}$$

and the set of vertices V is associated to the vector of binary random variables

$$\mathbf{X} = (X_0, \ldots, X_{N-1})^T. \tag{16}$$

Let the vector of the probability distributions of the opinions at step n be

$$\mathbf{Y}^n = (Y_0^n, \ldots, Y_{N-1}^n). \tag{17}$$

From Eq. 13, it suffices to study the behaviour of the process for initial opinion vector $\mathbf{e}_i$, in which case we have

$$\begin{cases} p(Y_i^0 = 0) = 0 & p(Y_i^0 = 1) = 1 \\ p(Y_j^0 = 0) = 1 & p(Y_i^j = 1) = 0 \text{ for } i \neq j \end{cases} \tag{18}$$

and the initial expectation vector is $E(\mathbf{X}) = \mathbf{e}_i$. Assuming that the clocks at the vertices of the graph ring in the order $t_0, t_1, \ldots, t_n$, we have

$$\mathbf{Y}^n = L_{t_n} \cdot \ldots \cdot L_{t_0} \cdot \mathbf{Y}^0 \tag{19}$$

and by linearity, the expectation vector at step n is

$$E(\mathbf{X}) = L_{t_n} \cdot \ldots \cdot L_{t_0} \cdot \mathbf{e}_i. \tag{20}$$

For any graph, including bipartite graphs, the above voter process reaches consensus with high probability [6]. That is, for any $\epsilon > 0$, after a sufficiently large number of steps, all the vertices will have the same opinion with probability at least $1 - \epsilon$. Moreover, the probability for the consensus being at opinion 1 is equal to the sum of the valences of the vertices with initial opinion 1 divided by the sum of all valences, which is twice the number of edges $|E|$. For the initial opinions corresponding to the basis vector $\mathbf{e}_i$, the expectation vector under the condition that consensus has been reached is

$$E_{cns}^i = \frac{d_i}{2|E|}(1, 1, \ldots, 1)^T. \tag{21}$$

Remark: Hassin and Peleg [9] studies a voter process where all vertices update simultaneously, i.e., in the fashion of L_{sync} rather than L_{exp}. The main result is that consensus is reached with high probability for all non-bipartite graphs. In [13], these two voter processes are compared and the main result is that the asynchronous process reaches consensus faster than the synchronous. The latter result is reflected in our experiments regarding the speed of the smoothing processes in Sect. 3.

The L_{perm} Smoothing Process: In-between L_{sync} and L_{exp} we define a third process L_{perm} where the graph vertices are updated one by one, according to a random permutation of them. Since there is no much literature on the theoretical properties of L_{perm}, we will only study it experimentally, generally expecting a behavior similar to L_{exp}.

The motivation for including L_{perm} in our investigations is two-fold. Firstly, as L_{exp} is different from L_{exp} in that all vertices are updated the same number of times, a comparison between the two methods can reveal the affect on L_{exp} of the fact that some vertices may be updated significantly fewer times than the average. Secondly, L_{perm} is similar to L_{sync} in that a single iteration of the process is clearly defined, i.e. when all vertices have been updated once. This is particularly convenient when we alternate between different steps of Laplacian smoothing, as for example in the case of Taubin smoothing.

3   Experimental Results

In our examples we use two graph types. First, graphs with the connectivity of a regular 2D grid and a real-valued function $f : V \rightarrow \mathbb{R}$ defined over its vertices. Secondly, irregular graphs with a vector valued function $\mathbf{f} : V \rightarrow \mathbb{R}^3$ defined on the vertex set. In the first case, f would typically be a spatially regular sample of measurements of a physical quantity and will often be visualized as a 2.5D height map. In the second case, $\mathbf{f}$ would typically represent the spatial coordinates of the vertices of a polygonal mesh embedded in 3D and the Laplacian smoothing process will be applied on each coordinate separately.

In Fig. 1 we applied the three Laplacian operators on a 50×50 regular grid with values

$$f(x, y) = \sin(2\pi x) + \sin(2\pi y). \tag{22}$$

Random uniform noise from the interval $[-1, 1]$ was added on the interior vertices. In all three cases the boundary of the grid is fixed and we apply the smoothing operators on the interior vertices only. After 50 iterations of the L_{sync}, the artifact related to the non-convergence of the operator on bipartite graphs is clearly visible, even though the fixed boundary means that eventually the result will be smooth as the boundary values slowly propagate towards the interior. Next, we applied the operator L_{exp} until one vertex had been updated for 50 times and as expected, the result was smooth. Notice that applying L_{exp} until one vertex has been updated for 50 times means that all other vertices are updated fewer than 50 times. However, as L_{exp} updates vertices on the fly, it mixes vertex values faster than L_{sync}. Finally, we notice that, as expected, L_{perm} yielded results similar to L_{exp}.

The next figures show results on triangle meshes. The Fandisk model was chosen for its flat areas and sharp edges, while the Eight model was chosen for the absence of sharp edges and its non-trivial topology. The subdivided Dipyramid was chosen for its rotational symmetry and its highly regular connectivity which produce artifacts when Laplacian smoothing is applied.

Figure 2 shows the results of applying L_{exp} on noisy Fandisk and Eight models. We notice that while most of the noise has been removed from both models, the edges of the Fandisk have not been preserved and there is more residual noise compared to L_{sync} and L_{perm}, see Figs. 3 and 4. The smoothed Eight model exhibits again some residual noise, but avoids the medium frequency artifacts created by L_{sync} and L_{perm} for the given edge preserving choice of parameters w_1 and w_2, see Fig. 3. We note that the larger amount of residual noise in L_{exp} is due to the fact that some mesh vertices are smoothed a few times only and that it can be a serious limitation in certain graphics applications.

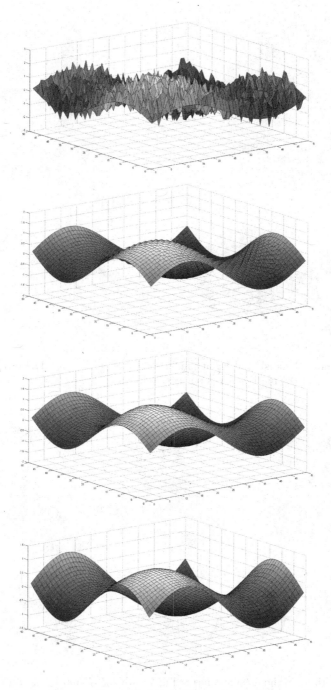

Fig. 1. Top to bottom: noisy grid data, 50 iterations of L_{sync}, L_{exp} and L_{perm} smoothing.

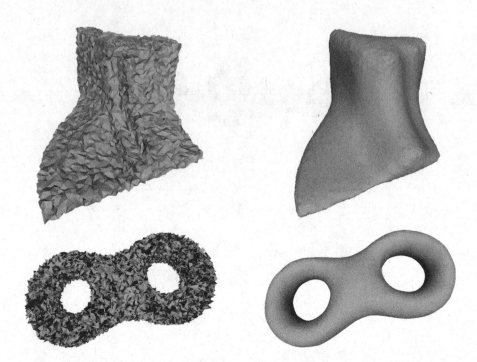

Fig. 2. The noisy Fandisk and Eight model smoothed with L_{exp}. The process termi-
nates when one of the mesh vertices reaches 50 iterations.

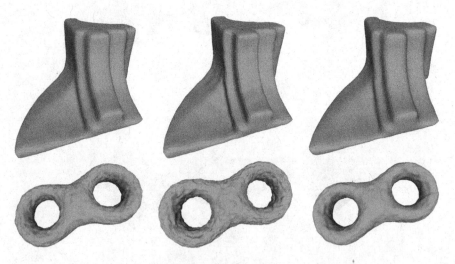

Fig. 3. Smoothing of the noisy meshes in Fig. 2 with L_{perm} and L_{sync}. **Left to right:**
50 iterations of L_{perm}, 50 iterations of L_{sync} and 100 iterations of L_{sync}. In all cases
the Taubin smoothing weights are $w_1 = 0.25$ and $w_2 = -0.20$.

Figure 3 compares L_{perm} and L_{sync} on Taubin smoothing. We notice that while L_{sync} and L_{perm} can produce results that visually are very similar to each other, the values of certain algorithmic variables such as the number of iterations, or the weights of the Taubin smoothing, are not directly comparable. In particular, for given Taubin smoothing weights, L_{perm} requires fewer iterations than L_{sync}. That was expected since L_{perm} updates the mesh vertices on the fly and thus, the new value of a vertex starts propagating immediately, rather than after the iteration is complete. We also note that in our basic Matlab implementation, a single iteration of L_{perm} runs faster than a single iteration of L_{sync}. This is because L_{sync} stores the updated vertex values in a temporary array which has to be copied at the end of each iteration.

Figure 4 compares L_{perm} and L_{sync} on the range of acceptable Taubin smoothing weights. We notice that the L_{perm} accepts a wider range of weights than L_{sync} and in particular, the anti-smoothing weight w_2 can be significantly larger in absolute value than the smoothing weight w_1.

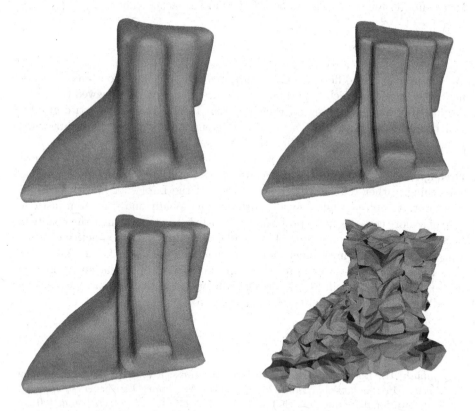

Fig. 4. Smoothing of the noisy meshes in Fig. 2 with 200 iterations of L_{perm} (left) and L_{sync} (right). **Top:** Taubin weights $w_1 = 0.25$ and $w_2 = 0.25$. **Bottom:** Taubin weights $w_1 = 0.25$ and $w_2 = 0.3$.

Fig. 5. From left to right: (a) The original mesh. (b) 200 iterations of L_{sync} with weight $w = 0.1$. (c) 200 iterations of L_{perm} Taubin smoothing with $w_1 = 0.25$ and $w_2 = 0.20$. (d) 200 iterations of L_{sync} Taubin smoothing with $w_1 = 0.25$ and $w_2 = 0.20$.

Finally, in Fig. 5 we apply the three processes on a linearly subdivided dipyramid. We notice that surface artifacts, as have been studied in [14], appear in all three cases and are very similar to subdivision artifacts. A possible explanation for this is that several subdivision schemes can be described as a combination a linear subdivision step followed by a weighted Laplacian smoothing step.

4   Conclusions

We compared three different implementations of Laplacian smoothing, depending on the order in which the graph vertices are updated. We reviewed their theoretical properties, focusing on their different behaviors over bipartite graphs. Our tests showed that these theoretical differences are visually significant when Laplacian smoothing is applied on regular grids.

As we noticed in the literature review in Sect. 1, Laplacian smoothing is most often applied in the form of several consecutive iterations of weighted Laplacian smoothing, rather than as direct application of the Laplacian matrix in Eq. 3. Moreover, it is often applied locally rather than globally and it is often just one step of a more complex mesh processing algorithm. In such settings, the results in Sect. 3 do not provide any compelling evidence against current practice of using L_{sync} as the default implementation of Laplacian smoothing. In the future, we plan to compare these three different implementations of Laplacian smoothing in the context of a specific 3D modeling problem, in particular, machine learning based surface reconstruction as, for example, in [1].

References

1. Annuth, H., Bohn, C.: Growing surface structures: a topology focused learning scheme. In: Madani, K., Dourado, A., Rosa, A., Filipe, J., Kacprzyk, J. (eds.) Computational Intelligence. SCI, vol. 613, pp. 401–417. Springer, Cham (2016). doi:10.1007/978-3-319-23392-5_22
2. Brouwer, A.E., Haemers, W.H.: Spectra of Graphs. Springer, New York (2011). doi:10.1007/978-1-4614-1939-6
3. Cashman, T.J., Hormann, K., Reif, U.: Generalized Lane-Riesenfeld algorithms. Comput. Aided Geom. Des. **30**(4), 398–409 (2013)

4. Cox, T., Griffeath, D.: Diffusive clustering in the two dimensional voter model. Ann. Probab. **14**(2), 347–370 (1986)
5. Desbrun, M., Meyer, M., Schröder, P., Barr, A.H.: Implicit fairing of irregular meshes using diffusion and curvature flow. In: SIGGRAPH, pp. 317–324 (1999)
6. Donnelly, P., Welsh, D.: Finite particle systems and infection models. Math. Proc. Cambridge Philos. Soc. **94**(01), 167–182 (1983)
7. Fleishman, S., Drori, I., Cohen-Or, D.: Bilateral mesh denoising. ACM Trans. Graph. **22**(3), 950–953 (2003)
8. Gadde, A., Narang, S.K., Ortega, A.: Bilateral filter: graph spectral interpretation and extensions. In: 2013 IEEE International Conference on Image Processing, pp. 1222–1226. IEEE (2013)
9. Hassin, Y., Peleg, D.: Distributed probabilistic polling and applications to proportionate agreement. Inf. Comput. **171**(2), 248–268 (2001)
10. Kobbelt, L., Campagna, S., Vorsatz, J., Seidel, H.-P.: Interactive multi-resolution modeling on arbitrary meshes. In: SIGGRAPH, pp. 105–114. ACM (1998)
11. Lane, J.M., Riesenfeld, R.F.: A theoretical development for the computer generation and display of piecewise polynomial surfaces. IEEE Trans. Pattern Anal. Mach. Intell. **2**(1), 35 (1980)
12. Li, Y., Chen, W., Wang, Y., Zhang, Z.-L.: Influence diffusion dynamics and influence maximization in social networks with friend and foe relationships. In: Proceedings of the Sixth ACM International Conference on Web Search and Data Mining, pp. 657–666. ACM (2013)
13. Nakata, T., Imahayashi, H., Yamashita, M.: A probabilistic local majority polling game on weighted directed graphs with an application to the distributed agreement problem. Networks **35**(4), 266–273 (2000)
14. Sabin, M.A., Barthe, L.: Artifacts in recursive subdivision surfaces. In: Curve and Surface Fitting: Saint-Malo, pp. 353–362 (2002)
15. Taubin, G.: A signal processing approach to fair surface design. In: SIGGRAPH, pp. 351–358 (1995)
16. Tomasi, C., Manduchi, R.: Bilateral filtering for gray and color images. In: Sixth International Conference on Computer Vision, 1998, pp. 839–846. IEEE (1998)
17. Ullah, B., Trevelyan, J., Ivrissimtzis, I.: A three-dimensional implementation of the boundary element and level set based structural optimisation. Eng. Anal. Boundary Elem. **58**, 176–194 (2015)
18. Vollmer, J., Mencl, R., Mueller, H.: Improved laplacian smoothing of noisy surface meshes. Comput. Graph. Forum **18**(3), 131–138 (1999)
19. Zhou, C., Zhang, P., Zang, W., Guo, L.: On the upper bounds of spread for greedy algorithms in social network influence maximization. IEEE Trans. Knowl. Data Eng. **27**(10), 2770–2783 (2015)

Author Index

Printed in the United States
By Bookmasters